T0192546

# DATA ANALYSIS USING HIERARCHICAL GENERALIZED LINEAR MODELS WITH R

# DATA ANALYSIS USING HIERARCHICAL GENERALIZED LINEAR MODELS WITH R

**Youngjo Lee**
**Lars Rönnegård**
**Maengseok Noh**

CRC Press is an imprint of the
Taylor & Francis Group, an **informa** business
A CHAPMAN & HALL BOOK

CRC Press
Taylor & Francis Group
6000 Broken Sound Parkway NW, Suite 300
Boca Raton, FL 33487-2742

First issued in paperback 2020



ISBN-13: 978-1-138-62782-6 (hbk)
ISBN-13: 978-0-367-65792-5 (pbk)



**Visit the Taylor & Francis Web site at**
**http://www.taylorandfrancis.com**

**and the CRC Press Web site at**
**http://www.crcpress.com**

# Contents

# List of notations

| Symbol | Description |
|---|---|
| $\boldsymbol{y}$ | Response vector |
| $\boldsymbol{X}$ | Model matrix for fixed effects |
| $\boldsymbol{Z}$ | Model matrix for random effects |
| $\boldsymbol{\beta}$ | Fixed effects |
| $\boldsymbol{v}$ | Random effect on canonical scale |
| $\boldsymbol{u}$ | Random effect on the original scale |
| $n$ | Number of observations |
| $m$ | Number of levels in the random effect |
| $h(\cdot)$ | Hierarchical log-likelihood |
| $L(\cdot;\cdot)$ | Likelihood with notation parameters; data |
| $f_\theta(\boldsymbol{y})$ | Density function for $\boldsymbol{y}$ having parameters $\theta$ |
| $\theta$ | A generic parameter indicating<br>any fixed effect to be estimated |
| $\phi$ | Dispersion component for the mean model |
| $\lambda$ | Dispersion component for the random effects |
| $g(\cdot)$ | Link function for the linear predictor |
| $r(\cdot)$ | Link function for random effects |
| $\boldsymbol{\eta}$ | Linear predictor in a GLM |
| $\boldsymbol{\mu}$ | Expectation of $\boldsymbol{y}$ |
| $s$ | Linearized working response in IWLS |
| $\boldsymbol{V}$ | Marginal variance matrix used in linear mixed models |
| $V(\cdot)$ | GLM variance function |
| $\boldsymbol{I}(.)$ | Information matrix |
| $p_d$ | Estimated number of parameters |
| $^T$ | Transpose |
| $\delta$ | Augmented effect vector |
| $\gamma$ | Regression coefficient for dispersion |

# List of notations

# Preface

Since the first paper on hierarchical generalized linear models (HGLMs) in 1996, interest in the topic grew to produce a monograph in 2006 (Lee, Nelder, and Pawitan, 2006). Ten years later this rather advanced monograph has been developed in a second edition (Lee, Nelder, and Pawitan, 2017) and two separate books on survival analysis (Ha, Jeong, and Lee, 2017) and this book, which shows how wide and deep the subject is. We have seen a need to write a short monograph as a guide for both students and researchers in different fields to help them grasp the basic ideas about how to model and how to make inferences using the h-likelihood. With data examples, we illustrate how to analyze various kinds of data using R. This book is aimed primarily toward senior undergraduates and first-year graduates, especially those searching for a bridge between Bayesian and frequentist statistics.

We are convinced that the h-likelihood can be of great practical use in data analysis and have therefore developed R packages to enhance the use of h-likelihood methods. This book aims to demonstrate its merits. The book includes several chapters divided into three parts. The first 5 chapters present various examples of data analysis using HGLM classes of models followed by the h-likelihood theory. For the examples in Chapters 2–5, R codes are presented after examples in each chapter. Most of these examples use the dhglm package, which is a very flexible package. Since there are numerous options, the code might seem technical at first sight, but we introduce it using a few simple examples and the details on how to use the dhglm package are found in Chapter 6 where we explain the code through additional examples.

In Chapters 6–9, the R packages dhglm, mdhglm, frailtyHL and jointdhglm are introduced. We explain how to use these packages by using example data sets, and the R code is given within the main text. The dhglm package fits several classes of models including: generalized linear models (GLMs), joint GLMs, GLMs with random effects (known as HGLMs) and HGLMs including models for the dispersion parameters, including double HGLMs (DHGLMs) introduced later. The mdhglm package fits multivariate DHGLMs where the response variables follow different distributions, which can also fit factor and structural equation models. The frailtyHL package is used for survival analysis using frailty models, which is an extension of Cox's proportional hazards model to allow random effects. The jointdhglm package allows joint models for HGLMs and survival time and competing risk models. In Chapter 10, we introduce variable selection methods via random-effect models. Furthermore, in Chapter 10 we study the random-effect models with discrete random effects and show that hypothesis testing can be expressed in terms of prediction of discrete random effects (e.g., null or alternative) and show how h-likelihood gives a general extension of likelihood-ratio test to multiple testing. Model-checking techniques and model-selection tools by using the h-likelihood modeling approach add further insight to the data analysis.

It is an advantage to have studied linear mixed models and GLMs before reading this book. Nevertheless, GLMs are briefly introduced in Chapter 2 together with a short review of GLM theory, for a reader who wishes to freshen up on the topic. The majority of data sets used in the book are available at URL

$$\texttt{http://cran.r-project.org/package=mdhglm}$$

for the R package mdhglm (Lee et al., 2016b). Several different examples are presented throughout the book, while the longitudinal epilepsy data presented by Thall and Vail (1990) is an example dataset used recurrently throughout the book from Chapter 1 to Chapter 6 allowing the reader to follow the model development from a basic GLM to a more advanced DHGLM.

We are grateful to Prof. Emmanuel Lesaffre, Dr. Smart Sarpong, Dr. Ildo Ha, Dr. Moudud Alam, Dr. Xia Shen, Mr. Jengseop Han, Mr. Daehan Kim, Mr. Hyunseong Park and the late Dr. Marek Molas for their numerous useful comments and suggestions.

*Youngjo Lee, Lars Rönnegård and Maengseok Noh*
*Seoul, Dalarna, and Busan*

CHAPTER 1

# Introduction

The objective of statistical inference is to draw conclusions about the study population following the sampling of observations. Different study problems involve specific sampling techniques and a statistical model to describe the analyzed situation. In this book we present HGLMs, a class of statistical models that allow flexible modeling of data from a wide range of applications and we describe the theory for their inferences. Further we present the methods of statistical testing based on HGLMs and prediction problems which can be tackled by this class. We also present extensions of classical HGLMs in various ways with a focus on dispersion modeling.

The advantage of the HGLM framework for specific statistical problems will be shown through examples including data and R code. The statistical inference and estimation are based on the Lee and Nelder (1996) hierarchical likelihood (*h-likelihood*). The h-likelihood approach is different from both classical frequentist and Bayesian methods, but at the same time unites the two (Figure 1.5) because it includes inference of both fixed and random unknowns. An advantage compared to classical frequentist methods is that inference is possible for unobservables, such as random effects, and consequently subject-specific predictions can be made. Once a statistical model is decided for the analysis of the data at hand, the likelihood leads to a way of statistical inferences. This book covers statistical models and likelihood-based inferences for various problems.

Lee and Nelder (1996) introduced inference for models including unobservable random variables, which include future outcomes, missing data, latent variables, factors, potential outcomes, etc. There are three major benefits of using the h-likelihood:

i) we can develop computationally fast algorithms for fitting advanced models,

ii) we can make inferences for unobservables and thereby make predictions of future outcomes from models including unobservables, and

iii) we can use model-checking tools for linear regression and generalized linear models (GLMs), making assumptions in all parts of an HGLM checkable.

The marginal likelihood is used for inference on the fixed effects both in classical frequentist and h-likelihood approaches, but the marginal likelihood involves multiple integration over the random effects that are most often not feasible to compute. For such cases the *adjusted profile h-likelihood*, a Laplace approximation of the marginal likelihood, is used in the h-likelihood approach. Because the random effects are integrated out in a marginal likelihood, classical frequentist method does not allow any direct inference of random effects.

Bayesians assume prior for parameters and for inference they often rely on Markov Chain Monte Carlo (MCMC) computations (Lesaffre and Lawson, 2012). The h-likelihood allows complex models to be fitted by maximizing likelihoods for fixed unknown parameters. So for a person who does not wish to express prior beliefs, there are both philosophical and computational advantages of using HGLMs. However, this book does not focus on the philosophical advantages of using the h-likelihood but rather on the practical advantages for applied users, having a reasonable statistical background in linear models and GLMs, to enhance their data analysis skills for more general types of data.

In this chapter we introduce a few examples to show the strength of the h-likelihood approach. Table 1.1 contains classes of models and the chapters where they are first introduced, and available R packages. Various packages have been developed to cover these model classes. From Table 1.1, we see that the dhglm package has been developed to cover a wider class of models from GLMs to DHGLMs . Detailed descriptions of dhglm package are presented in Chapter 7, where full structure of model classes are described.

Figure 1.1 shows the evolution of the model classes presented in this book together with their acronyms. Figure 1.1 also shows the building-block structure of the h-likelihood; once you have captured the ideas at one level you can go to a deeper level of modeling. This book aims to show how complicated statistical model classes can be built by combining interconnected GLMs and augmented GLMs, and inference can be made in a single framework of the h-likelihood. For readers who want a more detailed description of theories and algorithms on HGLMs and on survival analysis, we suggest the monographs by Lee, Nelder, and Pawitan (2017) and Ha, Jeong, and Lee (2017). This book shows how to analyze examples in these two books using available R-packages and we have also new examples.

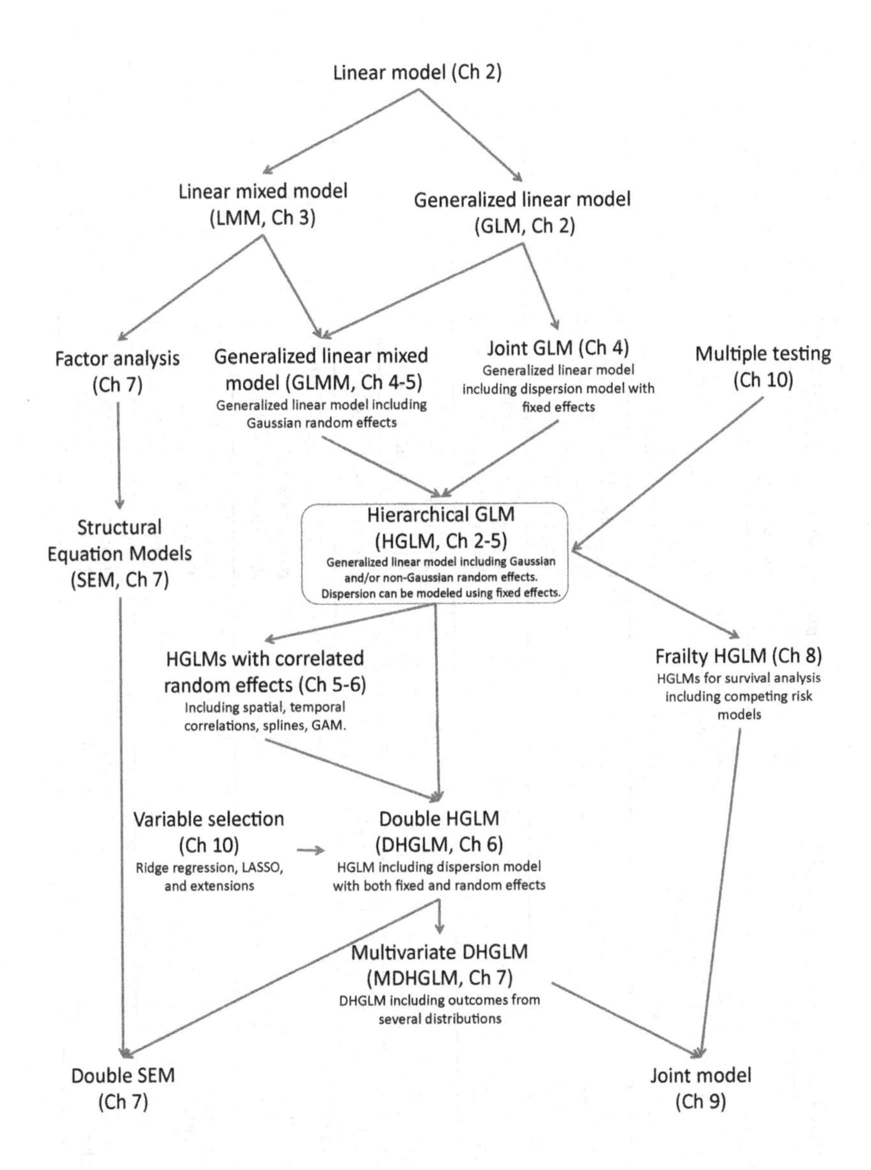


Figure 1.1 *A map describing the development of HGLMs.*

Table 1.1 *Model classes presented in the book including chapter number and available R packages*

| Model Class | R package | Developer | Chapter |
|---|---|---|---|
| GLM<br>Nelder and Wedderburn (1972) | `glm()` function<br>dhglm | <br>Lee and Noh (2016) | 2 |
| Joint GLM (Nelder and Lee, 1991) | dhglm | Lee and Noh (2016) | 4 |
| GLMM<br>Breslow and Clayton (1993) | dhglm<br>lme4<br>hglm | Lee and Noh (2016)<br>Bates and Maechler (2009)<br>Alam et al. (2015) | 4, 5 |
| HGLM<br>Lee and Nelder (1996) | dhglm<br>hglm | Lee and Noh (2016)<br>Alam et al. (2015) | 2,3,4,5 |
| Spatial HGLM<br>Lee and Nelder (2001b) | dhglm<br>spaMM | Lee and Noh (2016)<br>Rousset et al. (2016) | 5 |
| DHGLM (Lee and Nelder, 2006; Noh and Lee, 2017) | dhglm | Lee and Noh (2016) | 6 |
| Multivariate DHGLM<br>Lee, Molas, and Noh (2016a)<br>Lee, Nelder, and Pawitan (2017) | mdhglm<br>mixAK<br>mmm | Lee, Molas, and Noh (2016b)<br>Komarek (2015)<br>Asar and Ilk (2014) | 7 |
| Frailty HGLM<br>Ha, Lee, and Song (2001) | frailtyHL<br>coxme<br>survival | Ha et al. (2012)<br>Therneau (2015)<br>Therneau and Lumley (2015) | 8 |
| Joint DHGLM<br>Henderson et al. (2000) | jointdhglm<br>JM | Ha, Lee, and Noh (2015)<br>Rizopoulos (2015) | 8 |

## 1.1 Motivating examples

GLMs, introduced in Chapter 2, have been widely used in practice, based on classical likelihood theory. However, these models cannot handle repeatedly observed data, therefore various multivariate models have been suggested. Among others, HGLMs are useful for the analysis of such data. Using the following data examples from Lee, Nelder, and Pawitan (2017) it is shown that the computationally intractable classical marginal likelihood estimators can be obtained by using the Laplace approximation based on the h-likelihood for a Poisson model with random effects. With the h-likelihood approach, inferences can be made for random effects and the analysis can be further developed by fitting a wider range of models, which is not possible using only a classical marginal likelihood.

### *1.1.1 Epilepsy data*

Thall and Vail (1990) presented longitudinal data from a clinical trial of 59 epileptics, who were randomized to a new drug or a placebo (T=1 or T=0). Baseline data were available at the start of the trial; the trial included the logarithm of the average number of epileptic seizures recorded in the 8-week period preceding the trial (B), the logarithm of age (A), and number of clinic visit (V: a linear trend, coded (-3,-1,1,3)). A multivariate response variable (y) consists of the seizure counts during 2-week periods before each of four visits to the clinic.

The data can be retrieved from the R package dhglm (see R code at the end of the chapter). It is a good idea at this stage to have a look at and get acquainted with the data. From the boxplot of the number of seizures (Figure 1.2) there is no clear difference between the two treatment groups. In Figure 1.3 the number of seizures per visit are plotted for each patient, where the lines in this spaghetti plot indicate longitudinal patient effects. Investigate the data further before running the models below.

A simple first preliminary analysis could be to analyze the data (ignoring that there are repeated measurements on each patient) with a GLM having a Poisson distributed response using the R function *glm*. Let $y_{ij}$ be the corresponding response variable for patient $i(=1,\cdots,59)$ and visit $j(=1,\cdots,4)$. We consider a Poisson GLM with log-link function modeled as

$$\log(\mu_{ij}) = \beta_0 + x_{B_i}\beta_B + x_{T_i}\beta_T + x_{A_i}\beta_A + x_{V_j}\beta_V + x_{B_iT_i}\beta_{BT}, \quad (1.1)$$

where $\beta_0$, $\beta_B$, $\beta_T$, $\beta_A$, $\beta_V$, and $\beta_{BT}$ are fixed effects for the intercept,

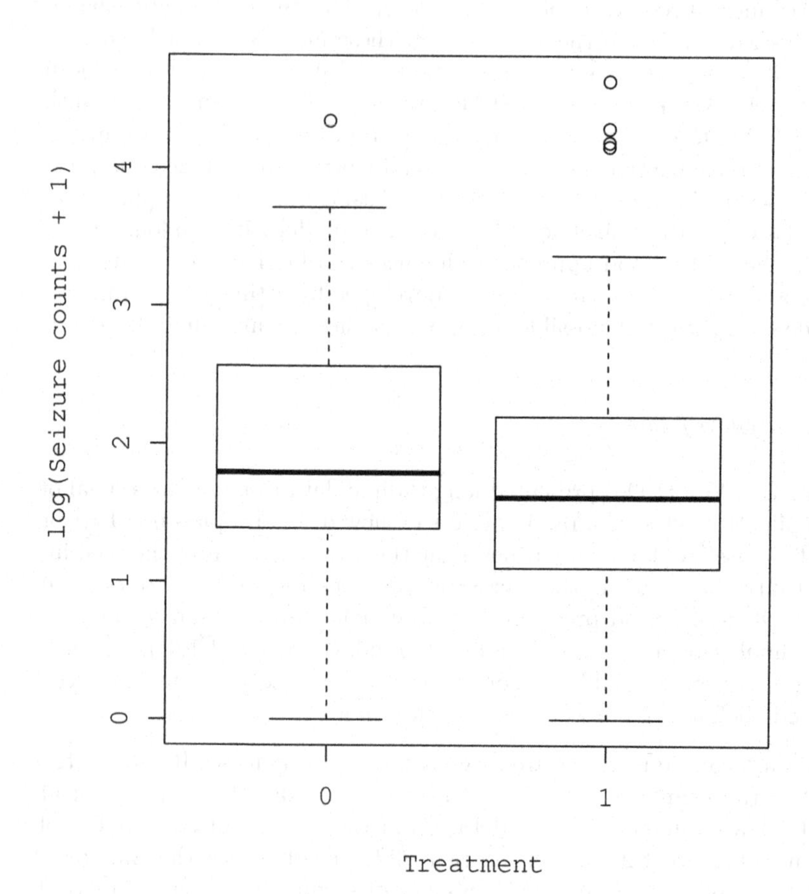


Figure 1.2 *Boxplot of the logarithm of seizure counts (new drug = 1, placebo = 0).*

logarithm of number of seizures preceding the trial (B), treatment (drug T=1, placebo T=0), logarithm of patient age (A), order of visit (V) and interaction effect (B:T). The *glm* function returns the following output

```
Call:
glm(formula=y~B+T+A+B:T+V,family=poisson(link=log),
    data=epilepsy)
```

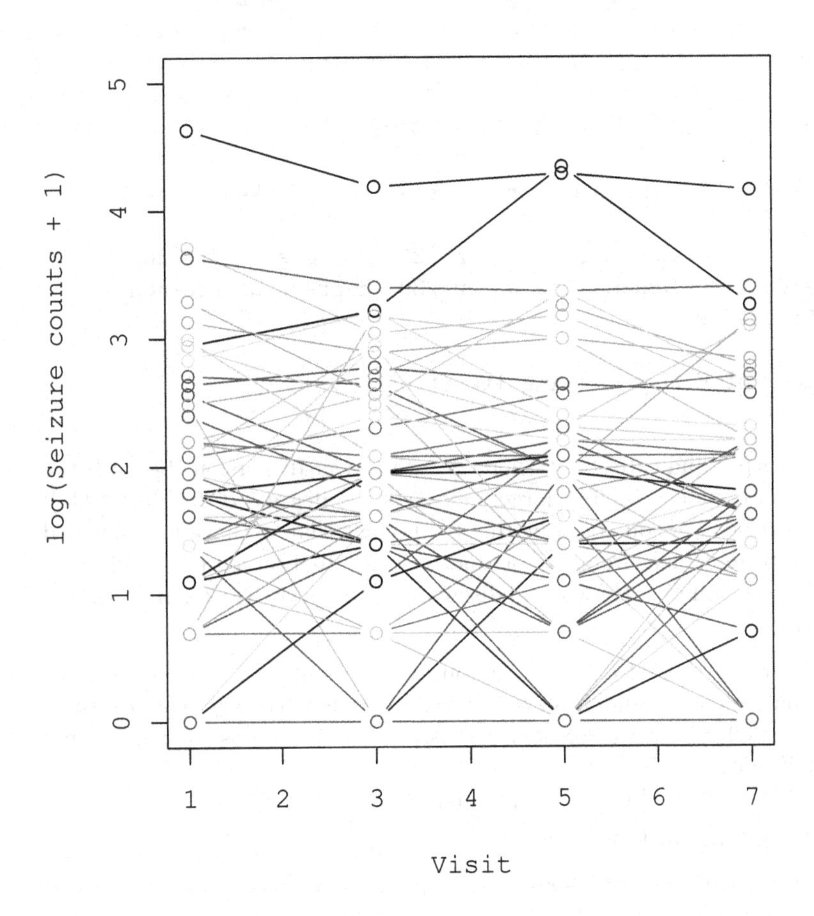


Figure 1.3 *Number of seizures per visit for each patient. There are 59 patients and each line shows the logarithm of seizure counts + 1 for each patient.*

```
Deviance Residuals:
    Min       1Q   Median       3Q      Max
-5.0677  -1.4468  -0.2655   0.8164  11.1387

Coefficients:
            Estimate Std. Error z value Pr(>|z|)
(Intercept) -2.79763    0.40729  -6.869 6.47e-12 ***
```

```
B             0.94952    0.04356  21.797  < 2e-16 ***
T            -1.34112    0.15674  -8.556  < 2e-16 ***
A             0.89705    0.11644   7.704 1.32e-14 ***
V            -0.02936    0.01014  -2.895  0.00379 **
B:T           0.56223    0.06350   8.855  < 2e-16 ***
---
Signif. codes:  0 *** 0.001 ** 0.01 * 0.05 . 0.1   1

(Dispersion parameter for poisson family taken to be 1)

    Null deviance: 2521.8  on 235  degrees of freedom
Residual deviance:  869.9  on 230  degrees of freedom
AIC: 1647.9

Number of Fisher Scoring iterations: 5
```

The output indicates that there is severe over-dispersion, because the residual deviance (869.9) largely exceeds the degrees of freedom (230). The over-dispersion could be due to the fact that we have not modeled the repeated measurements or due to some unobserved effect affecting the dispersion directly. We will discuss more on over-dispersion in Section 4.1.1.

There are repeated observations on each patient and therefore including patient as a random effect in a GLMM is necessary. For $y_{ij}$ given random effects $v_i$, we consider a Poisson distribution as the conditional distribution:

$$E(y_{ij}|v_i) = \mu_{ij} \text{ and } \mathrm{var}(y_{ij}|v_i) = \mu_{ij}.$$

with log-link function modeled as

$$\log(\mu_{ij}) = \beta_0 + x_{B_i}\beta_B + x_{T_i}\beta_T + x_{A_i}\beta_A + x_{V_j}\beta_V + x_{B_iT_i}\beta_{BT} + v_i, \quad (1.2)$$

where $v_i \sim \mathrm{N}(0, \lambda)$.

Likelihood inferences are possible in a classical frequentist framework using the glmer function in the R package lme4.

```
 Family: poisson  ( log )
Formula: y ~ B + T + A + B:T + V + (1 | patient)
   Data: epilepsy

     AIC      BIC   logLik deviance df.resid
  1345.3   1369.5   -665.6   1331.3      229
```

```
Scaled residuals:
    Min      1Q  Median      3Q     Max
-3.2832 -0.8875 -0.0842  0.6415  7.2819

Random effects:
 Groups  Name        Variance Std.Dev.
 patient (Intercept) 0.2515   0.5015
Number of obs: 236, groups:  patient, 59

Fixed effects:
            Estimate Std. Error z value Pr(>|z|)
(Intercept) -1.37817    1.17746  -1.170  0.24181
B            0.88442    0.13075   6.764 1.34e-11 ***
T           -0.93291    0.39883  -2.339  0.01933 *
A            0.48450    0.34584   1.401  0.16123
V           -0.02936    0.01009  -2.910  0.00362 **
B:T          0.33827    0.20247   1.671  0.09477 .
```

The output gives a variance for the random patient effect equal to 0.25. In GLMMs the random effects are always assumed normally distributed. The same model can be fitted within the h-likelihood framework using the R package hglm, which gives the output (essential parts of the output shown):

```
Call:
hglm2.formula(meanmodel=y~B+T+A+B:T+V+(1|patient),
    data=epilepsy,family=poisson(link = log), fix.disp = 1)


----------
MEAN MODEL
----------


Summary of the fixed effects estimates:

            Estimate Std. Error t-value Pr(>|t|)
(Intercept) -1.29517    1.22155  -1.060  0.29040
B            0.87197    0.13590   6.416 1.13e-09 ***
T           -0.91685    0.41292  -2.220  0.02760 *
A            0.47184    0.35878   1.315  0.19008
V           -0.02936    0.01014  -2.895  0.00425 **
B:T          0.33137    0.21015   1.577  0.11654
---
Signif. codes:  0 *** 0.001 ** 0.01 * 0.05 . 0.1   1
Note: P-values are based on 186 degrees of freedom
```

```
----------------
DISPERSION MODEL
----------------

Dispersion parameter for the random effects:
[1] 0.2747
```

The output gives a variance for the random patient effect equal to 0.27, similar to the lme4 package. In Chapter 4, we compare similarities and difference between HGLM estimates and lme4. Unlike the lme4 package, however, it is also possible to add non-Gaussian random effects to further model the over-dispersion. However, with the marginal-likelihood inferences, subject-specific inferences cannot be made. For subject-specific inferences, we need to estimate random patient effects $v_i$ via the h-likelihood.

With dhglm package it is possible to allow different distribution for different random effects. For example, in the previous GLMM, a gamma distributed random effect can be included for each observation, which gives a conditional distribution of the response that can be shown to be negative binomial, while patient effects are modeled as normally distributed:

$$E(y_{ij}|v_i, v_{ij}) = \mu_{ij} \text{ and } \text{var}(y_{ij}|v_i) = \mu_{ij}.$$

with log-link function modeled as

$$\log(\mu_{ij}) = \beta_0 + x_{B_i}\beta_B + x_{T_i}\beta_T + x_{A_i}\beta_A + x_{V_j}\beta_V + x_{B_iT_i}\beta_{BT} + v_i + v_{ij}, \tag{1.3}$$

where $v_i \sim \text{N}(0, \lambda_1)$, $u_{ij} = \exp(v_{ij}) \sim G(\lambda_2)$ and $G(\lambda_2)$ is a gamma distribution with $E(u_{ij}) = 1$ and $\text{var}(u_{ij}) = \lambda_2$. Then, this model is equivalent to the negative binomial HGLM such that the conditional distribution of $y_{ij}|v_i$ is the negative binomial distribution with the probability function

$$\begin{pmatrix} y_{ij} + 1/\lambda_2 - 1 \\ 1/\lambda_2 - 1 \end{pmatrix} \left(\frac{\lambda_2}{1+\mu^*_{ij}\lambda_2}\right)^{y_{ij}} \left(\frac{1}{1+\mu^*_{ij}\lambda_2}\right)^{1/\lambda_2} \mu_{ij}^{*y_{ij}},$$

where $\mu^*_{ij} = E(y_{ij}|v_i) = \mu_{0ij}u_i$ and $u_i = \exp(v_i)$.

```
Call:
hglm2.formula(meanmodel=y~B+T+A+B:T+V+(1|id)+
    (1|patient),data=epilepsy,family=poisson(link=log),
    rand.family=list(Gamma(link=log),gaussian()),
    fix.disp = 1)
```

```
----------
MEAN MODEL
----------

Summary of the fixed effects estimates:

             Estimate Std. Error t-value Pr(>|t|)
(Intercept) -1.27757    1.21959  -1.048   0.2971
B            0.87262    0.13560   6.435 3.01e-09 ***
T           -0.91389    0.41202  -2.218   0.0285 *
A            0.46674    0.35827   1.303   0.1953
V           -0.02652    0.01633  -1.624   0.1072
B:T          0.32983    0.20966   1.573   0.1184
---
Signif. codes:  0 *** 0.001 ** 0.01 * 0.05 . 0.1   1
Note: P-values are based on 114 degrees of freedom

----------------
DISPERSION MODEL
----------------

Dispersion parameter for the random effects:
[1] 0.1258 0.2430
```

The variance component for the random patient effect is now 0.24 with additional saturated random effects picking up some of the over-dispersion. By adding gamma saturated random effects we can model over-dispersion (extra Poisson variation). The example shows how modeling with GLMMs can be further extended using HGLMs. This can be viewed as an extension of Poisson GLMMs to negative-binomial GLMMs, where repeated observations on each patient follow the negative-binomial distribution rather than the Poisson distribution.

Dispersion modeling is an important, but often challenging task in statistics. Even for simple models without random effects, iterative algorithms are needed to compute maximum likelihood (ML) estimates. An example is the heteroscedastic linear model

$$\boldsymbol{y} \sim \mathrm{N}(\boldsymbol{X\beta}, \exp(\boldsymbol{X}_d\boldsymbol{\beta}_d))$$

which requires the restricted maximum likelihood (REML) for unbiased estimation of fixed effects in the variance, $\boldsymbol{\beta}_d$. Not surprisingly, estimation becomes even more challenging for models that include random

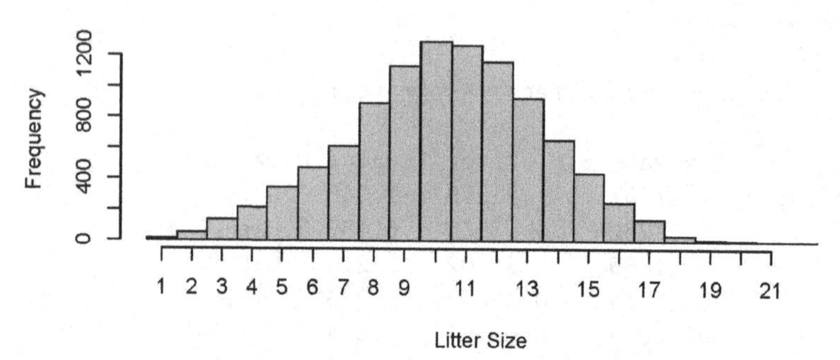


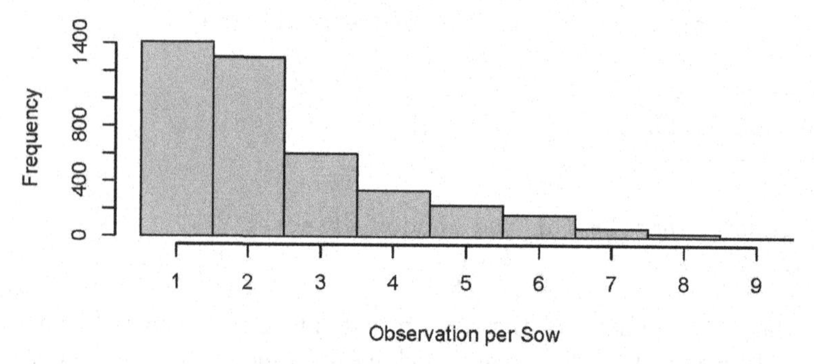


Figure 1.4 *Distributions of observed litter sizes and the number of observations per sow.*

effects, especially correlated random effects, but with double hierarchical generalized linear models (DHGLM) the problem becomes rather straightforward (Lee and Nelder, 2006; Lee, Nelder, and Pawitan, 2017).

### *1.1.2 An animal breeding study*

An important field of application for random effect estimation is animal breeding where the animals are ranked by their genetic potential given by the estimated random effects in a linear mixed model

$$\boldsymbol{y} = \boldsymbol{X\beta} + \boldsymbol{Za} + \boldsymbol{e}.$$

The outcome $\boldsymbol{y}$ can for instance be litter size in pigs or milk yield in dairy cows, and the random effects $\boldsymbol{a} \sim \mathrm{N}(0, \sigma_a^2 \boldsymbol{A})$ have a correlation matrix $\boldsymbol{A}$ computed from pedigree information (see e.g., Chapter 17 of Pawitan (2001)). The ranking is based on the best linear unbiased predictor (BLUP) $\hat{\boldsymbol{a}}$, referred to as "estimated breeding values" (for the mean), and by selecting animals with high estimated breeding values the mean of the response variable is increased for each generation of selection.

The model traditionally also assumes homoscedastic residuals,

$$\boldsymbol{e} \sim \mathrm{N}(0, \sigma_e^2 \boldsymbol{I}),$$

but there is a concern from the animal breeding industry that the residual variance might increase with selection. Therefore, a model including a heterogeneous residual variance has been suggested (Sorensen and Waagepetersen, 2003) with residuals distributed as

$$\boldsymbol{e} \sim \mathrm{N}(0, \exp(\boldsymbol{X}_d \boldsymbol{\beta}_d + \boldsymbol{Z}_d \boldsymbol{a}_d)) \tag{1.4}$$

Here $\boldsymbol{a}_d \sim \mathrm{N}(0, \sigma_d^2 \boldsymbol{A})$, and $\hat{\boldsymbol{a}}_d$ are estimated breeding values for the variance. The uniformity of a trait can be increased in a population by selecting animals with low estimated breeding values for the variance. Hence, it is desirable to have animals with large $\boldsymbol{a}$ but small $\boldsymbol{a}_d$. To implement this we often estimate $\mathrm{cor}(\boldsymbol{a}, \boldsymbol{a}_d)$. If it is negative we can select on animals with large $\boldsymbol{a}$ to reduce $\boldsymbol{a}_d$ and estimation of $\boldsymbol{a}_d$ can be ignored. However, if the correlation is positive we need to select specific animals that have large $\hat{\boldsymbol{a}}$ but small $\hat{\boldsymbol{a}}_d$. Thus, we need estimation methods for both $\boldsymbol{a}$ and $\boldsymbol{a}_d$.

By implementing a DHGLM using sparse matrix techniques (Rönnegård et al., 2010; Felleki et al., 2012) the estimation for this model becomes very fast even for a large number of observations and a large number of levels in the random effects. In Felleki et al. (2012), the model was fitted on data from 4,149 related sows having in total over 10,000 observations on litter size; see Figure 1.4. The computation time was reduced from days using MCMC, where several papers have been devoted to the computational issues of this problem (Waagepetersen et al., 2008; Ibanez et al., 2010) to a few minutes using DHGLM. In Rönnegård et al. (2013), a DHGLM was also used to estimate variance components and breeding values in a very large dataset for 177,411 related Swedish Holstein cows having a total of 1,693,154 observations on milk yield. Hence, the h-likelihood framework opens up completely new possibilities for analysis of large data. □

A possible future Bayesian alternative for this kind of application might be the use of Integrated Nested Laplace Approximations (INLA) (Rue et al., 2009) for fast deterministic computation of posterior distributions in a Bayesian context. At the time of writing, however, this possibility has not been implemented in the INLA software (www.r-inla.org) due to the complexity of extending the computations in the INLA software to include advanced dispersion modeling, which further highlights the simplicity and power of the h-likelihood approach.

## 1.2 Regarding criticisms of the h-likelihood

Likelihood is used both in the frequentist and Bayesian worlds and it has been the central concept for almost a centry in statistical modeling and inference. However, likelihood and frequentists cannot make inference for random unknowns whereas Bayesian does not make inference of fixed unknowns. The h-likelihood aims to allow inferences for both fixed and random unknowns and could cover both worlds (Figure 1.5).

The concept of the h-likelihood has received criticism since the first publication of Lee and Nelder (1996). Partly the criticism has been motivated because the theory in Lee and Nelder (1996) was not fully developed. However, these question marks have been clarified in later papers by Lee, Nelder and co-workers. One of the main concerns in the 1990's was the similarity with penalized quasi-likelihood (PQL) for GLMMs of Breslow and Clayton (1993), which has large biases for binary data. PQL estimation for GLMMs is implemented in e.g., the R package glmmPQL. Early h-likelihood methodology was criticized because of non-ignorable biases in binary data. These biases in binary data can be eliminated by improved approximations of the marginal likelihood through higher-order Laplace approximations (Lee, Nelder, and Pawitan, 2017).

Meng (2009, 2010) established Bartlett-like identities for h-likelihood. That is, the score for parameters and unobservables has zero expectation, and the variance of the score is the expected negative Hessian under easily verifiable conditions. However, Meng noted difficulties in inferences about unobservables: neither the consistency nor the asymptotic normality for parameter estimation generally holds for unobservables. Thus, Meng (2009, 2010) conjectured that an attempt to make probability statements about unobservables without using a prior would be in vain. Paik et al. (2015) studied the summarizability of h-likelihood estimators and Lee and Kim (2016) showed how to make probability statements about unobservables in general without assuming a prior as we shall see.

The h-likelihood approach is a genuine approach based on the extended likelihood principle. This likelihood principle is mathematical theory so there should be no controversy on its validity. However, it does not tell how to use the extended likelihood for statistical inference. Another important question, that has been asked and answered, is: For which family of models can the joint maximization of the h-likelihood be applied on? This is a motivated question since there are numerous examples where joint maximization of an extended likelihood containing both fixed effects $\boldsymbol{\beta}$ and random effects $\boldsymbol{v}$ gives nonsense estimates (see Lee and Nelder (2009). Such examples use the extended likelihood for joint maximization of both $\boldsymbol{\beta}$ and $\boldsymbol{v}$. If the h-likelihood $h$, defined in Chapter 2, is jointly maximized for estimating both $\boldsymbol{\beta}$ and $\boldsymbol{v}$, such nonsense estimates disappear. However, consistent estimates for the fixed effect can only be guaranteed for a rather limited class of models, including linear mixed models and Poisson HGLMs having gamma random effects on a log scale. Thus, as long as the marginal likelihood is used to estimate $\boldsymbol{\beta}$ and the h-likelihood $h$ is maximized to estimate the random effects these examples do not give contradictory results.

In close connection to the development of the h-likelihood, terminology has been used where a number of different *likelihoods* have been referred to. Some are objective functions for approximate estimation of GLM and GLMM, e.g., quasi-likelihood and extended quasi-likelihood, whereas some are used to explain the connection to classical frequentist inference and Bayesian inference, e.g., marginal likelihood and predictive probability. Other terms are joint likelihood, extended likelihood, and adjusted profile likelihood. For initiated statisticians acquainted with GLM and mixed model terminology, these terms make sense. For an uninitiated student, or researcher, the h-likelihood might seem simply another addition to this long list of likelihoods, but the central part that the h-likelihood plays in statistics is presented in the book. In later chapters we also show that the h-likelihood is the fundamental likelihood which the marginal and REML likelihoods, and predictive probabilities are derived from.

Lee and Nelder have developed a series of papers to show the iterative weighted least squares (IWLS) algorithm for GLMs can be extended to a general class of models including HGLMs. It is computationally efficient and therefore potentially very useful in statistical applications, to allow analysis of more and more complex models (Figure 1.1). For linear mixed models, there is nothing controversial about this algorithm because it can be shown to give BLUP. Neither is it controversial for GLMs with random effects in general, because the adjusted profile h-likelihoods (defined in Chapter 2) simply are approximations of marginal and restricted likelihoods for estimating fixed effects and variance components, and as such

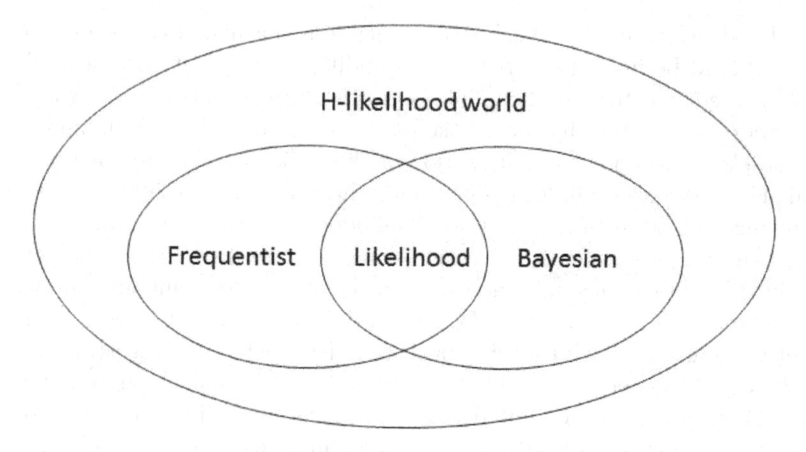


Figure 1.5 *The h-likelihood world.*

are easily acceptable for a frequentist. The h-likelihood is proportional to a Bayesian posterior distribution for models including random effects only with a flat prior, and as such is not controversial for a Bayesian statistician. The concept of predictive probability can easily be accepted both by Bayesians and frequentists (Lee and Kim, 2016). It allows both Bayesian credible interval and frequentist confidence interval interpretations. Consequently, the h-likelihood approach attempts to combine the two worlds of frequentist and Bayesian statistics, which might be controversial to some. The aim of this book, however, is to put these controversies aside and to highlight the computational and inferential advantages that the h-likelihood method can give.

## 1.3 R code

```
library(dhglm)
data(epilepsy)
model1 <- glm(y~B+T+A+B:T+V,family=poisson(link=log),
       data=epilepsy)
summary(model1)

library(lme4)
model2 <- glmer(y~B+T+A+B:T+V+(1|patient),
       family=poisson(link=log),data=epilepsy)
summary(model2)
```

```
library(hglm)
model3 <- hglm2(y~B+T+A+B:T+V+(1|patient),
        family=poisson(link=log),
        fix.disp=1, data=epilepsy)

# The option fix.disp=1
# holds the dispersion parameter constant,
# since the dispersion parameter is 1
# for a Poisson distribution.
# In later examples,
# we will see that the dispersion can be modeled
# in the hglm function,
# which is not possible in lme4 functions.

model4 <- hglm2(y~B+T+A+B:T+V+(1|id)+(1|patient),
        family=poisson(link=log),
        rand.family=list(Gamma(link=log),gaussian()),
        fix.disp=1, data=epilepsy)
summary(model4)
```

## 1.4 Exercises

1. Get acquainted with the model checking plots available in the hglm package by fitting a Gaussian distribution to the epileptic seizure count data. (This is highly inappropriate but we do it for the sake of the exercise.)

The model checking plots are produced with the command `plot(hglm.object)` where `hglm.object` is the fitted model using hglm. The QQ-plot for the residuals are found in the figure with heading *Mean Model Deviances* and the QQ-plot for the random effects are found in the figure with heading *Random1 Deviances.*

Compare these two QQ-plots to the ones from the Poisson GLMM (i.e., `model3` in the text). Which model seems to be the most suitable one?

CHAPTER 2

# GLMs via iterative weighted least squares

Before moving on to the HGLM class of models the basic concepts of GLMs are introduced. Five components specify a GLM (Box 1):

i) the response $\boldsymbol{y}$,

ii) the linear predictor $\boldsymbol{\eta} = \boldsymbol{X}\boldsymbol{\beta}$,

iii) the distribution of $\boldsymbol{y}$,

iv) the link function $g(\boldsymbol{\mu}) = \boldsymbol{\eta}$ with $\boldsymbol{\mu} = E(\boldsymbol{y})$, and

v) a prior weight $1/\phi$ (used in IWLS and will be explained later).

Before the 1970s numerous regression methods were used, including Poisson regression, logistic regression and probit regression. Each regression model required a unique estimation algorithm by maximizing the specific likelihood for that model.

Nelder and Wedderburn (1972) showed that these regression methods could be unified under the GLM framework and that IWLS could be used for estimation. GLMs capture all regression models where $f_{\theta}(\boldsymbol{y})$ belongs to the exponential family of distributions, and the user merely needs to specify the five components of a GLM.

In the following examples, model-checking plots are used and quantities such as deviance residuals and hat values are introduced (defined in Sections 2.5 and 2.6).

## 2.1 Examples

### *2.1.1 Snoring and heart disease data*

Agresti (2007) presented the data in Table 2.1 on an epidemiological survey of 2,484 subjects to investigate snoring as a possible risk factor for heart disease. The subjects were classified according to their snoring level $\boldsymbol{x}$, as reported by their spouses. We use scores (0, 2, 4, 5) for $\boldsymbol{x}$. The response variable "yes" is the number of heart disease cases and the variable "no" is the number of non-case.

**Box 1: GLM class of models**
In a GLM, the response $\boldsymbol{y}$ follows a distribution from the exponential family of distributions (including normal, binomial, Poisson, and gamma), and its expectation is modeled as

$$E(\boldsymbol{y}) = \boldsymbol{\mu}.$$

There is a link function $g(\cdot)$ connecting $\boldsymbol{\mu}$ with $\boldsymbol{X\beta}$ such that

$$g(\boldsymbol{\mu}) = \boldsymbol{X\beta}.$$

The variance of $\boldsymbol{y}$ is a function of $\boldsymbol{\mu}$. Take a Poisson distribution, for instance, the variance of $\boldsymbol{y}$ is equal to the mean $\boldsymbol{\mu}$. This relationship between the mean and the variance depends directly on the assumed distribution of $\boldsymbol{y}$. Thus, for all GLMs, the variance of $\boldsymbol{y}$ is the product of a *variance function* $V(\boldsymbol{\mu})$ and a *dispersion parameter* $\phi$. With $\boldsymbol{m}$ being the binomial denominator, we have the following variances and variance functions, $V(\boldsymbol{\mu})$.

| | Variance of $\boldsymbol{y}$ | $V(\boldsymbol{\mu})$ |
|---|---|---|
| Normal | $\phi$ | 1 |
| Poisson | $\boldsymbol{\mu}$ | $\boldsymbol{\mu}$ |
| Gamma | $\phi\boldsymbol{\mu}^2$ | $\boldsymbol{\mu}^2$ |
| Binomial | $\boldsymbol{\mu}(\boldsymbol{m}-\boldsymbol{\mu})/\boldsymbol{m}$ | $\boldsymbol{\mu}(\boldsymbol{m}-\boldsymbol{\mu})/\boldsymbol{m}$ |

For the Poisson and binomial distributions $\phi = 1$, whereas for normal and gamma $\phi$ is a parameter to be estimated. For the normal distribution $\phi$ is simply the residual variance.

For modeling proportion of heart disease cases ($\boldsymbol{p}$), we fit a logistic regression model as below,

$$\log\left(\frac{\boldsymbol{p}}{1-\boldsymbol{p}}\right) = \alpha + \beta\boldsymbol{x}.$$

We run the GLM by using the R function *glm* and from the output we see that snoring level is a significant risk factor for heart disease. In *glm*, the null deviance is the scaled deviance for intercept-only model and the scaled deviance for proposed model is called residual deviance.

The difference of two-scaled deviance 63.10 is very significant with 1 degree of freedom, so that the snoring level $x$ is significant. Since the scaled deviance of the current model is 2.8 with degrees of freedom 2 (p-value=0.25), we conclude that there is no lack of fit in using the logistic regression model.

Table 2.1 *Snoring and heart disease data*

| | | Heart Disease | |
|---|---|---|---|
| Snoring | x | yes | no |
| Never | 0 | 24 | 1355 |
| Occasional | 2 | 35 | 603 |
| Nearly every night | 4 | 21 | 192 |
| Every night | 5 | 30 | 224 |

```
Call:
glm(formula = cbind(yes, no) ~ x, family = binomial)

Deviance Residuals:
      1        2        3        4
-0.8346   1.2521   0.2758  -0.6845

Coefficients:
            Estimate Std. Error z value Pr(>|z|)
(Intercept) -3.86625    0.16621 -23.261  < 2e-16 ***
x            0.39734    0.05001   7.945 1.94e-15 ***
---
Signif. codes:  0 *** 0.001 ** 0.01 * 0.05 . 0.1   1

(Dispersion parameter for binomial family taken to be 1)

    Null deviance: 65.9045  on 3  degrees of freedom
Residual deviance:  2.8089  on 2  degrees of freedom
AIC: 27.061

Number of Fisher Scoring iterations: 4
```

### *2.1.2 Train accident data*

Consider the train accident data in the UK between 1975 and 2003 (Agresti, 2007). Let $\boldsymbol{\mu}$ denote the expected value of the number of train-related accidents $\boldsymbol{y}$ which are annual number of collisions between trains and road vehicles, for $\boldsymbol{t}$ million kilometers of train travel. The data is plotted in Figure 2.1.

To allow for a linear trend over time, we consider the following Poisson GLM for response $\boldsymbol{y}$ with covariate $\boldsymbol{x}$ (=number of years since 1975) and offset $\log(\boldsymbol{t})$.

$$\log(\boldsymbol{\mu}) = \log(\boldsymbol{t}) + \alpha + \beta\boldsymbol{x}.$$

Since the scaled deviance is 37.9 with 27 degrees of freedom (p-value=0.08; marginally significant), there is no lack of fit at a significant level 0.05.

```
Call:
glm(formula = y ~ x, family = poisson, offset = log(t))

Deviance Residuals:
    Min       1Q    Median       3Q       Max
-2.0580  -0.7825  -0.0826   0.3775   3.3873

Coefficients:
            Estimate Std. Error z value Pr(>|z|)
(Intercept) -4.21142    0.15892  -26.50  < 2e-16 ***
x           -0.03292    0.01076   -3.06  0.00222 **
---
Signif. codes:  0 *** 0.001 ** 0.01 * 0.05 . 0.1   1

(Dispersion parameter for poisson family taken to be 1)

    Null deviance: 47.376  on 28  degrees of freedom
Residual deviance: 37.853  on 27  degrees of freedom
AIC: 133.52

Number of Fisher Scoring iterations: 5
```

We apply the model-checking plots of Lee and Nelder (1998) for GLMs as shown in Figure 2.2. Two plots are used; the plot of studentized residuals against fitted values on the constant information scale (Nelder, 1990), and the plot of absolute residuals similarly. For a satisfactory model these two plots should show running means that are approximately straight

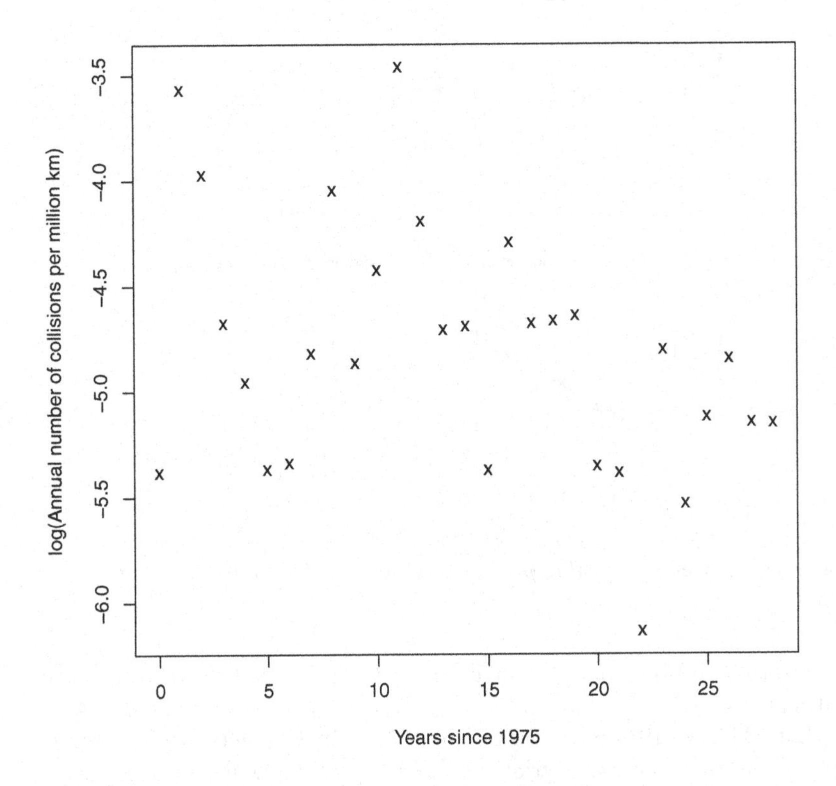


Figure 2.1 *Annual number of collisions between trains and road vehicles per million km for the years 1975–2003.*

and flat. If there is marked curvature in the first plot, this indicates either an unsatisfactory link function or missing terms in the linear predictor, or both. If the first plot is satisfactory, the second plot may be used to check the choice of variance function for the distributional assumption. If, for example, the second plot shows a marked downward trend, this implies that the residuals are falling in absolute value as the mean increases, i.e., that the assumed variance function is increasing too rapidly with the mean. In a normal probability plot, ordered values of residuals are plotted against the expected order statistics of the standard normal sample. In the absence of outliers this plot is approximately linear. We also use the histogram of residuals. If the distributional assumption is correct, it shows symmetry provided the deviance residual is the best normalizing transformation.

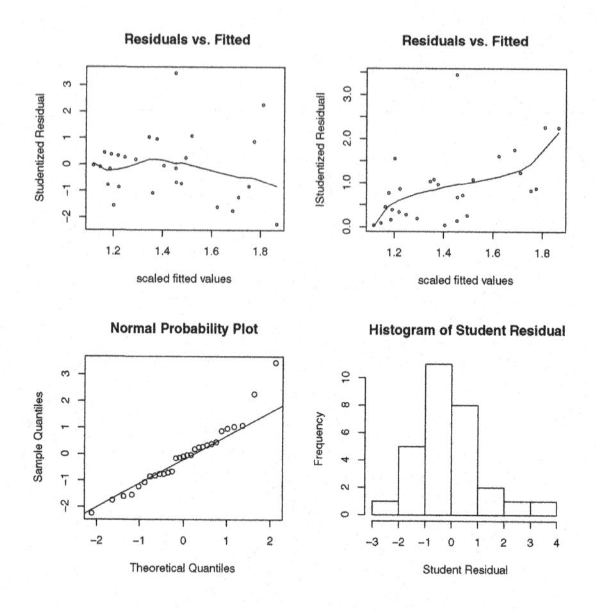


Figure 2.2 *Normal probability plot for the train accident data under a Poisson GLM.*

The normal probability plot in Figure 2.2 shows that there exist two outliers (y=12, 13 for year=1976, 1986). The average accident $\bar{y}$ is 4.2, so that 1976 and 1986 have large number of train accidents, which cannot be explained as a Poisson-variation. In Chapter 5, we will show how these data could be more appropriately modeled using an HGLM.

### *2.1.3 Epilepsy data continued*

As an example consider the epileptic seizure count data again. Fitting the data assuming a normal distribution ought to be an inappropriate model since we have count data, which we also can see from the normal QQ-plot of the deviance residuals (Figure 2.3).

Modeling the data as Poisson distributed is a better idea but still the residual plot indicates some problems (Figure 2.4), which is expected because there are repeated observations on each patient that needs to be modeled as well.

The hat values for the Poisson model can easily be plotted in R. Four observations show considerably higher values than the rest (Figure 2.5) and therefore should be investigated further.

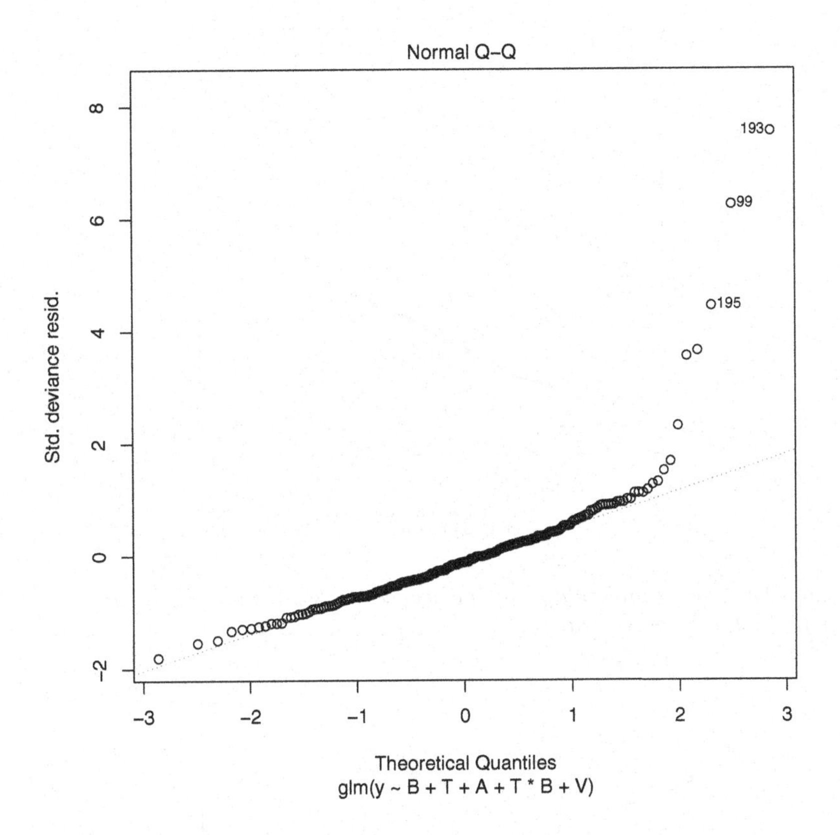


Figure 2.3 *Normal-probability plot of deviance residuals under a normal GLM for the epileptic seizure data.*

## 2.2 R code

*Snoring and heart data*

```
yes <- c(24, 35, 21, 30)
no <- c(1355, 603, 192, 224)
x <- c(0, 2, 4, 5)
snoring <- glm(cbind(yes,no) ~ x, family=binomial)
```

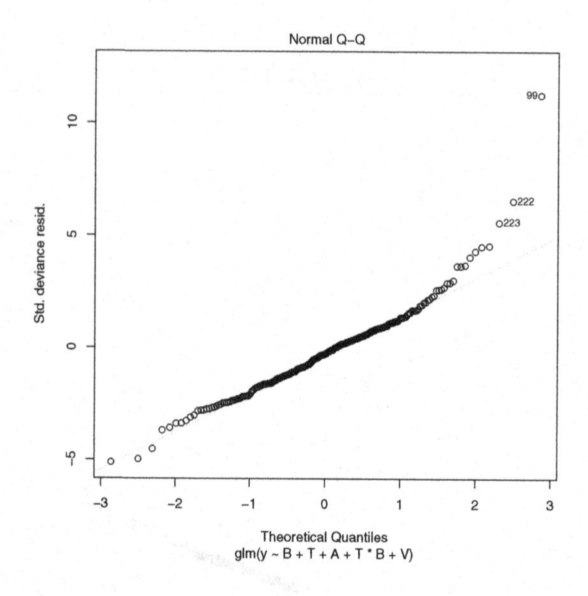


Figure 2.4 *Normal-probability plot of deviance residuals under a Poisson GLM for the epileptic seizure data.*

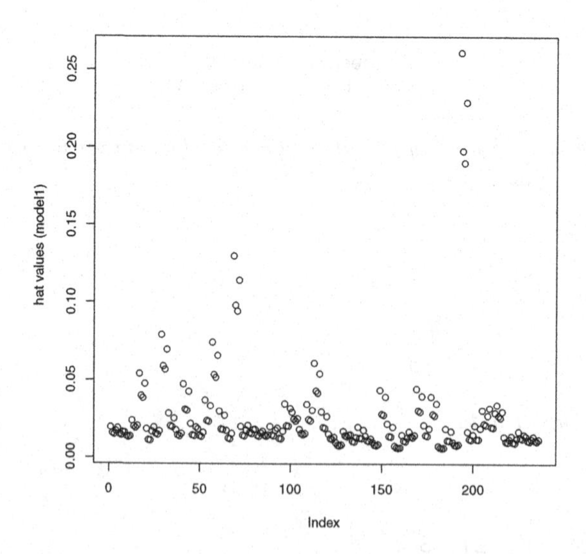


Figure 2.5 *Hat values under a Poisson GLM for the epileptic seizure count data.*

*Train accident data*

```
x <- rep(1975:2003)-1975
y<-c(2,12,8,4,3,2,2,3,7,3,5,13,
     6,4,4,2,6,4,4,4,2,2,1,4,2,3,4,3,3)
t<-c(436,426,425,430,426,430,
     417,372,401,389,418,414,397,443,436,
     431,439,430,425,415,423,437,463,487,
     505,503,508,516,518)
trainres1 <- glm( y~ x, family=poisson, offset=log(t))
## This can be available at mdhglm package
## data(train,package="mdhglm")

# model checking plot
train<-data.frame(cbind(x,y,t))

model_mu<-DHGLMMODELING(Model="mean",Link="log",
          LinPred=y~x, Offset=log(t))
model_phi<-DHGLMMODELING(Model="dispersion")

fit<-dhglmfit(RespDist="poisson",DataMain=train,
     MeanModel=model_mu,DispersionModel=model_phi)
plotdhglm(fit)

# Poisson GLM without two outliers
x1 <- x[c(-2,-12)]
y1 <- y[c(-2,-12)]
t1 <- t[c(-2,-12)]
trainres2 <- glm(y1~x1, family=poisson, offset=log(t1))

train2<-data.frame(cbind(x1,y1,t1))
model_mu<-DHGLMMODELING(Model="mean",Link="log",
          LinPred=y1~x1, Offset=log(t1))
model_phi<-DHGLMMODELING(Model="dispersion")

fit2<-dhglmfit(RespDist="poisson",DataMain=train2,
      MeanModel=model_mu,DispersionModel=model_phi)
plotdhglm(fit2)
```

*Epilepsy data (model-checking plots)*

```
data(epilepsy, package="dhglm")
model0 <- glm(y ~ B + T + A + B:T + V, family=gaussian,
```

```
          data=epilepsy)
plot(model0, which=2)


model1 <- glm(y~B+T+A+T*B+V, family=poisson(link=log),
              data=epilepsy)
plot(model1, which=2)
plot(hatvalues(model1))
```

## 2.3 Fisher's classical likelihood

Consider a simple linear model

$$y = X\beta + e \quad (2.1)$$

where $\boldsymbol{y}$ is the response, $\boldsymbol{X}$ is a design matrix, $\boldsymbol{\beta}$ is a vector of fixed effects and $\boldsymbol{e}$ is the residual assumed to be identically and independently distributed (i.i.d.) from a normal distribution with a common residual variance, say $\phi$.

Linear models and GLMs have two kinds of parameters: (1) mean structure parameters, $\boldsymbol{\beta}$, often called regression or fixed effects parameters and (2) dispersion parameters, denoted by $\boldsymbol{\phi}$.

The statistical model (2.1) describes how the responses $\boldsymbol{y}$ are generated and explains how the population of interest is constructed. For statistical inferences we want to draw useful scientific conclusions about unknowns, assuming the statistical model on the observed data $\boldsymbol{y}$ is true. Thus, model checking is important for the validity of inferences because the data generation model is assumed to be true. For the model checking the fitted residuals $\hat{\text{e}}$ can be used. The distribution of the residuals can be checked because $\hat{\text{e}}$ is a consistent estimator of $\boldsymbol{e}$.

### *2.3.1 Fisher's classical likelihood*

A solution to inference for any fixed unknowns, say $\theta$, was proposed by Fisher (1922). Fisher developed a likelihood theory, expressing the probability to observe the data as a function of the parameter value. To grasp Fisher's basic idea, consider a classical statistical model, consisting of two types of objects, data $\mathbf{y}$ and parameter $\theta$, and two related processes on them: a stochastic model describing the data generation and a statistical model for parameter inference. The two processes are defined as:

- **Stochastic Model:** Generate an instance of the data $\mathbf{y}$ from a probability function with fixed parameters $\theta$

$$f_\theta(\mathbf{y}).$$

- **Statistical Inference:** Given the data $\mathbf{y}$, make an inference about unknown fixed $\theta$ in the stochastic model by using the likelihood

$$L(\theta; \mathbf{y}).$$

The connection between these two processes is:

$$L(\theta; \mathbf{y}) \equiv f_\theta(\mathbf{y}),$$

where $L$ and $f$ are algebraically identical, but on the left-hand side $\mathbf{y}$ is fixed while $\theta$ varies and on the right-hand side $\theta$ is fixed while $\mathbf{y}$ varies.

The notation used for the likelihood here is that the parameters and data are separated by a semicolon, i.e., $L(parameter; data)$, where the data are an observed response $\boldsymbol{y}$. Furthermore, in the probability function $f_\theta(data)$, $\theta$ denotes the fixed unknown parameters and the data are an observed response $\mathbf{y}$. Here $\theta$ will be used generically denoting any kind of unknown fixed parameter.

### *2.3.2 Bayesian approach*

For the regression model 2.1, $\boldsymbol{\theta} = (\boldsymbol{\beta}, \phi)$ were fixed unknown parameters. A full Bayesian approach requires specification of a prior distribution $\pi(\boldsymbol{\theta})$ pretending all parameters are random. The advantage is that a full probabilistic framework can be used and inference can be directly made from the posterior distribution

$$p(\boldsymbol{\theta}|\boldsymbol{y}) \propto p(\boldsymbol{y}|\boldsymbol{\theta})\pi(\boldsymbol{\theta}), \tag{2.2}$$

where $p(\boldsymbol{y}|\boldsymbol{\theta}) \equiv f_\theta(y)$. Here $p(\boldsymbol{y}|\boldsymbol{\theta})$ is the model specification, while $\pi(\boldsymbol{\theta})$ is not part of the model specification but is necessary to obtain the posterior. Bayesian often wants to fit the entire distribution $p(\boldsymbol{\theta}|\boldsymbol{y})$ and use MCMC techniques. In machine learning, the penalty is a function of $\pi(\boldsymbol{\theta})$ and the resulting objective function is called a penalized likelihood. The mode estimator for (2.2) is called a *penalized least squares estimate* (more generally, *penalized maximum likelihood estimate*).

An informative prior will shrink the estimates $\hat{\boldsymbol{\beta}}$ toward the center of the prior distribution in a similar fashion as random effects in mixed linear models where the shrinkage depends on the specified distribution of the random effects. Note however that the prior distribution in the

Bayesian approach is not a part of the model while the distribution of random effects in linear mixed models is an essential part of the model specification to describe correlations among responses. Thus, we may view penalized ML estimators as the use of h-likelihood estimators but where the true model has fixed unknowns. Shrinkage of the h-likelihood estimation can be viewed as regularization of fixed parameter estimates by the penalty.

In the remaining chapter, we study how to have ML procedures for GLMs. We study penalized maximum likelihood estimation in Chapter 10.

## 2.4 Iterative weighted least squares

### *2.4.1 Linear model*

Consider a linear model

$$\mathbf{y} = \mathbf{X}\boldsymbol{\beta} + \mathbf{e}$$

$$\mathbf{e} = \boldsymbol{y} - \boldsymbol{X}\boldsymbol{\beta} \sim \mathrm{N}(0, \phi \boldsymbol{I})$$

By minimizing the least squares

$$\mathbf{e}^T\mathbf{e} = (\boldsymbol{y} - \boldsymbol{X}\boldsymbol{\beta})^T(\boldsymbol{y} - \boldsymbol{X}\boldsymbol{\beta})$$

we have the least square estimator

$$\hat{\boldsymbol{\beta}} = (\boldsymbol{X}^T\boldsymbol{X})^{-1}\boldsymbol{X}^T\boldsymbol{y}.$$

Unbiased estimator for $\phi$ is

$$\hat{\phi} = \frac{1}{n-p}\hat{\mathrm{e}}^T\hat{\mathrm{e}}$$

using residuals

$$\hat{\mathrm{e}} = (\boldsymbol{y} - \boldsymbol{X}\hat{\boldsymbol{\beta}})$$

and $p$ is the number of parameters in $\boldsymbol{\beta}$, which is the REML estimator. The variance estimator of $\hat{\boldsymbol{\beta}}$ is

$$\hat{\phi}(\boldsymbol{X}^T\boldsymbol{X})^{-1}$$

### *2.4.2 Generalized linear model*

GLM extends the linear model to the exponential family of distributions. Since explicit expressions for ML estimators are not usually available,

Table 2.2 *Deviance components for some common distributions*

| Distribution | Deviance component |
|---|---|
| Normal | $(y_i - \hat{\mu}_i)^2$ |
| Poisson | $2[y_i \log(y_i/\hat{\mu}_i) - (y_i - \hat{\mu}_i)]$ |
| Binomial | $2[y_i \log(y_i/\hat{\mu}_i) + (m_i - y_i) \log\{(m_i - y_i)/(m_i - \hat{\mu}_i\}]$ |
| Gamma | $2[-\log(y_i/\hat{\mu}_i) + (y_i - \hat{\mu}_i)/\hat{\mu}_i]$ |

$m_i$ is the number of trials for the $i$th observation in the binomial distribution.

they must be calculated iteratively, known as iterative weighted least squares (IWLS), which will be disussed later in this section..

The (unscaled) deviance $D$, a measure of model fit, is proportional to the log likelihood-ratio statistic

$$D = 2\phi[l(y; y) - l(\mu; y)]$$

where $\ell = l(\mu; y) = \log L(\mu; \mathbf{y})$ is the log likelihood for the fitted model and $l(y; y) = l(\mu; y)|_{\mu=y}$. The deviance can be expressed as a sum of unscaled deviance components

$$D = \sum d_i$$

one for each $i$th observation. The expression of these deviance components for the GLM family of distributions are found in Table 2.2.

Note that in a normal linear model $d_i = (y_i - \mu_i)^2 = e_i^2$ so that the sum of error squares becomes $\boldsymbol{e}^T\boldsymbol{e} = D$. This means the least square estimator in linear models is the maximum likelihood estimator in GLMs which maximizes $l(\mu; y)$. So, the deviance $D$ can be viewed as an extension of the error sum of squares to the class of GLM models.

The ML estimator of the regression parameters is

$$\hat{\boldsymbol{\beta}} = (\boldsymbol{X}^T\Sigma^{-1}\boldsymbol{X})^{-1}\boldsymbol{X}^T\Sigma^{-1}\mathbf{s},$$

where $\Sigma = \phi W^{-1}$, $W$ is a diagonal weight matrix, whose $i$th element is defined as

$$W_i = \left(\frac{d\mu_i}{d\eta_i}\right)^2 \frac{1}{V(\mu_i)},$$

$V(\mu)$ is the GLM variance function specifying the part of the variance of $\boldsymbol{y}$ that depends on $\mu$, and $\mathbf{s}$ is the adjusted dependent variate which is a linearization of $g(\boldsymbol{y})$ around $\mu$, with

$$\mathbf{s} = \boldsymbol{\eta} + (\boldsymbol{y} - \boldsymbol{\mu})\frac{\partial g(\boldsymbol{\mu})}{\partial \boldsymbol{\mu}}.$$

In an IWLS algorithm $1/\phi$ plays the part of prior weight. Furthermore, the variance of the estimator $\boldsymbol{\beta}$ is

$$(\boldsymbol{X}^T\Sigma^{-1}\boldsymbol{X})^{-1} = \phi(\boldsymbol{X}^TW\boldsymbol{X})^{-1}.$$

The linear model has an explicit solution for $\hat{\boldsymbol{\beta}}$ that does not require an iterative procedure, whereas GLMs in general require iteration. Take Poisson regression with a log link function for instance,

$$\log(\boldsymbol{\mu}) = \boldsymbol{X}\boldsymbol{\beta}.$$

The IWLS works as follows

1. Specify a starting value for $\boldsymbol{\beta}$, say $\boldsymbol{\beta}^{(0)}$.
2. Compute $\boldsymbol{\eta}^{(0)} = \boldsymbol{X}\boldsymbol{\beta}^{(0)}$ and $\boldsymbol{\mu}^{(0)} = \exp(\boldsymbol{\eta}^{(0)})$.
3. Compute s. For the log link we have

$$\mathrm{s} = \boldsymbol{\eta}^{(0)} + (\boldsymbol{y} - \boldsymbol{\mu}^{(0)})\frac{1}{\boldsymbol{\mu}^{(0)}}.$$

4. Fit the linear regression model

$$\mathrm{s} = \boldsymbol{X}\boldsymbol{\beta} + e.$$

   Here the residual variance is $1/\boldsymbol{\mu}$ because

$$\mathrm{var}(\mathrm{s}) = \mathrm{var}\left(\frac{\boldsymbol{y}-\boldsymbol{\mu}}{\boldsymbol{\mu}}\right) = \frac{\mathrm{var}(\boldsymbol{y})}{\boldsymbol{\mu}^2}.$$

   Thus, the model can be fitted as a weighted least squares (WLS)

$$\boldsymbol{X}^TW\boldsymbol{X}\hat{\boldsymbol{\beta}} = \boldsymbol{X}^TW\mathrm{s}$$

   where $W = \mathrm{diag}(\boldsymbol{\mu})$ for the Poisson model with log link. Let $\boldsymbol{\beta}^{(1)}$ be the estimates of $\boldsymbol{\beta}$ from this WLS.
5. Go to 2. replacing $\boldsymbol{\beta}^{(0)}$ by $\boldsymbol{\beta}^{(1)}$ and repeat until convergence.
6. After convergence report $\hat{\boldsymbol{\beta}}$ and

$$\mathrm{var}(\hat{\boldsymbol{\beta}}) = (\boldsymbol{X}^TW\boldsymbol{X})^{-1};$$

   $\phi$ does not appear in these computations because $\phi = 1$ in Poisson models. □

Nelder and Wedderburn (1972) showed that this algorithm is identical to the ML estimation procedure of Fisher scoring. Furthermore, the curvature of the log-likelihood around the ML estimator is equal to the log-likelihood curvature of the WLS model and consequently the standard errors are identical to those obtained by ML.

## 2.5 Model checking using residual plots

Birnbaum (1962) proved that the classical likelihood function contains all the information in the observed data about the fixed parameter, provided that the assumed stochastic model is right. This means that to make inferences about the true value of the fixed parameter using the information only in the data, we need the likelihood function and nothing else. However, this likelihood principle implies that the likelihood captures all the information in the data for analyses, *if the model is correct.* Thus, we should check model assumptions to verify our analysis. We can check various model assumptions by treating computed residuals as i.i.d. samples from a common distribution. Hence, their first two moments should be the same if the assumed model is correct.

In linear models, the raw residuals $\hat{e} = y - X\hat{\beta}$ can be used for model checking. However, for GLM, in general, the variance of the raw residuals is not constant over the range of response values and alternative residuals are required for model checking. Consider standardized deviance residuals defined as

$$r_{D,i} = \text{sign}(y_i - \hat{\mu}_i)\sqrt{d_i/\phi}$$

and alternatively, Pearson residuals

$$r_{P,i} = \frac{y_i - \hat{\mu}_i}{\sqrt{\phi V(\hat{\mu}_i)}}.$$

In linear models, both residuals are identical. The sum of squared deviance residuals $D^* = \sum r_{D,i}^2$ is the log likelihood-ratio statistic (the scaled deviance in the dhglm package and the residual deviance in the glm() function), and $P^* = \sum r_{P,i}^2$ is the Pearson chi-squared statistic. In linear models, $D = P = \mathbf{e}^T\mathbf{e}$, so that both $D$ and $P$ can be considered an extension of the sum of error squares in linear models to GLMs. However, $D$ gives an ML estimator for $\mu$, while $P$ cannot be used for estimation of $\mu$ due to severe biases. If the model is appropriate, the (scaled) deviance $D^*$ or Pearson chi-squared statistic are asymptotically Chi-squared distributed with $n - p$ degrees of freedom. Thus, these statistics have been used for the lack of fit test in GLMs.

## 2.6 Hat values

The hat matrix $\mathbf{H}$ is a matrix transforming the observations $\boldsymbol{y}$ to their fitted values $\hat{\boldsymbol{y}}$

$$\hat{\boldsymbol{y}} = \mathbf{H}\boldsymbol{y}.$$

For a linear model $\boldsymbol{y} = \boldsymbol{X\beta} + \boldsymbol{e}$, we have

$$\mathbf{H} = \boldsymbol{X}(\boldsymbol{X}^T\boldsymbol{X})^{-1}\boldsymbol{X}^T.$$

The diagonal elements of $\mathbf{H}$ are the *hat values* here denoted by $q_i$. They take values from 0 to 1 and indicate how much information there is in the model for each observation, where low values show high information.

In general, GLMs have the hat matrix

$$\mathbf{H} = \boldsymbol{X}(\boldsymbol{X}^T W\boldsymbol{X})^{-1}\boldsymbol{X}^T W.$$

Studentized residuals adjust for the hat values and are obtained as

$$\frac{r_{D,i}}{\sqrt{1-q_i}} \text{ or } \frac{r_{P,i}}{\sqrt{1-q_i}}.$$

Deviance residuals are the best normalization transformation under the GLM class of models (Pierce and Schafer, 2008). For the model checking in this book we use studentized deviance residuals and it is reasonable to assume that they are i.i.d. samples from the standard normal distribution if the model is correctly specified.

The deviance residuals can be used to model the dispersion parameter $\phi$. Take a linear model for instance where

$$E(\sum \hat{e}_i^2) = \phi \sum (1 - q_i)$$

so that the unbiased estimator of the residual variance is

$$\hat{\phi} = \frac{\sum \hat{e}_i^2}{\sum(1-q_i)} = \frac{\sum \hat{e}_i^2}{n-p}.$$

Furthermore, in GLMs, we use the unscaled deviance to estimate the dispersion parameter

$$\hat{\phi} = \frac{\sum d_i}{n-p} = \frac{D}{n-p}.$$

## 2.7 Exercises

1. Implement the IWLS algorithm given for the Poisson regression model in the beginning of the chapter (Section 2.4.2) using the following response, y, and explanatory variable, x.

```
y <- c(18,17,15,20,10,20,25,13,12)
x <- c(0.5, 0.5, 1, 0, 1, 0, 0, 1, 1)
```

2. a) Compute the deviance components for the fitted model in Exercise 1 (using the formula for the Poisson distribution in Table 2.2.

b) Compute the standardized deviance residuals.

c) Check the distributional assumption of the model by plotting the standardized deviance residuals in a QQ-plot.

3. Implement an IWLS algorithm for a gamma regression model with a log link function and $\phi = 1$. (You might find the table in Box 1 useful.) Fit the same data as above.

4. Using the general health questionnaire score data available in the SMIR package, fit a logistic GLM with the glm function in R (see Section 2.4.3 of Lee, Nelder, and Pawitan (2017) to check your results). The response variable is case vs. non-case and the linear predictor is Constant + sex + ghq. Check whether it is reasonable to assume ghq to have a linear effect or whether the ghq should be fitted as a factor.

```
data(ghq, package="SMIR")
# sex : men, women
# c : case
# nc: non-case
# ghq : general health questionair score
```

5. The ozone data (analyzed in Section 2.4.4 of Lee, Nelder, and Pawitan (2017)) contains nine meteorological variables. Find a suitable distribution for ozone concentration as response and a suitable linear predictor by comparing model checking plots and AIC for different GLMs.

```
data(ozone, package="mdhglm")
# y : ozone concentration
# x1 - x9 : nine meteorological variables
```

CHAPTER 3

# Inference for models with unobservables

In Chapter 1, we introduced several HGLM examples and showed some of the strengths of the h-likelihood approach. We illustrated that the h-likelihood method (together with the associated software) can fit rather complex models in an elegant manner. In contrast, classical likelihood software may not be as flexible, whereas Bayesian MCMC approaches allow fitting these models but at the expense of more computation time and requires to assume priors for fixed parameters.

In this chapter we define the h-likelihood and provide insight to inference and predictions based on the h-likelihood. We introduce the *extended likelihood principle* underlying the h-likelihood framework and show how it is related both to classical likelihood and Bayesian inference. Five important points are made:

i) inference about random effects can be made using h-likelihood, while classical likelihood cannot give any information about random effects,

ii) h-likelihood inference of random effects takes into account the uncertainty in estimating the fixed effects, whereas empirical Bayes (EB) estimation of random effects assumes known values of the fixed effects,

iii) model checking is possible for all parts of the model,

iv) all necessary inferential tools can be derived from the h-likelihood,

and

v) the h-likelihood can be used for predictions of unobserved random variables such as future outcomes.

The use of model-checking plots and model selection for HGLMs are introduced using the epilepsy seizure data and thereafter the theoretical details are presented in the following sections.

## 3.1 Examples

### *3.1.1 Epilepsy data continued*

Model selection in HGLMs is exemplified using the epileptic seizure data presented in Section 1.1.1. The random patient effect is here modeled to be Gaussian, and in this example the significance of the interaction effect B:T is tested (R code is in the next section).

Consider the Poisson GLMM (1.2) again.

$$\log(\mu_{ij}) = \beta_0 + x_{B_i}\beta_B + x_{T_i}\beta_T + x_{A_i}\beta_A + x_{V_j}\beta_V + x_{B_iT_i}\beta_{BT} + v_i,$$

where $v_i \sim \mathrm{N}(0, \lambda)$. Suppose that we wish to test $\mathrm{H}_0 : \boldsymbol{\beta}_{BT} = 0$.

The essential part of the *hglm* output is

```
Summary of the fixed effects estimates:

            Estimate Std. Error t-value Pr(>|t|)
(Intercept) -1.29517    1.22155  -1.060  0.29040
B            0.87197    0.13590   6.416 1.13e-09 ***
T           -0.91685    0.41292  -2.220  0.02760 *
A            0.47184    0.35878   1.315  0.19008
V           -0.02936    0.01014  -2.895  0.00425 **
B:T          0.33137    0.21015   1.577  0.11654


Dispersion parameter for the random effects:
[1] 0.2747


---------------
LOG-LIKELIHOODS
---------------
h-likelihood: -624.3322
Adjusted profile likelihood
   Profiled over random effects: -665.7751
   Profiled over fixed and random effects: -674.242
Conditional AIC: 1272.685

EQL estimation converged in 5 iterations.
```

The standard errors for the fixed effects B, T, A, V, and B:T are computed from the curvature of the marginal likelihood, in the same way as in classical likelihood theory. However, the Laplace approximation in (3.3) is generally used instead of the computationally intractable marginal likelihood to allow fast and simple computations for all HGLMs.

For model comparisons and testing Lee and Nelder (1996) proposed to use the h-likelihood for random effects $f_\theta(\boldsymbol{y}, \boldsymbol{v})$, the marginal likelihood for fixed effects $f_\theta(\boldsymbol{y})$ and for the dispersion parameters $f_\theta(\boldsymbol{y}|\hat{\beta})$. When the Laplace approximation is used $\log(f_\theta(\boldsymbol{y}))$ is replaced by $p_v(h)$ and $\log(f_\theta(\boldsymbol{y}|\hat{\beta}))$ by $p_{\boldsymbol{\beta},\boldsymbol{v}}(h) = (h - \frac{1}{2}\log(|\mathbf{I}(\boldsymbol{v}, \boldsymbol{\beta})|/(2\pi)))|_{v=\hat{v},\beta=\hat{\beta}}$. In the epilepsy example, $p_{\boldsymbol{v}}(h) = -665.8$ and $p_{\boldsymbol{\beta},\boldsymbol{v}}(h) = -674.2$, which can be used to test model differences similarly as in classical likelihood theory.

Suppose for instance we wish to test the interaction effect B:T. A Wald test gives $p = 0.11654$ ($\hat{\boldsymbol{\beta}}_{BT} = 0.331$, s.e.=0.210). A likelihood-ratio test (LRT) can be computed by fitting a model without the interaction effect, which gives $p_v(h) = -667.0998$. The LRT is computed as $-2$ times the difference in $p_v(h)$, which is here 2.65. The LRT is $\chi^2(1)$ distributed giving $p = 0.103$, and consequently both the Wald test and LRT give similar P-values.

For the nested models as above, we can use the likelihood-ratio tests for a model selection. However, for the non-nested models, we may use the Akaike information criterion (AIC). The conditional AIC (cAIC) is defined as

$$-2\log f_\theta(\boldsymbol{y}|\boldsymbol{v}) + 2p_D$$

where $p_D$ are the estimated degrees of freedom for fixed parameters and random effects (Lee and Nelder, 1996). For this example the estimated degrees of freedom is 185.6. Note that degrees of freedom for fixed parameters are integer, while those for random effects are positive real numbers. It can be computed from the hglm object by taking the number of observations, 236, minus `model1$df`, or equivalently by summing up the first 236 hat values (see Section 4.1.1 for more detail). For dispersion models, we consider the AIC based on the restricted likelihood (rAIC) defined as

$$p_{\boldsymbol{\beta},v}(h) + 2p_r$$

where $p_r$ is the number of dispersion parameters. The cAIC and rAIC can be used for model selection among non-nested models: for more detailed discussion see Ha et al. (2007). If the AIC difference is larger than 1 it is considered to be significant, and if the difference is less than 1 it is not (Sakamoto et al., 1986).

For the epilepsy data, we consider following six models:

i) Poisson GLM with log-link function (1.1),

ii) Poisson-normal HGLM with log-link function (1.2),

iii) negative binomial (NB)-normal HGLM with log-link function (1.3),

iv) NB-gamma HGLM with $u_i \sim G(\lambda_1)$ in (1.3).

As we shall see, large outliers are mainly caused by zeros in observations, which indicates an extra-Poisson variation among repeated measures. Thus, we fit over-dispersed Poisson models with $\text{var}(y_{ij}|v_i) = \phi\mu_{ij}$ in Section 4.1.3:

v) over-dispersed Poisson GLM with log-link function (1.1),

vi) over-dispersed Poisson-normal HGLM with log-link function (1.2).

From outputs of the dhglm package, we can have model selection criteria based on likelihood values, such as cAIC and rAIC. In the absence of random effects, the cAIC becomes AIC, so that we can directly compare cAIC in HGLMs with AIC in GLMs. In Table 3.1, we see that both cAIC and rAIC prefer NB-gamma HGLM. This means that correlation among repeated measures should be modeled via random effects, but the normal random effects for patient are not enough to describe over-dispersion among repeated measures. Over-dispersion can be modeled by over-dispersed quasi-Poisson models or negative-binomial HGLMs (corresponding some Poisson-gamma HGLMs). We prefer negative-binomial HGLM over quasi-Poisson models because the former are real statistical models, while latter are quasi models (Lee and Nelder, 2004) where there is no agreement in estimation of dispersion parameters (Nelder and Lee, 1992). In this dataset, we need to model not only the over-dispersion among repeated measures but also the correlation caused by patients. Between over-dispersed Poisson models and negative binomial HGLMs, we see negative-binomial HGLMs are preferred according to cAIC and rAIC. Furthermore, the gamma distribution of $u_i$ for patients is preferred to normal distribution of $v_i = \log u_i$.

The outputs from the dghlm package for the NB-gamma HGLM are as follows and model-checking plot based on residuals is in Figure 3.1. From the figures we see that many large negative outliers are caused by zero observations, which is explained by negative-binomial model for the conditional distribution. Model checking plots for $v_i$ ($\lambda_1$) and $v_{ij}$ ($\lambda_2$) are in Figure 3.2. The largest negative outlier of $v_i$ corresponds to patient 58. The largest positive outlier of $v_{ij}$ is from the third observation on the patient 25 having the largest 76 seizure count, while the smallest outlier from the third observation on the patient 58 having 0 seizure count. Thus, the distibutional assumptions for random effects are in suspect. The distributional assumptions on $v_i$ and $v_{ij}$ are further investigated in Section 6.2.9.

Table 3.1 *cAIC and rAIC from Poisson models for the epilepsy data*

| model | cAIC | rAIC |
|---|---|---|
| Poisson GLM | 1,647.9 | 1,664.7 |
| Poisson-normal HGLM | 1,272.7 | 1,350.5 |
| NB-normal HGLM | 1,201.1 | 1,310.5 |
| NB-gamma HGLM | 1,163.9 | 1,274.8 |
| over-dispersed Poisson GLM | 1,321.9 | 1,332.8 |
| over-dispersed Poisson-normal HGLM | 1,219.4 | 1,320.9 |

```
Distribution of Main Response :
                          "poisson"
[1] "Estimates from the model(mu)"
y ~ B + T + A + B:T + V + (1 | patient) + (1 | id)
[1] "log"
            Estimate Std. Error t-value
(Intercept) -1.30406    1.34162  -0.972
B            0.90046    0.14317   6.290
T           -0.83068    0.42867  -1.938
A            0.50809    0.39911   1.273
V           -0.02807    0.01667  -1.684
B:T          0.32455    0.21928   1.480
[1] "Estimates for logarithm of lambda=var(u_mu)"
[1] "gamma" "gamma"
        Estimate Std. Error t-value
patient   -1.290     0.2215  -5.824
id        -1.989     0.1612 -12.336
[1] "== Likelihood Function Values and Condition AIC =="
                                      [,1]
-2 log(likelihood)            :   1255.5890
-2 log(restricted likelihood):   1270.8333
cAIC                          :   1163.9246

Scaled Deviance : 142.5631 on 108.3168 degrees of freedom
```

## 3.2 R code

*Epilepsy data*

```
# testing the interaction effect
```

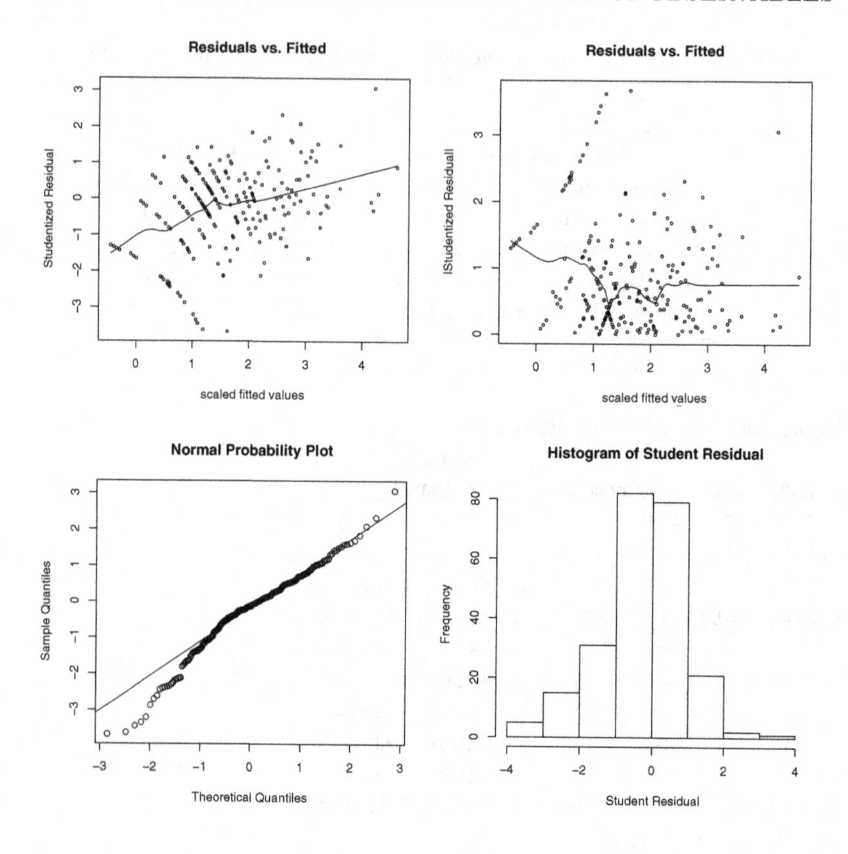


Figure 3.1 *Model-checking plot based on residuals for the NB-gamma HGLM on the epilepsy data.*

```
library(hglm)
data(epilepsy, package="dhglm")
model3 <- hglm(fixed=y~B+T+A+B:T+V, random=~1|patient,
            family=poisson(link=log),
            fix.disp=1,data=epilepsy,
            calc.like = TRUE)

#Likelihoods for model without interaction effect
logLik(hglm(fixed=y~B+T+A+V, random=~1|patient,
          family=poisson(link=log), fix.disp=1,
          data=epilepsy, calc.like=TRUE))
```

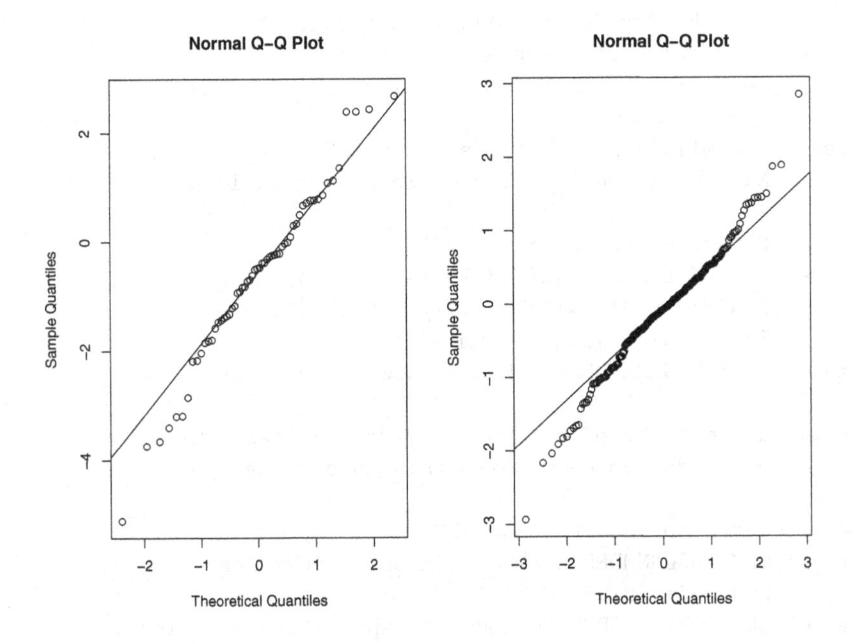


Figure 3.2 *Model-checking plot for $v_i$ (left) and $v_{ij}$ (right) for the NB-gamma HGLM on the epilepsy data.*

```
# i) Poisson GLM
model_mu<-DHGLMMODELING(Model="mean", Link="log",
         LinPred=y~B+T+A+B:T+V)
model_phi<-DHGLMMODELING(Model="dispersion",Link="log")

res1<-dhglmfit(RespDist="poisson",DataMain=epilepsy,
      MeanModel=model_mu,DispersionModel=model_phi)

# ii) Poisson-Gaussian HGLM
model_mu<-DHGLMMODELING(Model="mean", Link="log",
         LinPred=y~B+T+A+B:T+V+(1|patient),
         RandDist=c("gaussian"))
model_phi<-DHGLMMODELING(Model="dispersion",Link="log")

res2<-dhglmfit(RespDist="poisson",DataMain=epilepsy,
      MeanModel=model_mu,DispersionModel=model_phi)

# iii) Netative-binomial-Gaussian HGLM
model_mu<-DHGLMMODELING(Model="mean", Link="log",
```

```
        LinPred=y~B+T+A+B:T+V+(1|patient)+(1|id),
        RandDist=c("gaussian","gamma"))
model_phi<-DHGLMMODELING(Model="dispersion",Link="log")

res3<-dhglmfit(RespDist="poisson",DataMain=epilepsy,
      MeanModel=model_mu,DispersionModel=model_phi)

# iv) Netativebinomial-gamma HGLM
model_mu<-DHGLMMODELING(Model="mean", Link="log",
      LinPred=y~B+T+A+B:T+V+(1|patient)+(1|id),
      RandDist=c("gamma","gamma"))
model_phi<-DHGLMMODELING(Model="dispersion",Link="log")

res4<-dhglmfit(RespDist="poisson",DataMain=epilepsy,
      MeanModel=model_mu,DispersionModel=model_phi)

# v) over-dispersed Poisson GLM
model_mu<-DHGLMMODELING(Model="mean", Link="log",
      LinPred=y~B+T+A+B:T+V)
model_phi<-DHGLMMODELING(Model="dispersion",Link="log",
      LinPred=phi~1)

res5<-dhglmfit(RespDist="poisson",DataMain=epilepsy,
     MeanModel=model_mu,DispersionModel=model_phi)

# vi) over-dispersed Poisson-Gaussian HGLM
model_mu<-DHGLMMODELING(Model="mean", Link="log",
      LinPred=y~B+T+A+B:T+V+(1|patient),
      RandDist=c("gaussian"))
model_phi<-DHGLMMODELING(Model="dispersion",Link="log",
      LinPred=phi~1)

res6<-dhglmfit(RespDist="poisson",DataMain=epilepsy,
      MeanModel=model_mu,DispersionModel=model_phi)
```

## 3.3 Likelihood inference for random effects

In this book we consider extended statistical models that consist of three types of objects, data $\mathbf{y}$, parameter (fixed unknowns) $\theta$ and unobservables (random unknowns) $\boldsymbol{v}$. Statistical inferences need to be made for both unknowns $\theta$ and $\boldsymbol{v}$, based upon the observed data $\mathbf{y}$. However, the classical likelihood $L(\theta; \mathbf{y})$ in Chapter 2 cannot be used for such inferences because it does not have unobservables $\boldsymbol{v}$.

Suppose for instance that we have repeated observations such that the responses can no longer be assumed to be independent. In such cases, random effects are often included to model the correlation structure between responses.

Consider a linear mixed model, for $i = 1, \cdots, m$ and $j = 1, \cdots, n_i$

$$y_{ij} = x_{ij}\beta + v_i + e_{ij}, \tag{3.1}$$

where $y_{ij}$ is the observed random variable (response), $\beta$ is the vector of fixed effects and $v_i \sim \mathcal{N}(\mathbf{0}, \lambda)$ are i.i.d. random effects, and $e_{ij} \sim \mathcal{N}(0, \phi)$ is an i.i.d residual random error or measurement error. Parameters $\phi$ and $\lambda$ are variance components. In this model, there are two types of unknowns; fixed unknowns $\theta = (\beta, \phi, \lambda)$ and random unknowns $\boldsymbol{v} = (v_1, \cdots, v_m)^T$.

The linear mixed model (3.1) may be written in matrix form as

$$\boldsymbol{y} = \boldsymbol{X\beta} + \boldsymbol{Zv} + \boldsymbol{e}.$$

In the classical likelihood setting the model for the data generation process $f_\theta(y)$ is given by the density function of a multivariate normal distribution

$$\mathrm{N}(\boldsymbol{X\beta}, \lambda \boldsymbol{ZZ}^T + \phi \boldsymbol{I})$$

with the corresponding marginal likelihood

$$L(\boldsymbol{\beta}, \lambda, \phi; \mathbf{y}) = (2\pi|\boldsymbol{V}|)^{-\frac{1}{2}}\exp\{-\frac{1}{2}(\boldsymbol{y} - \boldsymbol{X\beta})^T \boldsymbol{V}^{-1}(\boldsymbol{y} - \boldsymbol{X\beta})\},$$

where $\boldsymbol{V}$ is the marginal covariance matrix of $\mathbf{y}$, $\lambda \boldsymbol{ZZ}^T + \phi \boldsymbol{I}$. This marginal likelihood can be used to estimate, and make inference about, the fixed parameters $\boldsymbol{\beta}$, $\lambda$ and $\phi$. Note, however, that the random effect $\boldsymbol{v}$ is not included and that the classical likelihood does not directly give estimates of, nor inference about, the random effects.

In summary, the classical likelihood is given by the density function of the observed data only, $f_\theta(\boldsymbol{y})$. ML estimates are derived by maximizing $f_\theta(\boldsymbol{y})$ or equivalently $\log(f_\theta(\boldsymbol{y}))$. But it does not give any guidelines for inference on random effects because the random effects have been integrated out in the marginal likelihood. The hierarchical likelihood is a remedy to this problem.

### *3.3.1 The hierarchical likelihood*

Lee and Nelder (1996) proposed the use of the hierarchical likelihood

$$H(\boldsymbol{\theta}, \boldsymbol{v}; \boldsymbol{y}) = f_\theta(\boldsymbol{y}|\boldsymbol{v}) f_\theta(\boldsymbol{v}) = f_\theta(\boldsymbol{v}, \boldsymbol{y}),$$

where $f_\theta(\boldsymbol{v}, \boldsymbol{y})$ is the joint density of $\boldsymbol{v}$ and $\boldsymbol{y}$. Furthermore, it is related to the conditional distribution of $\boldsymbol{v}$ given $\boldsymbol{y}$ as

$$H(\boldsymbol{\theta}, \boldsymbol{v}; \boldsymbol{y}) = f_\theta(\boldsymbol{v}, \boldsymbol{y}) = f_\theta(\boldsymbol{y}) f_\theta(\boldsymbol{v}|\boldsymbol{y}).$$

The notation $f_\theta(\cdot)$ is used to denote the density function of random variables with parameters $\boldsymbol{\theta}$; the arguments within the brackets can be either conditional or unconditional. Thus, $f_\theta(\boldsymbol{v}, \boldsymbol{y})$ and $f_\theta(\boldsymbol{v}|\boldsymbol{y})$ have different functional forms even though the same $f_\theta(\cdot)$ is used to indicate density functions with parameters $\boldsymbol{\theta}$. Two types of notations $f(\cdot|\boldsymbol{\theta}) = f_\theta(\cdot)$ and $f(\cdot|\boldsymbol{\theta}, \cdot) = f_\theta(\cdot|\cdot)$ are used interchangeably, where $f(\cdot|\boldsymbol{\theta})$ and $f(\cdot|\boldsymbol{\theta}, \cdot)$ are often used in Bayesian literature and $f_\theta(\cdot)$ and $f_\theta(\cdot|\cdot)$ in frequentist literature.

Bjørnstad (1996) introduced the extended likelihood principle where all information in the observed data for parameters $\boldsymbol{\theta}$ and unobservables $\boldsymbol{v}$ are in the extended likelihood, such as the hierarchical likelihood. Lee and Nelder (1996) found that the scale of $\boldsymbol{v}$ is important for meaningful statistical inference; they called the extended likelihood in a particular scale the *hierarchical likelihood.* The logarithm of the hierarchical likelihood is referred to as the *h-likelihood.*

For the h-likelihood, there is a close connection both to classical frequentist inference and Bayesian inference. In the absence of random effects the hierarchical likelihood is the same as the classical likelihood, i.e., $H = f_\theta(\boldsymbol{y})$. For example, in GLMs, the h-likelihood is the Fisher likelihood because there is no random effect. Furthermore, in the absence of fixed parameters $\boldsymbol{\theta}$

$$H(\boldsymbol{v}; \boldsymbol{y}) = f(\boldsymbol{y}) f(\boldsymbol{v}|\boldsymbol{y}).$$

When $\boldsymbol{v}$ is a fixed unknown and $f(\boldsymbol{v})$ is a prior then the hierarchical likelihood is proportional to the posterior distribution $f(\boldsymbol{v}|\boldsymbol{y})$ used for inference in Bayesian statistics. In hierarchical models such as linear mixed models, $\boldsymbol{v}$ is random and $f(\boldsymbol{v})$ is part of the model. Thus, to make this distinction clear we call $f(\boldsymbol{v}|\boldsymbol{y})$ the predictive density (or probability) for random effect $\boldsymbol{v}$, not posterior of $\boldsymbol{v}$. For prediction of future $\boldsymbol{v}$, one may want to compute the whole predictive density $f(\boldsymbol{v}|\boldsymbol{y})$, as we shall see, by using the bootstrap method.

Following the work of Henderson (1975) on linear mixed models, Lee and Nelder (1996) proposed that the random effects could be estimated by finding the mode of the joint density $f_\theta(\boldsymbol{y}, \boldsymbol{v})$. Robinson (1991) refer to this principle as the "method of most likely unobservable." Using the mode of $H$ to estimate $\boldsymbol{v}$ and the curvature (second derivatives) of $H$

around $\hat{\boldsymbol{v}}$ for inference simplifies the computations drastically compared to MCMC. However, it requires an appropriate scale of $\boldsymbol{v}$ because the joint density will depend upon the transformation of $\boldsymbol{v}$ and an inappropriate scale may give misleading conclusions. For example, the mode of the joint likelihood is not invariant to transformation of $\boldsymbol{v}$ and different conclusions will be drawn depending on the scale of $\boldsymbol{v}$ chosen when the mode is used for inference about the random effects. The novelty of Lee and Nelder's (1996) method was to limit the possible joint likelihoods to a given scale of $\boldsymbol{v}$, resolving the invariance problem.

An interesting aspect of using the mode of $H$ for estimating $\boldsymbol{v}$ is that it opens up the possibility to develop variable selection models using the hierarchical likelihood (Lee and Oh, 2014). It gives a statistical interpretation in terms of estimated random effects for penalized likelihood methods applied in machine learning (Hastie et al., 2009). But this will be covered in a later chapter (Chapter 10).

*Using the h-likelihood for inference on random effects, fixed effects and dispersion parameters*

The marginal likelihood, for inference of fixed effects $\theta$, is derived from $f_\theta(\boldsymbol{y}, \boldsymbol{v})$ by integrating out the random effects

$$f_\theta(\mathrm{y}) = \int f_\theta(\boldsymbol{y}, \boldsymbol{v}) d\boldsymbol{v}.$$

This is a classical Fisher likelihood and for the linear mixed model example the integral has an explicit solution given in Box 2.

For estimating variance components Patterson and Thompson (1971) suggested that a REML approach should be used to improve the estimation properties with reduced bias. REML for linear models can be extended to GLMs through a more general specification as a conditional likelihood $f_\theta(\boldsymbol{y}|\hat{\boldsymbol{\beta}})$ where $\hat{\boldsymbol{\beta}}$ is the estimator for the mean parameters (Smyth and Verbyla, 1996; Lee and Nelder, 2001a). For instance, by maximizing the REML likelihood for a simple linear model, the standard sample variance estimator is derived (Box 3).

In summary, the following basic ideas are used for the h-likelihood method.

- The hierarchical likelihood $f_\theta(\boldsymbol{y}, \boldsymbol{v})$ is used for inference on random effects. By maximizing $f_\theta(\boldsymbol{y}, \boldsymbol{v})$ we use the mode of $f_\theta(\boldsymbol{v}|\boldsymbol{y})$ as an estimate of $\boldsymbol{v}$. For the prediction of future $\boldsymbol{v}$ we use the predictive probability as we shall discuss.
- The classical (marginal) likelihood $f_\theta(\boldsymbol{y})$ is used for inference on fixed

effects. By maximizing $f_\theta(\boldsymbol{y})$ we obtain the ML estimator for $\boldsymbol{\theta}$. If the computation of the marginal likelihood is intractable one can use an approximation, such as the Laplace approximation.

- The REML likelihood $f_\theta(\boldsymbol{y}|\hat{\boldsymbol{\beta}})$ can be used for inference on variance components and dispersion parameters to reduce potential bias. By maximizing $f_\theta(\boldsymbol{y}|\hat{\boldsymbol{\beta}})$ we obtain the REML estimator for the dispersion parameters.

**Box 2: Marginal, REML and h-likelihoods for a linear mixed model** For a linear mixed model

$$\boldsymbol{y} = \boldsymbol{X\beta} + \boldsymbol{Zv} + \boldsymbol{e}$$
$$\boldsymbol{v} \sim \mathrm{N}(0, \lambda \boldsymbol{I})$$
$$\boldsymbol{e} \sim \mathrm{N}(0, \phi \boldsymbol{I})$$

the marginal, REML and h-likelihoods are

- $\log(f_\theta(\boldsymbol{y})) = \log \int H(\boldsymbol{\theta}, \boldsymbol{v}; \boldsymbol{y}) d\boldsymbol{v} =$

  $$-\tfrac{1}{2}\log(|2\pi \boldsymbol{V}|) - \tfrac{1}{2}(\boldsymbol{y} - \boldsymbol{X\beta})^T \boldsymbol{V}^{-1}(\boldsymbol{y} - \boldsymbol{X\beta})$$

  where $\boldsymbol{V} = \boldsymbol{ZZ}^T\lambda + \boldsymbol{I}\phi$. This gives the ML estimator $\hat{\boldsymbol{\beta}} = (\boldsymbol{X}^T\boldsymbol{V}^{-1}\boldsymbol{X})^{-1}\boldsymbol{X}^T\boldsymbol{V}^{-1}\boldsymbol{y}$.
- $\log(f_\theta(\boldsymbol{y}|\hat{\boldsymbol{\beta}})) =$

  $$-\tfrac{1}{2}\log(|2\pi \boldsymbol{V}|) - \tfrac{1}{2}(\boldsymbol{y} - \boldsymbol{X}\hat{\boldsymbol{\beta}})^T V^{-1}(\boldsymbol{y} - \boldsymbol{X}\hat{\boldsymbol{\beta}}) - \tfrac{1}{2}\log(|\boldsymbol{X}^T\boldsymbol{V}^{-1}\boldsymbol{X}|)$$

  which gives the REML estimator equations for variance components.
- $\log(f_\theta(\boldsymbol{y}, \boldsymbol{v})) = \log(f_\theta(\boldsymbol{y}|\boldsymbol{v})) + \log(f_\theta(\boldsymbol{v})) =$

  $$-\tfrac{n}{2}\log(2\pi\phi) - \tfrac{1}{2\phi}(\boldsymbol{y} - \boldsymbol{X\beta} - \boldsymbol{Zv})^T(\boldsymbol{y} - \boldsymbol{X\beta} - \boldsymbol{Zv}) - \tfrac{m}{2}\log(2\pi\lambda) - \tfrac{\boldsymbol{v}^T\boldsymbol{v}}{2\lambda}$$

  where $n$ is the number of observations and $m$ is the length of $v$. The joint maximization for $\boldsymbol{\beta}$ and $\boldsymbol{v}$ gives Henderson's mixed model equations

  $$\begin{pmatrix} \frac{1}{\phi}\boldsymbol{X}^T\boldsymbol{X} & \frac{1}{\phi}\boldsymbol{X}^T\boldsymbol{Z} \\ \frac{1}{\phi}\boldsymbol{Z}^T\boldsymbol{X} & \frac{1}{\phi}\boldsymbol{Z}^T\boldsymbol{Z} + \mathbf{I}\frac{1}{\lambda} \end{pmatrix} \begin{pmatrix} \hat{\beta} \\ \hat{v} \end{pmatrix} = \begin{pmatrix} \frac{1}{\phi}\boldsymbol{X}^T\boldsymbol{y} \\ \frac{1}{\phi}\boldsymbol{Z}^T\boldsymbol{y} \end{pmatrix},$$

  which give the BLUP for $\boldsymbol{v}$ and the ML estimator for $\boldsymbol{\beta}$.

**Box 3: Deriving sample variance from the REML likelihood $f(\boldsymbol{y}|\hat{\mu})$**

Suppose there are $n$ i.i.d. observation $y_1, y_2, ..., y_n$ from $\mathrm{N}(\mu, \sigma^2)$ where both parameters are unknown. The maximum likelihood estimator for $\mu$ is the sample mean $\hat{\mu} = \frac{1}{n}\sum_{i=1}^{n} y_i$, which is $\mathrm{N}(\mu, \sigma^2/n)$ distributed, whereas direct maximization of log$L$ gives the biased estimator $\hat{\sigma}^2 = \frac{1}{n}\sum(y_i - \bar{y})^2$. So, we may consider the REML likelihood given by

$$f(\boldsymbol{y}|\hat{\mu}) = \frac{f(\boldsymbol{y})}{f(\hat{\mu})} = \frac{\left(\frac{1}{\sqrt{2\pi\sigma^2}}\right)^n \exp(-\frac{1}{2\sigma^2}\sum_{i=1}^{n}(y_i - \mu)^2)}{\frac{1}{\sqrt{2\pi(\sigma^2/n)}}\exp(-\frac{1}{2(\sigma^2/n)}\left(\frac{1}{n}\sum_{i=1}^{n} y_i - \mu\right)^2)}$$

$$= \frac{1}{\sqrt{n}}\left(\frac{1}{\sqrt{2\pi\sigma^2}}\right)^{n-1} \times$$

$$\exp\left(-\frac{1}{2\sigma^2}\left[\left(\sum_{i=1}^{n}(y_i - \mu)^2\right) - \frac{1}{n}\left(\sum_{i=1}^{n} y_i - n\mu\right)^2\right]\right)$$

$$= \frac{1}{\sqrt{n}}\left(\frac{1}{\sqrt{2\pi\sigma^2}}\right)^{n-1} \exp(-\frac{1}{2\sigma^2}\left[\sum_{i=1}^{n}(y_i - \bar{y})^2\right])$$

Hence, the REML log-likelihood is (ignoring constant terms)

$$\log L_{REML} = -\frac{n-1}{2}\log(\sigma^2) - \frac{1}{2\sigma^2}\left[\sum_{i=1}^{n}(y_i - \bar{y})^2\right]$$

By maximizing log$L_{REML}$ the sample variance estimator is obtained: $\hat{\sigma}^2 = \frac{1}{n-1}\sum(y_i - \bar{y})^2$. Hence, the REML estimator adjusts for the degrees of freedom and for large $n$ the two estimators will be similar. However, when the number of mean parameters (i.e., the number of parameters included in the mean part of the model) grows with sample size, the two estimators can be very different.

For a linear mixed model, these three likelihoods are presented in Box 2. It will be shown in subsequent sections that both the marginal likelihood and REML likelihood are derived from the h-likelihood, and that approximations of the marginal likelihood can be derived directly from the h-likelihood. Thus, the h-likelihood is the basic quantity where all the necessary quantities such as marginal and REML likelihood for fixed parameters are derived. Furthermore, it allows inference about unobservables as well as function of both fixed and random unknowns.

## 3.4 Extended likelihood principle

Birnbaum (1962) proved that the classical likelihood function contains all the information in the observed data about the fixed parameter. Bjørnstad (1996) extended this principle and showed that all information in the data $\mathbf{y}$ for parameters $\theta$ and unobservables $\boldsymbol{v}$ is in the extended likelihood. This means that inference about fixed parameters and unobservables, using the information only in the data, requires the extended likelihood function and nothing else. However, these likelihood principles do not show how the information in the data can be retrieved from the likelihood. In the absence of $\boldsymbol{v}$, the extended likelihood becomes the marginal likelihood. Fisher advocated the use of ML estimation and established the underlying theory. In the absence of $\theta$, we see that the extended likelihood gives Bayesian posterior and its use has been advocated by Bayesian statisticians. This gives an insight on how to make inferences in at least these two extreme cases, so that we may develop a procedure which gives identical inferences to that using the marginal likelihood for $\theta$ and that exploiting the property of the predictive probability (posterior) for $\mathbf{v}$ in these two extreme cases. This was first attempted by Lee and Nelder (1996, 2005).

In the context of HGLMs, Lee and Nelder (1996, 2005) advocated the use of the h-likelihood and presented how information in the data for unobservables and parameters can be retrieved from it under the extended likelihood framework for models, consisting of all three types of objects ($\theta$, $\boldsymbol{v}$, and $\mathbf{y}$).

Similarly as for classical likelihood inference, we have a model for the data generation process and a corresponding likelihood.

- **Stochastic Model**: (i) Generate an instance of the random quantities $\boldsymbol{v}$ from a probability function $f_\theta(\boldsymbol{v})$ and then with $\boldsymbol{v}$ fixed, and (ii) generate an instance of the data $\mathbf{y}$ from a probability function $f_\theta(\mathbf{y}|\boldsymbol{v})$. The combined stochastic model is given by the product of the two probability functions

$$f_\theta(\boldsymbol{v})f_\theta(\mathbf{y}|\boldsymbol{v}).$$

- **Statistical Inference**: Given the data $\mathbf{y}$, we can (i) make inferences about $\theta$ by using the marginal likelihood $L(\theta;\mathbf{y}) \equiv f_\theta(\mathbf{y})$, and (ii) given $\theta$, we can make inferences about $\boldsymbol{v}$ by using a conditional likelihood (or predictive probability) of the form

$$L(\theta, \boldsymbol{v}; \boldsymbol{v}|\mathbf{y}) \equiv f_\theta(\boldsymbol{v}|\mathbf{y}).$$

Note that the likelihood $f_\theta(\boldsymbol{y})$ and the predictive probability $f_\theta(\boldsymbol{v}|\boldsymbol{y})$ share the common parameter $\boldsymbol{\theta}$. The extended likelihood, based on the joint density $f_\theta(\boldsymbol{v},\mathbf{y})$, for unknowns $(\boldsymbol{v}, \theta)$ is given by

$$L(\theta, \boldsymbol{v}; \boldsymbol{v}, \mathbf{y}) = L(\theta; \mathbf{y})L(\theta, \boldsymbol{v}; \boldsymbol{v}|\mathbf{y}), \tag{3.2}$$

where

$$\begin{aligned} L(\theta, \boldsymbol{v}; \boldsymbol{v}, \mathbf{y}) &\equiv f_\theta(\boldsymbol{v},\mathbf{y}), \\ L(\theta; \mathbf{y}) &\equiv f_\theta(\mathbf{y}), \\ L(\theta, \boldsymbol{v}; \boldsymbol{v}|\mathbf{y}) &\equiv f_\theta(\boldsymbol{v}|\mathbf{y}). \end{aligned}$$

The connection between these two processes is given by

$$f_\theta(\mathbf{y})f_\theta(\boldsymbol{v}|\mathbf{y}) \equiv L(\theta, \boldsymbol{v}; \boldsymbol{v}, \mathbf{y}) \equiv f_\theta(\boldsymbol{v}, \mathbf{y}) = f_\theta(\boldsymbol{v})f_\theta(\mathbf{y}|\boldsymbol{v}).$$

On the left-hand side $\mathbf{y}$ is fixed while $(\boldsymbol{v}, \theta)$ vary, while on the right-hand side $\theta$ is fixed while $(\boldsymbol{v}, \mathbf{y})$ vary. In the extended likelihood framework the $\boldsymbol{v}$ appear in the stochastic model as random instances, but in statistical inference as unknowns.

From (3.2), we see that the extended likelihood is the product of two likelihoods, the Fisher likelihood $f_\theta(\mathbf{y})$ and the conditional likelihood $f_\theta(\boldsymbol{v}|\boldsymbol{y})$. In likelihood theory the product of two likelihoods is a way of gathering information from the two independent sources of data (Lee, Nelder, and Pawitan (2017); Chapter 1). It is straightforward to note the close connection between the Fisher likelihood and the h-likelihood, because it uses the Fisher likelihood for inferences about $\theta$. In this book, the conditional likelihood $f_\theta(\boldsymbol{v}|\boldsymbol{y})$ is called the predictive probability to highlight its probability property

$$\int f_\theta(\boldsymbol{v}|\boldsymbol{y})d\boldsymbol{v} = 1,$$

which in earlier literature is called the predictive likelihood (Bjørnstad, 1990). The predictive probability is useful to give confidence (or predictive) intervals for random effects (Lee and Kim, 2016) and takes the form of the posterior probability in a Bayesian framework. Hence, there is a close connection between Bayesian inference and h-likelihood inference for random effects. A main difference is that the h-likelihood approach makes statistical inference about the unobservables $\boldsymbol{v}$ without assuming priors on the fixed parameters $\theta$ in the model. The prior $\pi(\theta)$ is often not checked by data because it is not part of the model, while the distribution of $f_\theta(\boldsymbol{v})$ needs to be checked to verify the model because it is a part of the model: see Chapter 5 of Lee, Nelder, and Pawitan (2017). In summary, the h-likelihood allows simultaneous inference for both fixed and random unknowns.

### *3.4.1 Definition of the h-likelihood*

For continuous $\boldsymbol{v}$, Lee and Nelder (1996) proposed the use of the hierarchical (h-)likelihood, an extended likelihood limited to a pre-defined scale of $\boldsymbol{v}$, for inferences for both fixed and random unknowns. To get an understanding of what this predefined scale of $\boldsymbol{v}$ is, consider the following example (Molas, 2012). Suppose we can choose to model the random effects in a linear predictor as

$$\eta_1 = \boldsymbol{X\beta} + \boldsymbol{v}$$

or as

$$\eta_2 = \boldsymbol{X\beta} + \exp(\boldsymbol{v}).$$

We would then have two alternative extended likelihoods based on two different scales of random effects:

$$L_1(\theta, \boldsymbol{v}; \boldsymbol{y}, \boldsymbol{v}) = f_\theta(\boldsymbol{y}|\eta_1) f_\theta(\boldsymbol{v}),$$

and

$$L_2(\theta, \boldsymbol{v}; \boldsymbol{y}, \boldsymbol{v}) = f_\theta(\boldsymbol{y}|\eta_2) f_\theta(\boldsymbol{v}).$$

The modes of these two likelihoods differ and the question is which scale of random effects to use for statistical inferences. The h-likelihood is defined as the extended likelihood having $\boldsymbol{v}$ on a canonical scale. The strong canonical scale is defined such that the random effects $\boldsymbol{v}$ do not interfere with the estimation of the fixed effects (Lee, Nelder, and Pawitan (2017); Chapter 4). This means that the marginal likelihood gives the same mode estimators about fixed effects as the h-likelihood. Hence, there is no conflict between classical likelihood inference and h-likelihood inference if the scale is canonical. For example, in linear mixed models of Box 2, $\boldsymbol{v}$ is on a canonical scale to $\boldsymbol{\beta}$ which implies that joint maximization of $h$ with respect to $\boldsymbol{\beta}$ and $\boldsymbol{v}$ gives the ML estimator for $\boldsymbol{\beta}$. However, in general the canonical scale may not exist. In HGLMs, the h-likelihood is defined under a weak canonical scale where the random effects combine additively with the fixed effects in a linear predictor. From the linear predictor $\eta_1$ above, we see that $L_1(\theta, \boldsymbol{v}|\boldsymbol{y}, \boldsymbol{v})$ is the h-likelihood, which gives a consistent inference framework (Lee, Nelder, and Pawitan, 2017).

The likelihood $L_1(\theta, \boldsymbol{v}; \boldsymbol{y}, \boldsymbol{v}) = f_\theta(\boldsymbol{y}|\eta_1) f_\theta(\boldsymbol{v})$ is called a hierarchical likelihood, as the random effects enter linearly in the linear predictor, and the h-likelihood

$$h = \log(L_1)$$

is used to denote the log hierarchical likelihood. It should also be noted that for $L_2(\theta, \boldsymbol{v}; \boldsymbol{y}, \boldsymbol{v})$ a reasonable mode estimator for $\boldsymbol{v}$ is not guaranteed but this is also a rather awkward model not commonly used. Nevertheless, this example shows the importance of having the random effects on the correct scale for joint maximization of fixed and random effects. An important difference between transforming fixed vs. random effects is that a transformation of random effects requires the need to multiply the density function for the random effects with a Jacobian. Suppose $u = r(v)$ for some function $r()$ then

$$f_\theta(u) = f_\theta(r(v)) \times |\frac{dv}{du}|.$$

Furthermore, when $\boldsymbol{v}$ is discrete there is no Jacobian involved so that all extended likelihoods are the h-likelihood (Lee and Bjørnstad, 2013): see chapter 10 for further discussions.

## 3.5 Laplace approximations for the integrals

For the linear mixed model both the marginal likelihood and REML likelihood are straightforward to derive (Box 2), but for most other distributions the integral for the marginal likelihood has no analytical form and some approximation is required. Numerical integration is infeasible if the number of integrands is large and MCMC algorithms are often too slow. An alternative, used in the h-likelihood approach, is to use Laplace approximations (Box 4).

For the marginal likelihood the Laplace approximation around the fitted random effects is

$$\begin{aligned}\int f_\theta(\boldsymbol{y}, \boldsymbol{v}) d\boldsymbol{v} &= \int exp(\log(f_\theta(\boldsymbol{y}, \boldsymbol{v})))d\boldsymbol{v} \\ &\approx \left.\left\{\left|-\frac{1}{2\pi}\frac{\partial^2 \log(f_\theta(\boldsymbol{y}, \boldsymbol{v}))}{\partial \boldsymbol{v}^2}\right|^{-\frac{1}{2}} f_\theta(\boldsymbol{y}, \boldsymbol{v})\right\}\right|_{\boldsymbol{v}=\hat{\boldsymbol{v}}}\end{aligned}$$

where $\hat{\boldsymbol{v}}$ is obtained from the mode of $f_\theta(\boldsymbol{y}, \boldsymbol{v})$.

Furthermore, applying a Laplace approximation to eliminate random effects together with a quadratic approximation around $\hat{\boldsymbol{\beta}}$ on the REML likelihood $f_\theta(\boldsymbol{y}|\hat{\boldsymbol{\beta}})$ to eliminate fixed effects we get (Lee and Nelder, 2001a; Molas and Lesaffre, 2011)

$$f_\theta(\boldsymbol{y}|\hat{\boldsymbol{\beta}}) \approx ... \approx \left\{ \left| \frac{1}{2\pi} \boldsymbol{I}(\boldsymbol{\beta}, \boldsymbol{v}) \right|^{-\frac{1}{2}} f_\theta(\boldsymbol{y}, \boldsymbol{v}) \right\} \Bigg|_{\boldsymbol{\beta}=\hat{\boldsymbol{\beta}}, \boldsymbol{v}=\hat{\boldsymbol{v}}}$$

where $\boldsymbol{I}(\boldsymbol{\beta}, \boldsymbol{v}) = -\begin{pmatrix} \partial^2 h/\partial\beta^2 & \partial^2 h/\partial\beta\partial v \\ \partial^2 h/\partial v\partial\beta & \partial^2 h/\partial v^2 \end{pmatrix}$ with $h \equiv \log(f_\theta(\boldsymbol{y}, \boldsymbol{v}))$.

Lee, Nelder, and Pawitan (2017) present these approximations in terms of adjusted profile h-likelihoods. An important idea that Lee and Nelder (2006) introduced was that these adjusted profile h-likelihoods not only gives computationally fast estimators but can also be used for joint inference of fixed and random effects. Thus, the uncertainty in estimating the fixed effects will also be part of the inference on random effects.

**Box 4: Laplace approximation**
Here the general mathematical definition of a Laplace approximation is given.

For some integral $\int \exp[f(x)]dx$ the (first-order) Laplace approximation is

$$\int \exp[f(x)]dx \approx \left\{ \left| -\frac{1}{2\pi} \frac{\partial^2 f(x)}{\partial x^2} \right|^{-\frac{1}{2}} \exp[f(x)] \right\} \Bigg|_{x=x_0}$$

where $x_0$ is a global maximum of some function $f(x)$.

### 3.5.1 *Adjusted profile h-likelihood*

Now we want to show inference tools based on the h-likelihood for an extended class of models, where the marginal and restricted likelihoods are hard to compute. To this end, the Laplace approximation for the log-marginal likelihood is specified as an adjusted profile h-likelihood (APHL)

$$p_{\boldsymbol{v}}(h) = [h - \frac{1}{2}\log(|\mathbf{I}(\boldsymbol{v})|/2\pi)]|_{\boldsymbol{v}=\hat{\boldsymbol{v}}} \tag{3.3}$$

where $\mathbf{I}(\boldsymbol{v})$ is the information matrix for the random effects, and $\hat{\boldsymbol{v}}$ is the maximum h-likelihood estimator of the random effects using $h$ as objective function.

Furthermore, the approximation for the REML log-likelihood $\log f_\theta(\boldsymbol{y}|\hat{\boldsymbol{\beta}})$ can also be expressed as an APHL:

$$p_{\boldsymbol{\beta},\boldsymbol{v}}(h) = [h - \frac{1}{2}\log(|\mathbf{I}(\boldsymbol{\beta}, \boldsymbol{v})|/2\pi)]|_{\boldsymbol{\beta}=\hat{\boldsymbol{\beta}},\boldsymbol{v}=\hat{\boldsymbol{v}}} \tag{3.4}$$

where $\mathbf{I}(\boldsymbol{\beta}, \boldsymbol{v})$ is the information matrix for the fixed and random effects. The estimates of fixed effects and dispersion parameters are computed by maximizing these two likelihoods.

The advantage of presenting these approximations as APHLs is that we then have two likelihoods $h$ and $p_{\boldsymbol{v}}(h)$ that can be maximized to obtain estimates for the random and fixed effects by putting $\partial h/\partial \boldsymbol{v} = 0$ and $\partial p_{\boldsymbol{\beta}}(h)/\partial \boldsymbol{v} = 0$. These likelihoods may also be used for joint inference on random and fixed effects, which was one of the revolutionizing ideas of Lee and Nelder (2006).

So, the standard method in the h-likelihood approach is to use $h$, which is equivalent to using $f_\theta(\boldsymbol{v}|\boldsymbol{y})$, for inference on the random effects and $p_{\boldsymbol{v}}(h)$ for inference of the fixed effects $\boldsymbol{\beta}$. The REML estimator for the dispersion parameters is derived from $p_{\boldsymbol{\beta},\boldsymbol{v}}(h)$. The method is referred to as the HL(1,1). One can also allow ML procedures where $p_{\boldsymbol{v}}(h)$, instead of $p_{\boldsymbol{\beta},\boldsymbol{v}}(h)$, is used to estimate the dispersion parameters. This may be suitable if the number of observations is large and the REML adjustment therefore can be expected to be low. Using ML instead of REML can reduce the computational cost of the estimation algorithm.

For some models $p_{\boldsymbol{v}}(h)$ can be approximated by the profile likelihood of $h$ for estimation of fixed effects and originally Lee and Nelder (1996) advocated to use the modes for fixed and random effects from $h$. It is referred to as HL(0,1). This gives an exact procedure if the scale of $\boldsymbol{v}$ is canonical but in general a canonical scale may not exist. For instance, if applied to binary data HL(0,1) gives non-ignorable bias and higher-order approximations may be required (Lee, Nelder, and Pawitan, 2017). In terms of the dhglm package, which will be introduced in some more detail in the Chapter 6, the logic in notation here is that in the dhglmfit function an estimator HL(a,b) uses the a$^{th}$-order Laplace approximation to the marginal likelihood for fixed effects in the linear predictor for the mean (`mord` = `a`) and the b$^{th}$-order approximation to the restricted likelihood for the dispersion parameters (`dord` = `b`).

HL(1,2) is useful for bias reduction in the analysis of binary data but is computationally expensive and applicable only to a restrictive class of models.

The second-order approximation to the log restricted likelihood is

$$s_{\beta,v}(h) = p_{\beta,v}(h) - \{F(h)/24\},$$

with $F(h) =$

$\mathrm{tr}[-\{3(\partial^4 h/\partial v^4) + 5(\partial^3 h_p/\partial v^3)D(h,v)^{-1}(\partial^3 h/\partial v^3)\}D(h,v)^{-2}]|_{v=\hat{v}}$,

where $D(h,v) = -\frac{\partial^2 h}{\partial^2 v}$.

In the dhglm package outputs in Chapter 6,
"$-2$ log(likelihood)" means $-2p_{\boldsymbol{v}}(h)$ and
"$-2$ log(restricted likelihood)" means $-2p_{\boldsymbol{\beta},\boldsymbol{v}}(h)$.

However, if we specify HL(1,2),
"$-2$ log(restricted likelihood)" represents
$-2s_{\beta,v}(h) = -2\{p_{\beta,v}(h) - F(h)/24\}$.

In GLMs (without a random effect)
"$-2$ log(likelihood)" means $-2\ell = -2\log\{f_\theta(\boldsymbol{y})\}$ and
"$-2$ log(restricted likelihood)" means $-2\log\{f_\theta(\boldsymbol{y}|\hat{\boldsymbol{\beta}})\}$.

### 3.6 Street magician

Here an example is presented to illustrate the fundamental idea of likelihood inference and how it may differ from Bayesian inference.

You walk down the street and come across a street magician. He has a small bag with a number of dice. There are two types of dice in the bag; white ones and blue ones. The white are numbered 1 to 6, while the blue have three sides with 1's and three sides with 2's. The magician draws a die at random from the bag without showing it to you and rolls the die. He claims that the number that turns up is a 2.

a) Which type of die would you guess he has rolled, a white or a blue one?

b) The entertainer lets you bet on the color of the dice. Which odds would you accept?

c) Now the entertainer informed you that there are 20 white and 10 blue dices in the bag. What is your guess on the color of the die that he rolled.

d) Instead of rolling a die, a magician generates a random number $Y$ from N($\mu$,1). If he drew a white die, he generates a random number with $\mu = 0$, otherwise he generate a random number with $\mu = 1$. From the observed random number $Y = y_0$, can we make an inference about the color of the die?

e) Instead of drawing a die, a magician generated a random variable $v$ from $N(0,1)$. Suppose that he observed $v = v_0$. Then without telling the value of realized value $v_0$, he generates random variable Y from $N(v_0, 1)$.

Then, he informs us the value $y_0$. Can we make an inference about $v_0$ given the observed data $y_o$?

### *3.6.1 Solution*

a) The likelihood for a white die is $1/6$ and for a blue die is $1/2$. Therefore, as a likelihoodist, the maximum likelihood guess is that the dice is blue. Let $Y$ be the number of dice and let $C$ be a color of dice and $c$ be a realized value of the color of dice. Then, the likelihood ratio is

$$\frac{P(Y=2|C=blue)}{P(Y=2|C=white)} = \frac{1/2}{1/6} = 3.$$

b) To be able to make a probability statement we need to know the distribution of the two types of dice in the bag. This is unknown however, which means that for a likelihoodist the odds cannot be computed. A Bayesian would guess the distribution and thereby compute the odds. Now the Bayesian odds become

$$\frac{P(Y=2|C=blue)\pi(C=blue)}{P(Y=2|C=white)\pi(C=white)} = 3\frac{\pi(C=blue)}{\pi(C=white)}.$$

c) The problem can be solved using a probabilistic argument, but here we also show that both a classical likelihood ratio and the ratio of extended likelihoods can be used to draw the same conclusion. Let $c$ be a realized value of the color of dice such that

$$L(c=blue) = P(C=blue) = 1/3,$$

and

$$L(c=white) = P(C=white) = 2/3.$$

Then, the ratio of extended likelihood is

$$\begin{aligned}\frac{L(c=blue, Y=2)}{L(c=white, Y=2)} &= \frac{P(Y=2|c=blue)L(c=blue)}{P(Y=2|c=white)L(c=white)} \\ &= \frac{1/2\times 1/3}{1/6\times 2/3} = 3/2.\end{aligned}$$

Thus, the maximum extended likelihood guess is that the die is blue. Furthermore, we can compute the conditional likelihood

$$
\begin{aligned}
L(c = blue|Y = 2) &= \frac{L(c = blue, Y = 2)}{L(Y = 2)} \\
&= \frac{P(Y = 2|c = blue)L(c = blue)}{P(Y = 2|c = blue)L(c = blue) + P(Y = 2|c = white)L(c = white)} \\
&= \frac{P(Y = 2|c = blue)}{P(Y = 2|c = blue) + P(Y = 2|c = white)L(c = white)/L(c = blue)} \\
&= \frac{1/2 \times 1/3}{1/2 \times 1/3 + 1/6 \times 2/3} = \frac{3}{5}
\end{aligned}
$$

and

$$L(c = white|Y = 2) = \frac{2}{5}.$$

Note that the conditional likelihood $L(c = blue|Y = 2)$ depends upon the likelihood ratio $L(c = white)/L(c = blue)$, so that it is invariant with respect to the transformation of data and parametrization. Furthermore,

$$\frac{L(c = blue, Y = 2)}{L(c = white, Y = 2)} = \frac{L(c = blue|Y = 2)}{L(c = white|Y = 2)} = \frac{3}{2},$$

i.e., the mode of the conditional likelihood $L(c|Y = 2)$ is the same as the mode of the extended likelihood $L(c, Y = 2)$.

In (c) we have information on $P(C)$ (part of the model), while in (b) no information is available on $P(C)$, so that we need a guess $\pi(C)$.

d) This problem is covered as a hypothesis testing in Chapter 10.

e) This problem is the subject of Chapters 3–9.

## 3.7 H-likelihood and empirical Bayes

Inference on random effects have important practical use in predictions. A typical example, is for instance, if there are repeated observations on patients' hospital visits and the lifetime of these patients are to be predicted. This would require a survival analysis including random effects for patients and the uncertainty in the predictions will include the uncertainty of the fitted random effects.

When $\theta$ is known, we can make inferences about $\boldsymbol{v}$ using $f_\theta(\boldsymbol{v}|\mathbf{y})$. However, $\theta$ is unknown, so that we may make inferences using $f_{\hat\theta}(\boldsymbol{v}|\mathbf{y})$ with $\hat\theta$ being the ML estimator. This is the so-called EB approach, which gives consistent estimation for predictive probability because $\hat\theta$ is consistent. However, in finite samples this approach often has a poor inferential performance because it cannot account for uncertainty, caused by estimating $\theta$; especially when the number of observations is low and the number of parameters in $\theta$ is large.

**Box 5: Inference for random effects in a linear mixed model derived from the h-likelihood**

For a linear mixed model

$$\boldsymbol{y} = \boldsymbol{X\beta} + \boldsymbol{Zv} + \boldsymbol{e}$$
$$\boldsymbol{v} \sim \mathrm{N}(0, \lambda \boldsymbol{I})$$
$$\boldsymbol{e} \sim \mathrm{N}(0, \phi \boldsymbol{I})$$

the h-likelihoods are $\log(f_\theta(\boldsymbol{y}, \boldsymbol{v})) = \log(f_\theta(\boldsymbol{y}|\boldsymbol{v})) + \log(f_\theta(\boldsymbol{v})) =$

$$-\tfrac{n}{2}\log(2\pi\phi) - \tfrac{1}{2\phi}(\boldsymbol{y} - \boldsymbol{X\beta} - \boldsymbol{Zv})^T(\boldsymbol{y} - \boldsymbol{X\beta} - \boldsymbol{Zv}) -$$
$$\tfrac{m}{2}\log(2\pi\lambda) - \tfrac{\boldsymbol{v}^T\boldsymbol{v}}{2\lambda}$$

where $n$ is the number of observations and $m$ is the length of $v$. The HL(0,1) and HL(1,1) methods are equivalent for a linear mixed model, so the estimates of both $\boldsymbol{\beta}$ and $\boldsymbol{v}$ can be computed by maximizing the h-likelihood. The score equations $\frac{\partial h}{\partial \beta} = 0$ and $\frac{\partial h}{\partial \boldsymbol{v}} = 0$ give Henderson's mixed model equations:

$$\begin{pmatrix} \frac{1}{\phi}\boldsymbol{X}^T\boldsymbol{X} & \frac{1}{\phi}\boldsymbol{X}^T\boldsymbol{Z} \\ \frac{1}{\phi}\boldsymbol{Z}^T\boldsymbol{X} & \frac{1}{\phi}\boldsymbol{Z}^T\boldsymbol{Z} + \mathbf{I}\frac{1}{\lambda} \end{pmatrix} \begin{pmatrix} \boldsymbol{\beta} \\ \boldsymbol{v} \end{pmatrix} = \begin{pmatrix} \frac{1}{\phi}\boldsymbol{X}^T\boldsymbol{y} \\ \frac{1}{\phi}\boldsymbol{Z}^T\boldsymbol{y} \end{pmatrix}$$

and the information matrix (computed from the second derivatives) is

$$\begin{pmatrix} \frac{1}{\phi}\boldsymbol{X}^T\boldsymbol{X} & \frac{1}{\phi}\boldsymbol{X}^T\boldsymbol{Z} \\ \frac{1}{\phi}\boldsymbol{Z}^T\boldsymbol{X} & \frac{1}{\phi}\boldsymbol{Z}^T\boldsymbol{Z} + \mathbf{I}\frac{1}{\lambda} \end{pmatrix}.$$

This is different from an EB approach $f_{\hat\theta}(\boldsymbol{v}|\boldsymbol{y})$ where the information matrix

$$\left(\frac{1}{\phi}\boldsymbol{Z}^T\boldsymbol{Z} + \mathbf{I}\frac{1}{\lambda}\right)$$

would typically be used for inference on the random effects ignoring the uncertainty in the estimates of $\hat{\boldsymbol{\beta}}$. A more thorough exposition is found in Section 5.4 of Lee, Nelder, and Pawitan (2017) showing that the h-likelihood gives correct inference.

Such an uncertainty about $\hat{\boldsymbol{\theta}}$ is included in $f_\theta(\mathbf{y})$, and can be used for inference on random effects (Lee and Nelder, 1996, 2001a). Thus, an important question is how to eliminate the nuisance parameter $\boldsymbol{\theta}$ from the predictive probability $f_{\boldsymbol{\theta}}(\boldsymbol{v}|\boldsymbol{y})$, using the information on $\boldsymbol{\theta}$ in the likelihood $f_{\boldsymbol{\theta}}(\boldsymbol{y})$. Box 5 illustrates the difference using a linear mixed model as an example.

### *3.7.1 Prediction of future outcomes*

Lee and Nelder (2009) showed that their h-likelihood approach can be used for inference about general models including unobservable random variables, which include future outcomes, missing data, latent variables, factors in factor analysis, potential outcomes, etc. An unobservable random variable $v$ is often of interest in making statistical inferences. Suppose that we have the number of epileptic seizures in an individual for five weeks, $\mathbf{y} = (3, 2, 5, 0, 4)$. Suppose also that these counts are i.i.d. from a Poisson distribution with mean $\theta$. Here, $\hat{\theta} = (3 + 2 + 5 + 0 + 4)/5 = 2.8$ is the ML estimator of $\theta$, which maximizes the Fisher likelihood $f_\theta(\mathbf{y})$.

The inferences about $\theta$ can be made by using the likelihood. Now we want to have a predictive probability function for the seizure counts for the next week $v$. Then, because $f_\theta(v = i|y) = f_\theta(v = i)$, the plug-in technique gives the predictive distribution for the seizure count $v$ of the next week:

$$f_{\hat{\theta}}(v = i|y) = f_{\hat{\theta}}(v = i) = \exp(-2.8)2.8^i/i!.$$

Pearson (1920) pointed out the limitation of this Fisher likelihood using the plug-in method because it cannot account for uncertainty in estimating $\theta$. This plug-in technique is a kind of EB method. With Jeffreys' prior, $\pi(\theta) \propto \theta^{-1/2}$, the resulting marginal posterior $\pi(v|y)$ gives a predictive probability with higher probabilities for larger $y$. Pearson (1920) pointed out that this Bayesian procedure handles uncertainty caused by estimating $\theta$. However, it depends upon the choice of a prior and it might be difficult to justify why the choice of Jeffreys' prior is the right choice. Here the h-likelihood including $v$ is proportional to

$$f_\theta(3, 2, 5, 0, 4, v) = \exp(-6\theta)\theta^{3+2+5+0+4+v}/(3!2!5!0!4!v!).$$

Now, $\hat{\theta}(v) = (3 + 2 + 5 + 0 + 4 + v)/6$ is the potential ML estimate if $v$ is observed. Then, the normalized profile likelihood $f_{\hat{\theta}(v)}(3, 2, 5, 0, 4, v)$ gives the predictive probability $p(\boldsymbol{v}|\boldsymbol{y})$, almost identical to Pearson's but without assuming a prior on $\theta$ (Figure 3.3). This is a method to eliminate $\boldsymbol{\theta}$ from the predictive probability $f_{\boldsymbol{\theta}}(\boldsymbol{v}|\boldsymbol{y})$. This example shows that standard methods for likelihood inferences, such as a profiling method, can be used for the prediction problem by using the h-likelihood without assuming a prior on $\theta$.

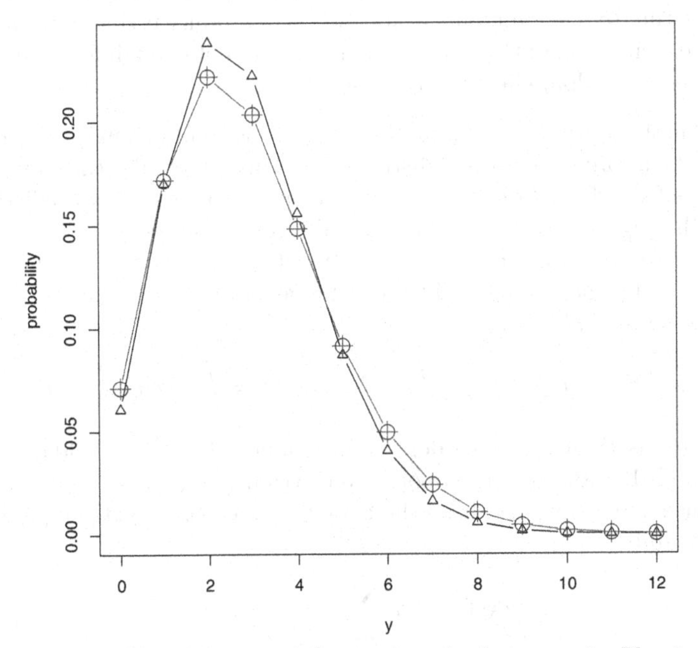


Figure 3.3 *Predictive density of the number of seizure counts:.Plug-in method (*$\triangle$*), Bayesian method (*$\circ$*) and h-likelihood method (+).*

Because the Fisher likelihood $f_\theta(y)$ does not involve $v$, the other component, the predictive probability, $f_\theta(v|y)$ carries all the information in the data about the unobservables. Thus, the prediction of random effects can be made via the EB method using the estimated predictive probability (or posterior)

$$p(v|y) = f_{\hat\theta}(v|y) = \pi(v|y, \hat\theta),$$

where $\hat\theta$ is the usual ML estimator (Carlin and Louis, 2000). However, using $f_{\hat\theta}(v|y)$ to make inferences about $v$ is naive and Bjørnstad (1990) has shown how badly it performs in measuring the true uncertainty in estimating $v$. Note that maximization of the h-likelihood

$$h = \log f_\theta(y|v) + \log f_\theta(v) = \log f_\theta(v|y) + \log f_\theta(y)$$

yields EB-mode estimators for $v$, without computing

$$f_\theta(v|y) = f_\theta(y, v)/f_\theta(y).$$

However, the Hessian matrix (i.e., matrix of second derivatives) based upon $f_\theta(v|y)$ gives a naive variance estimate for the prediction $\hat v$ because it does not properly account for the uncertainty caused by estimating $\theta$, that is in $f_\theta(y)$. The h-likelihood considers both components and gives proper estimators for random effects and their variance estimators. However, the estimation of the first two moments are not enough for accurate inferences of random effects if it is not normal.

The example above shows that $\hat{v}$ is neither a consistent estimator of $v$ nor follows the asymptotic normal distribution. Thus, interval estimations of random effects differ from those of fixed effects. Note that the predictive probability $f_{\hat{\theta}}(v|y)$ gives an asymptotically correct inference. Thus, it is necessary to have a finite sample adjustment to account for information loss caused by estimating $\theta$. This can be generally done. Lee and Kim (2016) showed that

$$p(v|y) = E_{\hat{\theta}}(f_{\hat{\theta}}(v|y)) \equiv \int f_t(v|y) f(\widehat{\theta} = t)dt = \int f_t(v|y) c(\theta = t)dt,$$

where $c(\theta)$ is the confidence density in Chapter 1 of Lee, Nelder, and Pawitan (2017). Because the bootstrap distribution gives an estimate of confidence density, we can have the bootstrap method to get the predictive probability

$$p(v|y) \equiv \frac{1}{B}\sum_{j=1}^{B} f_{\theta_j^*}(v|y),$$

where $\theta_1^*, \ldots, \theta_B^*$ are the bootstrap replicates of $\widehat{\theta}$. In complex models it may not be easy to design the bootstrap scheme, so that it is convenient to generate the bootstrap replicates of $\widehat{\theta}$ from the asymptotic normal distribution of $\widehat{\theta}$ or the normalized likelihood. Via simulation studies, Lee and Kim (2016) demonstrate that bootstrap methods provide excellent prediction intervals for random effects, including the prediction of future outcomes above. Cao et al. (2017) showed that the bootstrap method gives an excellent prediction interval for subject-specific function estimations.

## 3.8 Exercises

1. Fit the Poisson model (in Section 3.5) for the epileptic seizure data with and without random patient effects (i.e., with the hglm and glm functions). Compare the standard errors of estimated fixed effects for the two models. Why do the standard errors increase when random patient effects are added?

2. Run the model in Section 3.5 including a random patient effect and

compute a 95% confidence interval for the variance of the random patient effect.

3. In this exercise, we compare the lme4 and hglm packages by fitting a linear mixed model to the epileptic seizure data. Use `lmer` function in lme4 to fit the linear mixed model using the same linear predictor as in previous exercises. Check that the estimates are the same as in hglm. Is there a relationship between the REML likelihood value from `lmer` and the value of $p_{\beta,v}(h)$ computed by the hglm function?

4. Show that maximizing $h$ or $p_v(h)$ gives the same estimates for the fixed effects in a linear mixed model.

CHAPTER 4

# HGLMs: from method to algorithm

HGLMs (Box 6) extend GLMs by allowing random effects in the linear predictor. HGLMs allow regression models for the residual variance and the variance for random effects. In Lee, Nelder, and Pawitan (2017), both the h-likelihood method (and related theory) and IWLS algorithm were described. It is important to keep apart what is the model, the method and the estimation algorithm. For example, GLM describes a family of models where the response comes from the exponential family of distributions. The method used to fit and make inference for these models is ML. In the GLM family of models, an IWLS algorithm can be used to compute the ML estimates and their standard errors. Hence, the computational machinery developed for least squares estimation for linear models can be used to fit GLMs, but the statistical method used is ML.

In this chapter we show how HGLMs can be fitted using interconnected and augmented GLMs. The theory is presented at the end of the chapter but first we present a few examples illustrating how the dispersion parameters can be modeled in HGLMs.

## 4.1 Examples

In quality control experiments there is a need to model both the mean and the dispersion (Nelder and Lee, 1998). Suppose for instance that a factory produces 12 mm screws and a set of treatments are performed to improve the production. It is important that the screws are 12 mm on average with as small variation as possible, so treatment effects for both the mean and the variance should be estimated. This requires that we have a model for the mean and a model for the variance.

*4.1.1 Epilepsy data continued*

Consider the epileptic seizure count data in Section 1.1.1 again. Fitting the Poisson model with log-link in (1.1)

$$\log(\mu_{ij}) = \beta_0 + x_{B_i}\beta_B + x_{T_i}\beta_T + x_{A_i}\beta_A + x_{V_j}\beta_V + x_{B_iT_i}\beta_{BT},$$

gives a deviance of 869.9 with degrees of freedom 230, clearly indicating over-dispersion.

**Box 6: HGLM class of models**

An HGLM has two random components: a response $\boldsymbol{y}$ and an unobserved random effect $\boldsymbol{v}$, such that $\boldsymbol{y}|\boldsymbol{v}$ follows a distribution from the exponential family, namely normal, binomial, Poisson, or gamma. The expectation of $\boldsymbol{y}|\boldsymbol{u}$ is modeled as

$$E(\boldsymbol{y}|\boldsymbol{u}) = \boldsymbol{\mu}$$

$$g(\boldsymbol{\mu}) = \boldsymbol{X}\beta + \boldsymbol{Z}\boldsymbol{v}$$

$$\boldsymbol{v} = r(\boldsymbol{u})$$

where $g(\cdot)$ is a link function, $\boldsymbol{X}$ and $\boldsymbol{Z}$ are design matrices and $\boldsymbol{\beta}$ is a fixed effect. The distribution of the random effect $\boldsymbol{u}$ is one of the conjugate distributions of exponential family: normal, beta, gamma or inverse-gamma. The random effect $\boldsymbol{v}$ is given on an appropriate (weak canonical) scale through the link function $r(\cdot)$ transforming $\boldsymbol{u}$ to guarantee correct model estimator. Hence, an HGLM includes two dispersion parameters, one for the distribution of $\boldsymbol{y}|\boldsymbol{v}$ and one for the distribution of $\boldsymbol{v}$. In multi-component models,

$$\boldsymbol{Z}\boldsymbol{v} = Z_1 v_1 + ... + Z_k v_k,$$

$v_1, ..., v_k$ can have different distributions among the four conjugate distributions.

For a linear mixed model, both $g(\cdot)$ and $r(\cdot)$ are identity links and both $\boldsymbol{y}|\boldsymbol{v}$ and $\boldsymbol{v}$ are normally distributed. GLMM assumes $\boldsymbol{v} = \boldsymbol{u}$ normally distributed with identity link for $r(\cdot)$.

The Pearson chi-square statistic has 1,036.3, it also largely exceeds the degrees of freedom. To accommodate this, we may fit the over-dispersed Poisson model with $\mathrm{var}(y) = \phi\mu$ by using dhglm package.

For the parameter estimation of $\phi$, we may use the deviance or Pearson chi-squared statistic, as follows: From the deviance we have $\hat{\phi} = 3.8 = \exp(1.33) = 869.9/230$, while from the Pearson chi-squared statistic $\hat{\phi} = 4.5 = \exp(1.505) = 1036.3/230$. We can adjust the standard errors of parameter estimates $\hat{\beta}$ by multiplying $\sqrt{\hat{\phi}}$.

```
# based on deviance statistic
Distribution of Main Response :
                        "poisson"
[1] "Estimates from the model(mu)"
y ~ B + T + A + B:T + V
[1] "log"
             Estimate Std. Error t-value
(Intercept) -2.79763    0.79209  -3.532
B            0.94952    0.08472  11.208
T           -1.34112    0.30482  -4.400
A            0.89705    0.22645   3.961
V           -0.02936    0.01972  -1.489
B:T          0.56223    0.12349   4.553
[1] "Estimates from the model(phi)"
phi ~ 1
[1] "log"
             Estimate Std. Error t-value
(Intercept)     1.33    0.09325   14.27
[1] "== Likelihood Function Values and Condition AIC =="
                                     [,1]
-2 log(likelihood)               :  1309.943
-2 log(restricted likelihood):  1330.809
cAIC                             :  1321.943

Scaled Deviance : 230.000 on 230.000 degrees of freedom
# based on Pearson chi-squared statistic
Distribution of Main Response :
                        "poisson"
[1] "Estimates from the model(mu)"
y ~ B + T + A + B:T + V
[1] "log"
             Estimate Std. Error t-value
(Intercept) -2.79763    0.86452  -3.236
B            0.94952    0.09247  10.269
T           -1.34112    0.33270  -4.031
A            0.89705    0.24716   3.629
V           -0.02936    0.02153  -1.364
```

```
B:T              0.56223     0.13478    4.172
[1] "Estimates from the model(phi)"
phi ~ 1
[1] "log"
             Estimate Std. Error t-value
(Intercept)     1.505    0.09325   16.14
[1] "== Likelihood Function Values and Condition AIC =="
                                   [,1]
-2 log(likelihood)              :  1314.318
-2 log(restricted likelihood) :  1334.134
cAIC                            :  1326.318

Scaled Deviance : 230.000 on 230.000 degrees of freedom
```

Via extensive numerical studies Nelder and Lee (1992) studied the performance of deviance and Pearson chi-squared statistics in estimating $\phi$. Because the deviance residuals are the best normalizing transformation under the exponential family it gives an estimator with small variance, but it gives an inconsistent estimate. The Pearson chi-squared statistics gives a consistent estimation because it is based upon the method of moments. Hilbe (2014) recommended to use the Pearson chi-statistics because it gives a consistent estimator, while in finite samples the deviance often gives more efficient estimators (Nelder and Lee, 1992). Thus, it is recommended to use the deviance in small samples, but in large samples the Pearson chi-squared statistics to have a smaller mean-squared error. In this example it is recommended to use the Pearson chi-squared statistic for estimating the dispersion parameter because the degrees of freedom is large (230). However, for the data from quality-improvement experiments in the next section the sample size is often not large, so it is recommended to use the deviance in estimating dispersion parameters.

In epilepsy data, the over-dispersion is caused by repeated measuring of the same subject. However, the over-dispersed Poisson models can account for the over-dispersion, but not correlation among repeated measures. Thus, for the analysis of repeated measures, we should use HGLMs in Section 4.2, which account for not only over-dispersion but also correlation among repeated measures.

### *4.1.2 Heteroscedastic log-linear model for injection-molding data*

Engel (1992) presented data from an injection-molding experiment in Table 4.1. An industrial Taguchi experiment was performed to study the influence of several controllable factors on the mean value and the variation in the percentage of shrinkage of products made by injection molding. The responses $y$ were percentages of shrinkage of products made by injection molding. There are seven controllable factors (A-G), in a $2^{7-4}$ fractional factorial design. Table 4.2 gives 8 combinations of levels for seven controllable factors (A-G) to allow the estimates of seven main effects for controllable factors. At each setting of the controllable factors, 4 observations were obtained from a $2^{3-1}$ fractional factorial with three noise factors (M-O).

Table 4.1 *Factors in the experiment*

| Controllable factors | Noise factors |
|---|---|
| A: cycle time | M: percentage regrind |
| B: mould temperature | N: moisture content |
| C: cavity thickness | O: ambient temperature |
| D: holding pressure | |
| E: injection speed | |
| F: holding time | |
| G: gate size | |

This dataset has been attended by many researchers because the model checking plots were not satisfactory. Lee and Nelder (1997) gave extensive discussion on how to choose a good model and presented the mean model with log-normal distribution and the identity link $\eta = \mu$, where

$$\eta = \beta_0 + \beta_A A + \beta_C C + \beta_D D + \beta_E E + \beta_G G + \beta_N N + \beta_{C \cdot N} C \cdot N + \beta_{E \cdot N} E \cdot N$$

and dispersion model,

$$\log \phi = \gamma_0 + \gamma_A A + \gamma_F F.$$

The effect of the dispersion factor F is quite large so that the observations with the lower level of F have weights $\exp(2\hat{\gamma}_F) \approx 104$ times as large as those with the higher level; it is almost as if the number of observations is reduced by half, i.e., restricted to those at the lower level of F. In consequence A, B and C are almost aliased to G, D, and E respectively in the mean model because the number of observation is reduced by half due to smallest weight for the higher level. Thus, parameter estimates of factors appearing together with near-aliased factors are unstable, with larger standard errors.

Even though much improved, the residual plots still look far from satisfactory (Figure 4.1 for the mean model and Figure 4.2 for the dispersion model) caused by the large differences in weights.

Table 4.2 *Experimental data from the injection molding experiment*

| Controllable factors | | | | | | | Percentage shrinkage for the following noise factors (M,N,O): | | | |
|---|---|---|---|---|---|---|---|---|---|---|
| $A$ | $B$ | $C$ | $D$ | $E$ | $F$ | $G$ | $(-1,-1,-1)$ | $(-1,1,1)$ | $(1,-1,1)$ | $(1,1,-1)$ |
| −1 | −1 | −1 | −1 | −1 | −1 | −1 | 2.2 | 2.1 | 2.3 | 2.3 |
| −1 | −1 | −1 | 1 | 1 | 1 | 1 | 0.3 | 2.5 | 2.7 | 0.3 |
| −1 | 1 | 1 | −1 | −1 | 1 | 1 | 0.5 | 3.1 | 0.4 | 2.8 |
| −1 | 1 | 1 | 1 | 1 | −1 | −1 | 2.0 | 1.9 | 1.8 | 2.0 |
| 1 | −1 | 1 | −1 | 1 | −1 | 1 | 3.0 | 3.1 | 3.0 | 3.0 |
| 1 | −1 | 1 | 1 | −1 | 1 | −1 | 2.1 | 4.2 | 1.0 | 3.1 |
| 1 | 1 | −1 | −1 | 1 | 1 | −1 | 4.0 | 1.9 | 4.6 | 2.2 |
| 1 | 1 | −1 | 1 | −1 | −1 | 1 | 2.0 | 1.9 | 1.9 | 1.8 |

From the fitted values for $\mu$ and $\phi$,

$$\begin{aligned}\mu &= 0.800 + 0.149A + 0.069C - 0.152D - 0.074G + 0.012E \\ &- 0.006N + 0.189C \cdot N - 0.173E \cdot N\end{aligned}$$

and

$$\log \phi = -2.849 - 0.608A + 2.324F,$$

we first set A to the high level and F to the low level to minimize variation, giving

$$\hat{\phi} = \exp(-2.849 - 0.608 - 2.324) = \exp(-5.781) = 0.003$$

with 95% confidence interval

$$\exp(-5.781 \pm 1.96 \times 0.693) = (0.0008, 0.012).$$

Then, we use other factors in the mean model for $\mu$, but not those in the dispersion model for $\phi$ to adjust the mean to the target.

```
Distribution of Main Response :
                          "gaussian"
[1] "Estimates from the model(mu)"
y ~ a + c + d + e + g + n + c:n + e:n
[1] "log"
              Estimate Std. Error  t-value
(Intercept)  0.799731   0.009739  82.1156
a            0.148589   0.035390   4.1987
c            0.068703   0.030886   2.2244
d           -0.152098   0.009727 -15.6373
e            0.011748   0.031591   0.3719
g           -0.074357   0.034873  -2.1323
n           -0.006003   0.007506  -0.7997
c:n          0.188582   0.034184   5.5166
e:n         -0.173397   0.034201  -5.0699
[1] "Estimates from the model(phi)"
phi ~ a + f
[1] "log"
             Estimate Std. Error t-value
(Intercept)    -2.849     0.1609 -17.708
a              -0.608     0.1597  -3.807
f               2.324     0.1594  14.582
[1] "== Likelihood Function Values and Condition AIC =="
                                  [,1]
-2 log(likelihood)             :  -9.368466
-2 log(restricted likelihood) :  52.746052
cAIC                           :   8.631534
```

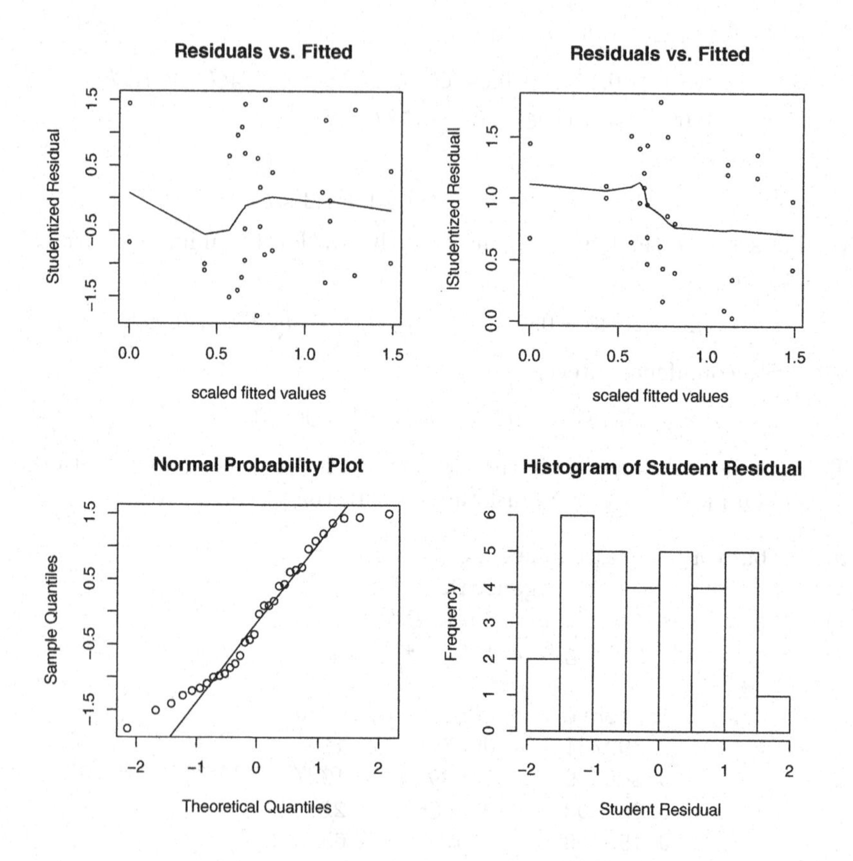


Figure 4.1 *Residual plot for the mean model from the injection molding data under a heteroscedastic log-linear model.*

```
Scaled deviance: 23.0000 on 23.0000 degrees of freedom
```

### *4.1.3 Gamma GLM with structured dispersion for crack growth data*

Hudak et al. (1978) presented crack-growth data (also listed in Lu and Meeker (1993)) from an experiment where crack lengths in inches are measured on a compact tension steel test operated in different laboratories. There are 21 metallic specimens, each subjected to 120,000 loading cycles, with the crack lengths recorded every $10^4$ cycles. We take $t =$no. cycles/$10^6$ here, so $t_j = j/100$ for $j = 1, ..., 12$. The crack increment sequences look rather irregular. Let $l_{ij}$ be the crack length of the $i$th

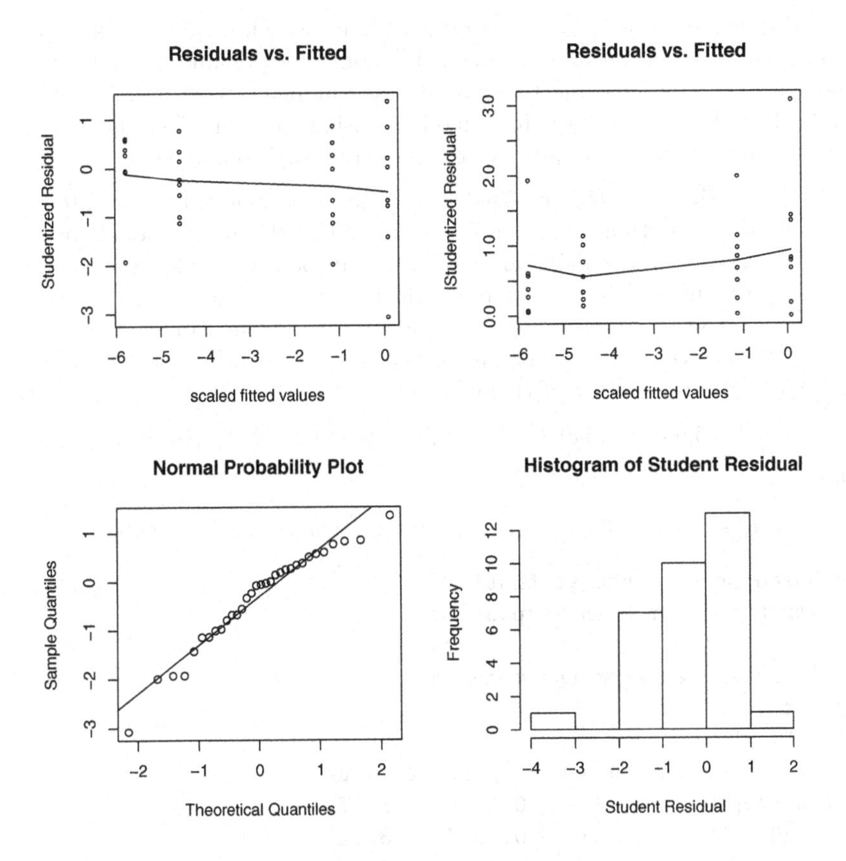


Figure 4.2 *Residual plot for the dispersion model from the injection molding data under a heteroscedastic log-linear model.*

specimen at the $j$th observation and let $y_{ij} = l_{ij} - l_{ij-1}$ be the corresponding increment of crack length, which has always a positive value. Models that describe the process of deterioration or degradation of units or systems are of interest and are also a key ingredient in processes that model failure events.

Consider a gamma GLM with response $y_{ij}$ and linear predictor for the mean part of the model

$$\log \mu_{ij} = \beta_0 + \beta_l l_{ij-1}.$$

The dispersion is modeled using a log link

$$\log \phi_{ij} = \gamma_0 + \gamma_1 t_j.$$

The estimated cycle effect ($\hat{\gamma}_1 = -10.5$) is negative, which means that the dispersion tends to decrease after each cycle. This model is unsatisfactory, however, because it does not account for possible dependence between observations due to repeated measurements (see also Figure 4.3). Therefore, the data will be modeled using random effects both in the mean and dispersion parts of the model in forthcoming chapters.

We may estimate $\phi$ either based on deviance or Pearson chi-squared statistic. Two outputs give similar results. For the mean model, parameter estimators are similar but for the dispersion model, they are somewhat different. In this example, degrees of freedom is large (239). We may prefer the Pearson chi-squared statistic in estimating $\phi$. We can compute 95% confidence intervals for $\mu_{ij}$ and $\phi_{ij}$ when $l_{ij-1} = 1.12$ and $t_j = 0.06$ at mean values of them: For $\mu_{ij}$,

$$\exp(-5.968 + 2.649 \times 1.12 \pm 1.96 \times 0.210) = (0.033, 0.075),$$

for $\phi_{ij}$,

$$\exp(-1.966 - 13.605 \times 0.06 \pm 1.96 \times 0.203) = (0.042, 0.092).$$

```
# based on deviance statistic
Distribution of Main Response :
                              "gamma"
[1] "Estimates from the model(mu)"
y ~ crack0
[1] "log"
            Estimate Std. Error t-value
(Intercept)   -5.934    0.10268  -57.79
crack0         2.626    0.08672   30.28
[1] "Estimates from the model(phi)"
phi ~ cycle
[1] "log"
            Estimate Std. Error t-value
(Intercept)   -2.127     0.1929 -11.025
cycle        -10.419     2.7240  -3.825
[1] "== Likelihood Function Values and Condition AIC =="
                                  [,1]
-2 log(likelihood)            :  -1435.545
-2 log(restricted likelihood) :  -1426.044
cAIC                          :  -1431.545

Scaled deviance: 239.0000 on 239.0000 degrees of freedom
# based on Pearson chi-squared statistic
Distribution of Main Response :
                              "gamma"
[1] "Estimates from the model(mu)"
```

```
y ~ crack0
[1] "log"
            Estimate Std. Error t-value
(Intercept)   -5.968    0.10273  -58.09
crack0         2.649    0.08564   30.94
[1] "Estimates from the model(phi)"
phi ~ cycle
[1] "log"
            Estimate Std. Error t-value
(Intercept)   -1.966     0.1927 -10.202
cycle        -13.605     2.7237  -4.995
[1] "== Likelihood Function Values and Condition AIC =="
                                  [,1]
-2 log(likelihood)            :  -1434.218
-2 log(restricted likelihood) :  -1424.562
cAIC                          :  -1430.218

Scaled deviance: 239.0000 on 239.0000 degrees of freedom
```

### *4.1.4 Similarities and differences between h-likelihood and PQL estimates: bacteria data example*

It is well known that the PQL estimator is seriously biased especially for binary data. Here, we show that the h-likelihood differs from PQL, providing efficient ML and REML estimators.

The bacteria dataset is available in the MASS library of R. There are 220 observations from 50 individuals. The response $y_{ij}$ (=1: presence of bacteria, 0: absence of bacteria) for $i$th individual and $j$th visit is binary. A fixed effect of treatment (drug, drug+, placebo) was included in the model together with a random ID effect ($v_i$).

$$\log(p_{ij}/(1-p_{ij})) = b_0 + b_1 I(i = drug) + b_2 I(i = drug+) + v_i$$
$$v_i \sim \mathrm{N}(0, \lambda)$$
$$p_{ij} = P(y_{ij} = 1|v_i)$$

Table 4.3 and Table 4.4 show that the REML and ML estimates for the dhglm package and the `glmer` function in the lme4 package. The method of the first-order Laplace approximation (HL(1,1)) to ML (i.e., using $p_v(h)$ for estimating the variance component $\lambda$) is identical to `glmer` with `nAGQ=1` where `nAGQ` is the number of points per axis for evaluating the adaptive Gauss-Hermite approximation to ML (Wolfinger, 1993). We see that estimates of HL(1,2) for ML are similar to those of

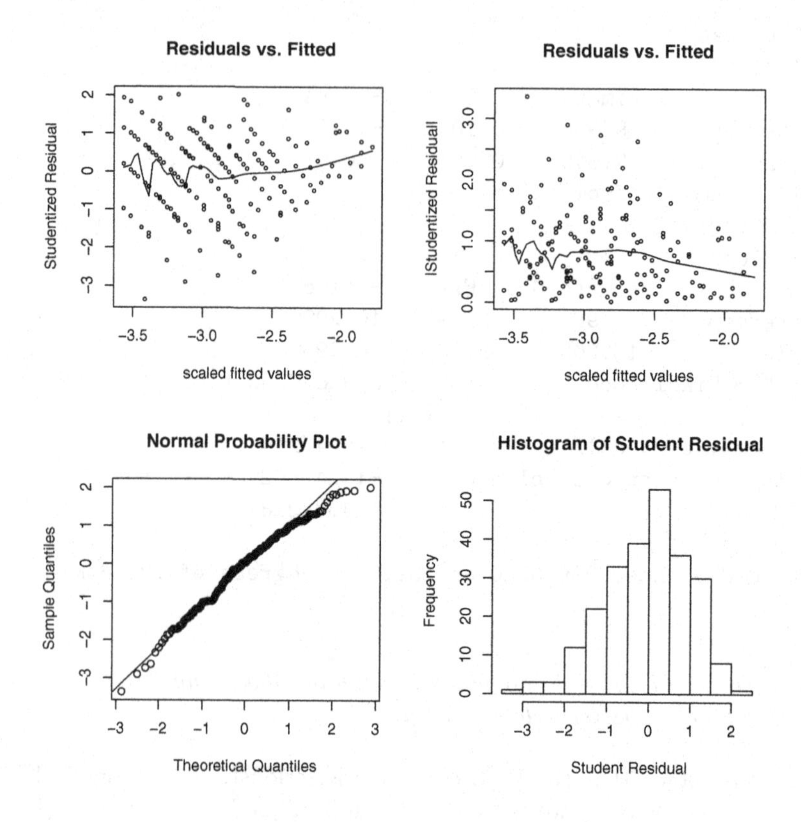


Figure 4.3 *Residual plot for the crack growth data under a Gamma GLM model with structured dispersion.*

**glmer** with **nAGQ=25** which gives exact, or at least highly accurate, ML estimators. Unlike the functions in lme4, the h-likelihood methods can also give REML estimators. The REML estimates give larger variance component estimates because they correct the degrees of freedom due to the uncertainty in the fixed effect estimates (intercept and treatment effects). See R code used to produce the estimates in Tables 4.3 and 4.4 in the next section.

### *4.1.5 Similarity between h-likelihood and Bayesian posterior mode estimates: HGLM vs. INLA using epilepsy data*

Here, we compare h-likelihood estimates to the ones produced by INLA (mentioned in Chapter 1).The h-likelihood $f_\theta(\boldsymbol{y}|\boldsymbol{v})f_\theta(\boldsymbol{v})$ can be viewed

Table 4.3 *Comparison of estimates (SE) by using ML methods*

| | 1st Laplace | |
|---|---|---|
| covariates | glmer(nAGQ=1) | dhglm |
| Intercept | 2.296(0.408) | 2.296(0.408) |
| trt(drug) | −1.202(0.571) | −1.202(0.571) |
| trt(drug+) | −0.710(0.588) | −0.710(0.588) |
| $\lambda$ | 0.966 | 0.966 |
| | GHQ | HL(1,2) |
| covariates | glmer(nAGQ=25) | dhglm |
| Intercept | 2.308(0.471) | 2.338(0.420) |
| trt(drug) | −1.209(0.596) | −1.222(0.590) |
| trt(drug+) | −0.720(0.608) | −0.725(0.606) |
| $\lambda$ | 1.061 | 1.097 |

Table 4.4 *Comparison of estimates (SE) by using REML methods*

| | PQL | HL(0,1) |
|---|---|---|
| covariates | glmer(nAGQ=0) | dhglm |
| Intercept | 2.032(0.378) | 2.051(0.389) |
| trt(drug) | −1.067(0.537) | −1.073(0.555) |
| trt(drug+) | −0.635(0.552) | −0.642(0.569) |
| $\lambda$ | 0.826 | 0.956 |
| | HL(1,1) | HL(1,2) |
| covariates | dhglm | dhglm |
| Intercept | 2.412(0.441) | 2.470(0.458) |
| trt(drug) | −1.256(0.622) | −1.283(0.647) |
| trt(drug+) | −0.752(0.636) | −0.775(0.661) |
| $\lambda$ | 1.335 | 1.533 |

as a function proportional to a Bayesian posterior distribution having flat priors for the hyperparameters $\theta$, $\pi(\boldsymbol{\theta}) = 1$. The h-likelihood method produces maximum h-likelihood estimates that should closely correspond to the mode of the posterior unless the posterior is sensitive to the specification of the priors.

The INLA and h-likelihood estimates for the epileptic seizure data example illustrate the similarity between the two. Take the treatment effect,

CTrt, for instance. The INLA mode estimate is -0.331 (with 95% CI [-0.633, -0.036] ) and the estimate from the hglm package is -0.333 (with 95% CI [-0.631, -0.036]). The variance component for the random effects from INLA are 0.244 and 0.130, compared to 0.237 and 0.127 from hglm. So, the estimates are similar and we can conclude that the Bayesian results are not prior sensitive.

The algorithm implemented in the hglm package allows both HL(0,1) and HL(1,1) estimation with a minor simplification to increase computational speed. It ignores that the random effects estimates used in $p_v(h)$ are functions of the dispersion parameters, whereas other packages for HGLM estimation (e.g., dhglm) include these derivatives giving improved estimates especially for binomial outcomes. Consequently, the algorithms implemented in the hglm package have been named `EQL` and `EQL1` for HL(0,1) and HL(1,1), respectively, to emphasize this approximation. For Poisson models, the difference between `EQL1` and HL(1,1) is hardly detectable though and it can be used for comparison with INLA. From the epileptic seizure data example we can see that INLA gives similar results as HL(1,1), which is expected since INLA uses Laplace approximations in a similar manner as HL(1,1). There may be differences for other examples, however, if the results from INLA are prior sensitive.

```
Estimates from hglm
Summary of the fixed effects estimates:
             Estimate Std. Error t-value Pr(>|t|)
(Intercept)   1.57467    0.07529  20.915  < 2e-16 ***
ClBase4       0.87889    0.13460   6.529 1.82e-09 ***
CTrt         -0.33306    0.15158  -2.197   0.0300 *
CBT           0.35010    0.20844   1.680   0.0957 .
ClAge         0.48093    0.35565   1.352   0.1789
CV4          -0.10191    0.08597  -1.185   0.2383

Log of variance components
     Estimate Std. Error
Ind    -1.441     0.2277
rand   -2.062     0.1623


Estimates from INLA
Fixed effects:
              mean     sd  0.025q   0.5q  0.975q   mode
(Intercept) 1.5765 0.0761  1.4242 1.5774  1.7241 1.5791
ClBase4     0.8795 0.1344  0.6143 0.8795  1.1443 0.8795
```

```
CTrt        -0.3328 0.1514 -0.6333 -0.3322 -0.0364 -0.3308
CBT          0.3507 0.2079 -0.0587  0.3505  0.7606  0.3501
ClAge        0.4831 0.3552 -0.2208  0.4844  1.1793  0.4869
CV4         -0.1031 0.0853 -0.2707 -0.1032  0.0646 -0.1034


Model hyperparameters:
                    mean   sd  0.025q 0.5q  0.975q  mode
Precision for Ind  4.610 1.327 2.570 4.423  7.751 4.098
Precision for rand 8.536 2.161 5.207 8.230 13.654 7.679


Log of variance components (from posterior modes)
      Estimate
Ind   -1.410
rand  -2.038
```

## 4.2 R code

*Epilepsy data (extra-dispersed Poisson)*

```
model_mu<-DHGLMMODELING(Model="mean", Link="log",
  LinPred=y~B+T+A+B:T+V)
model_phi<-DHGLMMODELING(Model="dispersion",
  Link="log",LinPred=phi~1)

## based on deviance statistic
res_jglm1<-dhglmfit(RespDist="poisson",DataMain=epilepsy,
  MeanModel=model_mu,DispersionModel=model_phi,
  Dmethod="deviance")

## based on Pearson chi-squared stistic
res_jglm2<-dhglmfit(RespDist="poisson",DataMain=epilepsy,
  MeanModel=model_mu,DispersionModel=model_phi,
  Dmethod="Pearson")
```

*Heteroscedastic log-linear model for injection-molding data*

```
data(injection, package="mdhglm")
injection[,2:11] <- 2*(injection[,2:11] - 1.5)
```

```
#Make explanatory variables -1 and 1
model_mu4<-DHGLMMODELING(Model="mean",Link="log",
        LinPred=y~a+c+d+e+g+n+c:n+e:n)
model_phi4<-DHGLMMODELING(Model="dispersion",Link="log",
    LinPred=phi~a+f)

fit4<-dhglmfit(RespDist="gaussian",DataMain=injection,
        MeanModel=model_mu4,DispersionModel=model_phi4)

plotdhglm(fit4)
plotdhglm(fit4,type="phi")
```

*Gamma GLM with structured dispersion for crack growth data*

```
data(crack_growth, package="mdhglm")
model_mu<-DHGLMMODELING(Model="mean",Link="log",
          LinPred=y~crack0)
model_phi<-DHGLMMODELING(Model="dispersion",Link="log",
          LinPred=phi~cycle)

fit1<-dhglmfit(RespDist="gamma",DataMain=crack_growth,
          MeanModel=model_mu,DispersionModel=model_phi)

fit2<-dhglmfit(RespDist="gamma",DataMain=crack_growth,
          MeanModel=model_mu,DispersionModel=model_phi,
          Dmethod="Pearson")

plotdhglm(fit1)
```

*Bacteria data estimation: comparison lme4 vs. dhglm*

```
data(bacteria, package="MASS")
y1 <- 1*(bacteria$y=="y")
bacteria <- data.frame(cbind(bacteria, y1))

###### lme4 package #########
#Estimates using glmer
library(lme4)
glmer.est <- numeric(3)
nAGQ <- c(0,1,25)
```

```
 for (i in nAGQ) {
     model.glmer<-glmer(y~trt+(1|ID),family=binomial,
                   data=bacteria,nAGQ=i)
     print(model.glmer@theta^2)
 }

###### dhglm package estimates #########
library(dhglm)
model_mu<-DHGLMMODELING(Model="mean", Link="logit",
           LinPred=y1~trt+(1|ID),RandDist="gaussian")
model_phi<-DHGLMMODELING(Model="dispersion")

#ML estimates
dhglm11<-dhglmfit(RespDist="binomial", DataMain=bacteria,
 MeanModel=model_mu, DispersionModel=model_phi, PhiFix=1,
          mord=1, dord=1, REML=FALSE)

dhglm12<-dhglmfit(RespDist="binomial", DataMain=bacteria,
 MeanModel=model_mu, DispersionModel=model_phi, PhiFix=1,
          mord=1, dord=2, REML=FALSE)

#REML estimates
dhglm01<-dhglmfit(RespDist="binomial", DataMain=bacteria,
 MeanModel=model_mu, DispersionModel=model_phi, PhiFix=1,
         mord=0, dord=1)

dhglm11<-dhglmfit(RespDist="binomial", DataMain=bacteria,
 MeanModel=model_mu, DispersionModel=model_phi, PhiFix=1,
         mord=1, dord=1)

dhglm12<-dhglmfit(RespDist="binomial", DataMain=bacteria,
 MeanModel=model_mu, DispersionModel=model_phi, PhiFix=1,
         mord=1, dord=2)
```

*HGLM estimation for the epileptic seizure data using the INLA input*

```
source("http://www.math.ntnu.no/inla/givemeINLA.R")
library(INLA)
#Following code for centering copied from the INLA web page
data(Epil)
my.center = function(x) (x - mean(x))
## make centered covariates
```

```
Epil$CTrt     = my.center(Epil$Trt)
Epil$ClBase4 = my.center(log(Epil$Base/4))
Epil$CV4      = my.center(Epil$V4)
Epil$ClAge    = my.center(log(Epil$Age))
Epil$CBT      = my.center(Epil$Trt*Epil$ClBase4)
hglm.epi<-hglm2(y~ClBase4+CTrt+CBT+ClAge+CV4+(1|Ind)+(1|rand),
  family=poisson(link=log),fix.disp=1,data=Epil,method="EQL1")
```

## 4.3 IWLS algorithm for interconnected GLMs

The need to keep apart principles, methods, and algorithms applies to all fields in statistics (Table 4.5). The aim of this chapter is to present extensions of the IWLS algorithm for GLMs to HGLMs, in light of the h-likelihood.

The algorithm is described in Lee, Nelder, and Pawitan (2017) for HGLM. Later in this chapter we start by describing it for joint modeling of mean and dispersion in a linear model but the reader well acquainted with GLMs is advised to read Section 4.6 as well (where the IWLS algorithm for HL(1,1) estimation for a Poisson GLMM is derived). The linear and linear mixed model examples show, however, the basic features of the algorithm and also how HGLMs can be extended to more complex models. A set of figures is used to illustrate conceptually how the IWLS computations can be extended by using interconnected GLMs for various components of the HGLM as in Figure 4.1 and how it can increase in complexity as random effects are added to the mean and dispersion models.

### *4.3.1 Joint linear model for mean and dispersion*

Consider a linear model

$$\mathbf{y} = \mathbf{X}\boldsymbol{\beta} + \mathbf{e}$$

$$\mathbf{e} \sim \mathrm{N}(0, \Phi)$$

where $\Phi = \mathrm{diag}(\phi_i)$

The ML estimator for $\boldsymbol{\beta}$ is $(\boldsymbol{X}^T\Phi^{-1}\boldsymbol{X})^{-1}\boldsymbol{X}^T\Phi^{-1}\boldsymbol{y}$ and the variance of the estimator is $(\boldsymbol{X}^T\Phi^{-1}\boldsymbol{X})^{-1}$. Now suppose that we have a regression model for the dispersion $\phi_i$

$$g(\phi_i) = G_i\gamma$$

where $g(\cdot)$ is a link function and $G_i$ is the $i$th row in a design matrix $\mathbf{G}$.

Table 4.5 *Overview of statistical principles, methods, and algorithms.*

| Principle | Method | Algorithm |
|---|---|---|
| Bayesian | compute posterior | MCMC, INLA |
| Likelihood | maximum likelihood | IWLS covering N-R , Fisher scoring |
| Extended likelihood | maximum adjusted profile h-likelihood | Augmented and Interconnected IWLS |
| | and compute predictive probability | Bootstrap, Laplace approximation |

MCMC = Markov chain Monte Carlo, INLA = Integrated Nested Laplace Approximation

N-R = Newton-Raphson, IWLS = Iterative Weighted Least Squares

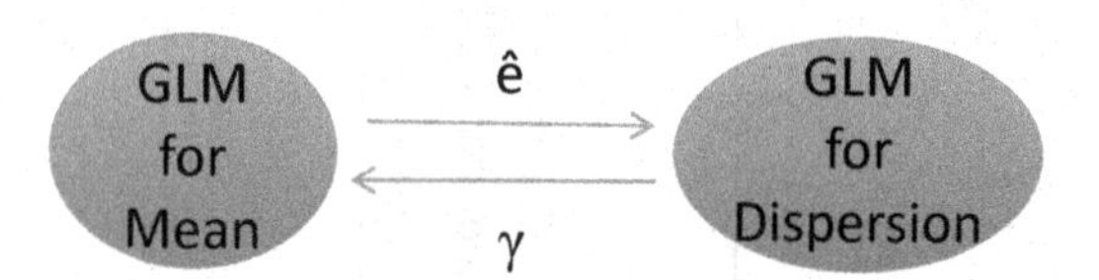


Figure 4.4 *Interconnected GLM for fitting joint GLMs.*

The ML estimate of the regression coefficient of the dispersion $\gamma$ can be computed by using $\hat{e}_i^2$ as response in a gamma GLM with mean $\phi_i$. If a log link is used it ensures that the estimated $\hat{\phi}_i$ is positive.

The REML estimate can also be computed by using $\hat{e}_i^2/(1-q_i)$ as response in a gamma GLM having a prior weight $(1-q_i)/2$. Here, $\mathbf{q}$ is a vector of hat values and $q_i$ is the $i$th diagonal element in the hat matrix

$$\mathbf{H} = (\boldsymbol{X}^T\Phi^{-1}\boldsymbol{X})^{-1}\boldsymbol{X}^T\Phi^{-1}.$$

The estimates of $\boldsymbol{\beta}$ and $\Phi$ need to be estimated iteratively, because the estimate of $\boldsymbol{\beta}$ depends on $\Phi$, and $\hat{\Phi}$ depends on the estimated residuals.

### *4.3.2 Joint GLMs for mean and dispersion*

Suppose $\boldsymbol{y}$ follows the GLM class of model with $\text{var}(y_i) = \phi_i V(\mu_i)$ and $\phi_i$ follows the regression model as above. Then given $\phi_i$, we see from Chapter 2 that the ML estimator $\hat{\boldsymbol{\beta}}$ can be obtained by using an IWLS algorithm for GLM model with prior weight $1/\phi_i$.

Given $\boldsymbol{\beta}$, the maximum likelihood estimate of the regression coefficient of the dispersion model $\gamma$ can be computed by using the deviance component $d_i$ as response in a gamma GLM with mean $\phi_i$. From Chapter 2, we can see that there is an intuitive explanation for estimating the dispersion parameters using the deviance components as response in a gamma GLM. The studentized deviance residuals are asymptotically standard normal and their squared values are gamma distributed.

The REML estimate can be computed by using $d_i/(1-q_i)$ as response in a gamma GLM having a prior weight $(1-q_i)/2$. Thus, ML estimator for joint GLMs can be computed by a see-saw algorithm between mean and dispersion (see Figure 4.4).

## 4.4 IWLS algorithm for augmented GLM

### 4.4.1 Linear mixed model

In this section we show that a linear mixed model can be written as a weighted least squares problem by augmenting the response vector. The weighted least squares estimator for this model is identical to Henderson's mixed model equations. So we can extend the estimation method for joint GLMs to joint GLMs including random effects by augmenting the response vector. We show how this is done for a linear mixed model and for GLMMs the reader is referred to Section 4.6.

Consider again the linear mixed model

$$\boldsymbol{y} = \mathbf{X}\boldsymbol{\beta} + \mathbf{Z}\boldsymbol{v} + \boldsymbol{e}$$

where $\boldsymbol{V} = \lambda \boldsymbol{Z}\boldsymbol{Z}^T + \phi \mathbf{I}_n$. Here $n$ is the number of observations and $\mathbf{I}_n$ is the identity matrix of size $n$. The length of $\boldsymbol{v}$ is $m$.

The model can be re-written as an augmented linear model

$$\boldsymbol{y}_a = \mathbf{X}_a \delta + e_a$$

where $\boldsymbol{y}_a = \begin{pmatrix} \boldsymbol{y} \\ \mathbf{0} \end{pmatrix}$, $\mathbf{X}_a = \begin{pmatrix} \mathbf{X} & \mathbf{Z} \\ \mathbf{0} & \mathbf{I}_m \end{pmatrix}$, $\delta = \begin{pmatrix} \boldsymbol{\beta} \\ \boldsymbol{v} \end{pmatrix}$, $e_a = \begin{pmatrix} \boldsymbol{e} \\ -\boldsymbol{v} \end{pmatrix}$

The variance-covariance matrix of the augmented residual vector is given by

$$\text{var}(e_a) \equiv W^{-1} = \begin{pmatrix} \phi \mathbf{I}_n & \mathbf{0} \\ \mathbf{0} & \lambda \mathbf{I}_m \end{pmatrix}.$$

The estimates from weighted least squares are given by

$$\mathbf{X}_a^T W \mathbf{X}_a \hat{\delta} = \mathbf{X}_a^T W y_a,$$

which is identical to Henderson's mixed model equations (the derivations are left as an exercise). The weight matrix $W$ may then be updated using the estimated variance components and the algorithm iterates until convergence.

### 4.4.2 HGLM class of models

Lee and Nelder (2001a) showed that the augmented linear model in the previous section can be extended to fit the HGLM class of models (Box 6). They further showed that the score functions from adjusted profile likelihoods $p_v(h)$ and $p_{\beta,v}(h)$, respectively, as well as the score function from the h-likelihood $h$ with respect to $\boldsymbol{v}$, can be obtained as the score functions of augmented GLMs. A GLM can be fitted using IWLS, and the entire HGLM (in Box 6) can be fitted using IWLS for augmented GLMs. An advantage of augmented IWLS algorithm is that we can make use of the computational machinery to fit least square problems (such as qr-factorization).

In Section 4.6, the algorithm for a Poisson GLMM is described in detail. These derivations are readily expandable to HGLMs and are described in (Lee, Nelder, and Pawitan, 2017) and the appendix of Molas and Lesaffre (2010).

*Correlated random effects.* In animal breeding applications for instance, the random effects are correlated so that $\boldsymbol{v} \sim \mathrm{N}(0, \lambda\mathbf{A})$ for a given correlation matrix $\mathbf{A}$. However, the model can be transformed into $\boldsymbol{u} \sim \mathrm{N}(0, \lambda\boldsymbol{I})$ by recomputing $\boldsymbol{Z}$ as $\boldsymbol{Z}\mathbf{A}^{\frac{1}{2}}$ where $\mathbf{A}^{\frac{1}{2}}$ is some square-root transformation of $\mathbf{A}$ (e.g., Cholesky decomposition). Thereby, the above augmented model having i.i.d. random effects can be applied. See Chapter 8 of Lee, Nelder, and Pawitan (2017) for further details. In some models $\mathbf{A}$ includes parameters. For instance, splines and the Intrinsic Autoregressive (IAR) model does not involve parameters in $\mathbf{A}$, whereas e.g., Markov Random Field (MRF) models in spatial statistics and factor models have parameters in $\mathbf{A}$, to be estimated. In factor analysis, $\boldsymbol{Z}\mathbf{A}^{\frac{1}{2}}$ is the factor loading and $\boldsymbol{u}$ is the factor. Thus, this generalizes factor models to count and proportional data.

## 4.5 IWLS algorithm for HGLMs

Consider the heteroscedastic linear mixed model

$$\boldsymbol{y} = \mathbf{X}\boldsymbol{\beta} + \mathbf{Z}\boldsymbol{v} + \boldsymbol{e}$$

with independent and heteroscedastic random effects $v_i \sim \mathrm{N}(0, \lambda_i)$ and residuals $e_i \sim \mathrm{N}(0, \phi_i)$. This model covers a homogeneous component model with $\lambda_i = \lambda$ and $\phi_i = \phi$. Now we can allow GLMs for the dispersion (residual variance) and random effect variance

$$g_1(\phi_i) = G_{1i}\gamma_1$$

$$g_2(\lambda_i) = G_{2i}\gamma_2.$$

By taking log link for these variance components, we avoid negative estimates for variance components. We consider a single component model,

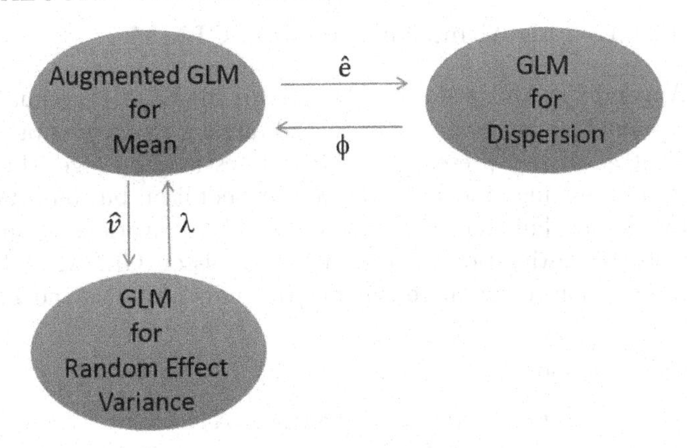


Figure 4.5 *Interconnected GLMs for HGLMs with structured dispersions.*

but extension to multi-component random-effect models is straightforward.

As in Figure 4.5 the REML estimates for $\gamma_1$ can be obtained by applying a gamma GLM to the response $\hat{e}_i^2/(1-q_i)$ with weights $(1-q_i)/2$, where the index $i$ goes from 1 to $n$. Those for $\gamma_1$ are computed by applying a gamma GLM to the response $\hat{v}_i^2/(1-q_i)$ with weights $(1-q_i)/2$, where the index $i$ goes from $n+1$ to $n+m$. The hat values $q_i$ are obtained from the hat matrix of the augmented model. In a k-component HGLM, the number of GLMs for random effect variances is $k$. Thus k-component HGLMs with structured dispersions can be composed of $k+2$ interconnected GLMs (i.e., one augmented GLM for the mean model, $k$ GLMs for variance components $\lambda$, and one GLM for the dispersion $\phi$). By replacing $\hat{e}_i^2$ and $\hat{v}_i^2$ by appropriate deviance components this algorithm can be extended to fit any HGLM including structured dispersion (Lee, Nelder, and Pawitan, 2017).

Furthermore, the same h-likelihood framework can be used to extend the fitting algorithm and inference for Double HGLMs, where random effects are allowed in dispersion models, and these can be further extended for outcomes from different DHGLMs with different distributions, i.e., multivariate DHGLM analysis. This will be presented in later chapters, but we can already now notice that the h-likelihood gives building blocks that are easily extendable to more and more complex models.

## 4.6 Estimation algorithm for a Poisson GLMM

In the Appendix of Molas and Lesaffre (2010) the HL(1,1) estimator for a Hurdle model having gamma distributed random effects was presented in excellent clarity. It presents the HL(1,1) estimation algorithm in a rather general setting, and here we use their notation but on a Poisson GLMM with one Gaussian random effect $\boldsymbol{v}$. The derivations should give insight into the estimation algorithm for a Poisson GLMM and serve as an introduction to the more general Appendix of Molas and Lesaffre (2010).

We consider the model

$$E(\boldsymbol{y}|\boldsymbol{v}) = \boldsymbol{\mu} \quad \text{and} \quad \log(\boldsymbol{\mu}) = \boldsymbol{X\beta} + \boldsymbol{Zv} \tag{4.1}$$

where the random effect is assumed i.i.d. normal, $\boldsymbol{v} \sim \mathrm{N}(0, \lambda \boldsymbol{I})$.

The following part of this section is divided into three parts: estimating equations for the random effects based on $h$, estimation of fixed effects based on $p_v(h)$ and estimation of the random effect variance component based on $p_{\boldsymbol{\beta},\boldsymbol{v}}(h)$.

### *4.6.1 Estimating equations for a Poisson GLMM*

The h-likelihood for the Poisson GLMM (4.1) has the following form

$$h = \sum_{ij} \log\left(\frac{\mu_{ij}^{y_{ij}}}{y_{ij}!} e^{-\mu_{ij}}\right) + \sum_{i} log\left(\frac{1}{\sqrt{2\pi\lambda}} e^{-\frac{v_i^2}{2\lambda}}\right)$$

and if we rewrite the Poisson part in standard GLM notation (McCullagh and Nelder, 1989), and drop constant terms in the Gaussian part, we have

$$h = \sum_{ij} (y_{ij}\theta_{ij} - b(\theta_{ij})) - \frac{m}{2}\log\lambda - \frac{1}{2\lambda}\sum_{i} v_i^2 \tag{4.2}$$

where $m$ is the number of levels in the random effect (i.e., the length of $\boldsymbol{v}$).

Now we compute the gradient and Hessian of (4.2)

$$\frac{\partial h}{\partial \beta_p} = \sum_{ij} (y_{ij} - \mu_{ij}) w_{ij} \frac{\partial \eta_{ij}}{\partial \mu_{ij}} \frac{\partial \eta_{ij}}{\partial \beta_p}$$

$$\frac{\partial h}{\partial v_i} = \sum_{ij} (y_{ij} - \mu_{ij}) w_{ij} \frac{\partial \eta_{ij}}{\partial \mu_{ij}} \frac{\partial \eta_{ij}}{\partial v_i} - \frac{v_i}{\lambda}$$

where $\eta_{ij} = \log(\mu_{ij}) = \boldsymbol{X}_{ij}\boldsymbol{\beta} + \boldsymbol{Z}_{ij,i}v_i$ is a linear predictor and $w_{ij}$ is a diagonal element of the GLM weight matrix

$$\mathbf{W} = \text{diag}\left[\frac{\partial\mu_{ij}}{\partial\eta_{ij}}\right] V^{-1}(\boldsymbol{\mu}).$$

Note that the log link is the canonical link for a Poisson distribution, so we have $\partial\mu_{ij}/\partial\eta_{ij} = w_{ij} = V(\mu_{ij})$ where $V(\mu)$ is the GLM variance function equal to $\mu$ for the Poisson distribution. Thus

$$\mathbf{W} = \text{diag}(\mu_{ij})$$

for the model considered here. Further, $\boldsymbol{Z}_{ij,i}$ is the $ij$th row of the $i$th column in $\boldsymbol{Z}$.

To make the notation a bit more general we now introduce the *pseudo-response* $\boldsymbol{\psi}$ (Lee, Nelder, and Pawitan, 2017), which is equal to a vector of zeros for a normally distributed random effect. The gradient may then be written in matrix notation as

$$\mathbf{D} = \frac{\partial h}{\partial(\beta, v)} = \begin{pmatrix} \boldsymbol{X}^T\mathbf{W}\frac{\partial\eta}{\partial\mu}(\boldsymbol{y}-\mu) \\ \boldsymbol{Z}^T\mathbf{W}\frac{\partial\eta}{\partial\mu}(\boldsymbol{y}-\mu) + \frac{\boldsymbol{\psi}}{\lambda} \end{pmatrix}$$

The negative expected Hessian matrix requires the second derivatives that are rather straight forward to derive, and we get:

$$\mathbf{H} = -E\left[\frac{\partial h}{\partial(\beta, v)\partial(\beta, v)}\right] = \begin{pmatrix} \boldsymbol{X}^T\mathbf{W}\boldsymbol{X} & \boldsymbol{X}^T\mathbf{W}\boldsymbol{Z} \\ \boldsymbol{Z}^T\mathbf{W}\boldsymbol{X} & \boldsymbol{Z}^T\mathbf{W}\boldsymbol{Z} + \frac{1}{\lambda}\boldsymbol{I} \end{pmatrix}$$

In the next step we derive an adjusted dependent variable $\mathbf{s}$ used in the estimation algorithm. The algorithm is iterative and applies Fisher scoring on the h-likelihood. It is based on the previously described matrices $\mathbf{H}$ and $\mathbf{D}$. The updating of the parameters $\delta = (\boldsymbol{\beta}, \boldsymbol{v})^T$ is performed as follows

$$\begin{aligned} \mathbf{H}(\delta^{+1} - \delta^0) &= \mathbf{D} \\ \Rightarrow \mathbf{H}\delta^{+1} &= \mathbf{D} + \mathbf{H}\delta^0 \end{aligned} \tag{4.3}$$

These are in our situation as follows:

$$\begin{pmatrix} \boldsymbol{X}^T\mathbf{W}\boldsymbol{X} & \boldsymbol{X}^T\mathbf{W}\boldsymbol{Z} \\ \boldsymbol{Z}^T\mathbf{W}\boldsymbol{X} & \boldsymbol{Z}^T\mathbf{W}\boldsymbol{Z} + \frac{1}{\lambda}\boldsymbol{I} \end{pmatrix} \begin{pmatrix} \boldsymbol{\beta}^{+1} - \boldsymbol{\beta}^0 \\ \boldsymbol{v}^{+1} - \boldsymbol{v}^0 \end{pmatrix} = \begin{pmatrix} \boldsymbol{X}^T\mathbf{W}\frac{\partial\eta}{\partial\mu}(\boldsymbol{y}-\boldsymbol{\mu}) \\ \boldsymbol{Z}^T\mathbf{W}\frac{\partial\eta}{\partial\mu}(\boldsymbol{y}-\boldsymbol{\mu}) + \frac{\boldsymbol{\psi}}{\lambda} \end{pmatrix}$$

$$\begin{pmatrix} \boldsymbol{X}^T\mathbf{W}\boldsymbol{X} & \boldsymbol{X}^T\mathbf{W}\boldsymbol{Z} \\ \boldsymbol{Z}^T\mathbf{W}\boldsymbol{X} & \boldsymbol{Z}^T\mathbf{W}\boldsymbol{Z} + \frac{1}{\lambda}\boldsymbol{I} \end{pmatrix} \begin{pmatrix} \boldsymbol{\beta}^{+1} \\ \boldsymbol{v}^{+1} \end{pmatrix} = \begin{pmatrix} \boldsymbol{X}^T\mathbf{W}\mathbf{s} \\ \boldsymbol{Z}^T\mathbf{W}\mathbf{s} + \frac{\boldsymbol{\psi}}{\lambda} \end{pmatrix}$$

where the vector $\mathbf{s}$ has elements

$$s_{ij} = \eta_{ij} + (y_{ij} - \mu_{ij})\frac{\partial \eta_{ij}}{\partial \mu_{ij}} = \eta_{ij} + (y_{ij} - \mu_{ij})\frac{1}{\mu_{ij}}.$$

This set of equations correspond to equations (4.3) of Lee and Nelder (1996).

Note also that the estimating equations correspond to those of an augmented weighted linear model

$$y_a = \mathbf{X}_a \delta + e_a$$

where $\boldsymbol{y}_a = \begin{pmatrix} \mathbf{s} \\ \boldsymbol{\psi} \end{pmatrix}$, $\mathbf{X}_a = \begin{pmatrix} \mathbf{X} & \mathbf{Z} \\ \mathbf{0} & \boldsymbol{I} \end{pmatrix}$, $\delta = \begin{pmatrix} \boldsymbol{\beta} \\ \boldsymbol{v} \end{pmatrix}$, $e_a = \begin{pmatrix} \boldsymbol{e} \\ -\boldsymbol{v} \end{pmatrix}$ with augmented weight matrix $\mathbf{W}_a = \begin{pmatrix} \mathbf{W} & \mathbf{0} \\ \mathbf{0} & \lambda \boldsymbol{I} \end{pmatrix}$.

Consequently, we can use IWLS for estimation.

### *4.6.2 Estimation of fixed effects using $p_{\boldsymbol{v}}(h)$*

Here we show how to compute adjustments ($\mathbf{m}$) for the response of the augmented model (above) necessary to use $p_{\boldsymbol{v}}(h)$ as an objective function used to estimate $\boldsymbol{\beta}$. By computing $\mathbf{m}$ we can adjust the augmented response and use the same IWLS as above. Detailed derivations are found in Lee and Lee (2012) and we follow the notation of Appendix B of Molas and Lesaffre (2010).

Suppose we have the augmented response:

$$\boldsymbol{y}_a = \begin{pmatrix} \boldsymbol{y} \\ \boldsymbol{\psi} \end{pmatrix}$$

Then the fixed effects estimates obtained by maximizing $p_{\boldsymbol{v}}(h)$ can be computed using Fisher scoring (4.3) with a modified augmented response

$$\begin{aligned} \boldsymbol{y}* &= \boldsymbol{y} - \mathbf{m} \\ \boldsymbol{\psi}* &= \boldsymbol{\psi} + \lambda \boldsymbol{Z}^T \mathbf{W} \tfrac{\partial \eta}{\partial \mu} \mathbf{m} \end{aligned}$$

where $\mathbf{W} = \text{diag}(\mu_{ij})$ and $\boldsymbol{\psi} = \mathbf{0}$ for the Poisson GLMM.

The technical details are as follows (see Supplementary Material of Lee and Lee (2012)). Let the $i$th element of $\mathbf{m}$ be $m_i = \frac{1}{2} k_i \frac{\partial \mu_i}{\partial \eta_i}$. Now we need to compute $k_i$, which is

$$k_i = \mathbf{P}_R[i,i]\frac{1}{w_{ii}} + \sum_{j=1}^{N} \mathbf{P}_R[j,j]\mathbf{A}[j,i]$$

where $\mathbf{P}_R = \mathbf{T}_R(\mathbf{T}_R^T\mathbf{\Sigma}_a^{-1}\mathbf{T}_R)^{-1}\mathbf{T}_R^T\mathbf{\Sigma}_a^{-1}$ with

$$\mathbf{T}_R = \begin{pmatrix} \boldsymbol{Z} \\ \boldsymbol{I} \end{pmatrix}, \mathbf{\Sigma}_a^{-1} = \begin{pmatrix} \mathbf{W} & \mathbf{0} \\ \mathbf{0} & \frac{1}{\lambda}\boldsymbol{I} \end{pmatrix}$$

and $\mathbf{A} = -\boldsymbol{Z}(\mathbf{T}_R^T\mathbf{\Sigma}_a^{-1}\mathbf{T}_R)^{-1}\boldsymbol{Z}^T$.

### *4.6.3 Estimation of the random effect variance using $p_{\beta,v}(h)$*

The score equation for the variance component is

$$\frac{\partial p_{\boldsymbol{\beta},\boldsymbol{v}}(h)}{\partial\lambda} = \frac{1}{2}\sum_i \frac{d_i - (1-q_i)\lambda}{\lambda^2} \tag{4.4}$$

where $d_i$ is the $i$th deviance component and the sum goes over the number of levels, $i$, in the random effect (i.e., the length of $\boldsymbol{v}$).

Equation (4.4) can be re-expressed as the score equation of a gamma distributed random variable $d_i^* = d_i/(1-q_i)$ with mean $\lambda$, variance $2\lambda/(1-q_i)$ and prior weight $(1-q_i)/2$, i.e.,

$$\frac{\partial p_{\boldsymbol{\beta},\boldsymbol{v}}(h)}{\partial\lambda} = \frac{1}{2}\sum_i \frac{(1-q_i)}{2}\frac{d_i^* - \lambda}{\lambda^2}.$$

For a normal random effect we have $d_i = v_i^2$, such that the variance component can be estimated using a gamma GLM with log-link, response $v_i^2/(1-q_i)$ and weights $(1-q_i)/2$, where $q_i$ is the $N+i$ diagonal element of the hat matrix:

$$\mathbf{P} = \mathbf{T}(\mathbf{T}^T\mathbf{\Sigma}_a^{-1}\mathbf{T})^{-1}\mathbf{T}^T\mathbf{\Sigma}_a^{-1}$$

with

$$\mathbf{T} = \begin{pmatrix} \boldsymbol{X} & \boldsymbol{Z} \\ \mathbf{0} & \boldsymbol{I} \end{pmatrix}.$$

## 4.7 Exercises

1. Run the model hglm.epi (see R code Section 4.2) setting method=EQL. How much do the estimates change? Summarize the theory for the HL(0,1) and HL(1,1) estimators.

2. Check, using an example, that
a) the following code computes the residual variance

```
model.mean <- lm(y~x)
```

```
e <- resid(model.mean)
hv <- hatvalues(model.mean)
model.variance<-glm((e^2)/(1-hv)~1,family=Gamma(link=log),
                    weights=(1-hv)/2 )
fitted(model.variance)[1]
```

b) Use the following code to check whether the residual variance depends on x.

```
model.variance<-glm((e^2)/(1-hv)~x,family=Gamma(link=log),
                    weights=(1-hv)/2 )
summary(model.variance)
```

3. The aim of this exercise is to give an understanding of the IWLS algorithm for a linear mixed model by inplementing the algorithm using the sleep study data (available in the lme4 package). The model is

$$\boldsymbol{y} = \boldsymbol{X\beta} + \boldsymbol{Zv} + \boldsymbol{e}$$

where $\boldsymbol{y}$ is the average reaction time per day (ms) for subjects in a sleep deprivation study, $\boldsymbol{\beta}$ includes an intercept term and effect of number of days of sleep deprivation, and $\boldsymbol{v}$ includes a random intercept term for each individual. It is a linear mixed model with $\boldsymbol{v} \sim \mathrm{N}(0, \boldsymbol{I}_m\sigma_v^2)$ and $\boldsymbol{e} \sim \mathrm{N}(0, \boldsymbol{I}_n\sigma_e^2)$ where $m$ is the number of individuals and $n$ is the total number of observations.

The model can be fitted in lme4 as

```
fm2 <- lmer(Reaction ~ Days + (1|Subject), sleepstudy)
```

a) Construct the augmented design matrix $\boldsymbol{X}_a = \begin{pmatrix} \boldsymbol{X} & \boldsymbol{Z} \\ \boldsymbol{0} & \boldsymbol{I}_m \end{pmatrix}$ and the augmented response $\boldsymbol{y}_a = \begin{pmatrix} \boldsymbol{y} \\ \boldsymbol{0} \end{pmatrix}$ in R.

b) Assuming $\sigma_v^2 = 1$ and $\sigma_e^2 = 2$, estimate the fixed and random effects (i.e., $\boldsymbol{\beta}$ and $\boldsymbol{v}$) using the `lm` function in R. Hint: An arbitrary design matrix, say `Xa`, can be specified in the `lm` function as

```
lm( y~.-1, data=data.frame(Xa) )
```

c) Given these fixed and random effects compute updated estimates of the variance component $\sigma_v^2$ and $\sigma_e^2$ using two gamma GLMs (one for each variance component).

d) Implement the IWLS algorithm by iterating between the functions in exercise b) and c).

4. Use the hglm package to fit the linear mixed model for the sleep study data including random slopes over Days. In lme4 the model is fitted as

```
fm2 <- lmer(Reaction~Days+(1|Subject)+(0+Days|Subject),
          sleepstudy)
```

Check whether the variance component for the random slopes is significant by applying a likelihood-ratio test on the REML likelihood $p_{\beta,v}$.

5. Show that the weighted least-square estimates from the augmented linear model in Section 4.4.1 give Henderson's mixed model equation.

CHAPTER 5

# HGLMs modeling in R

In this chapter a number of dataset are modeled using HGLMs and the R codes for running these examples are given at the end of the chapter, using either the hglm or dhglm package, or both. The hglm package is slightly more user friendly, especially for users acquainted with packages in R that fit linear mixed models (e.g., lme4). On the other hand, the dhglm package is much more flexible and the HL(1,1) and HL(1,2) methods are implemented making it especially useful for binary response variables.

In the first few examples we show analyses using normal, log-normal, gamma, Poisson, and binomial HGLMs. Binary models are notoriously difficult to fit using standard software, whereas the dhglm package is fast and produces reliable results. Thereafter, examples using HGLMs including structured dispersion are given. We also fit models with correlated random effects, including spatial models.

## 5.1 Examples

### *5.1.1 Linear mixed models for cake data*

In an experiment on the preparation of chocolate cakes, conducted at Iowa State College, 3 recipes for preparing the batter were compared (Cochran and Cox, 1957). Recipes I and II differed in that the chocolate was added at 40°C and 60°C, respectively, while recipe III contained extra sugar. In addition, 6 different baking temperatures were tested: these ranged in 10°C steps from 175°C to 225°C. For each mix, enough batter was prepared for 6 cakes, each of which was baked at a different temperature. Thus the recipes are the whole-unit treatments, while the baking temperatures are the sub-unit treatments. There were 15 replications, and it will be assumed that these were conducted serially according to a randomized blocks scheme: that is, one replication was completed before starting the next, so that differences among replicates represent time differences. A number of measurements were made on the cakes. The

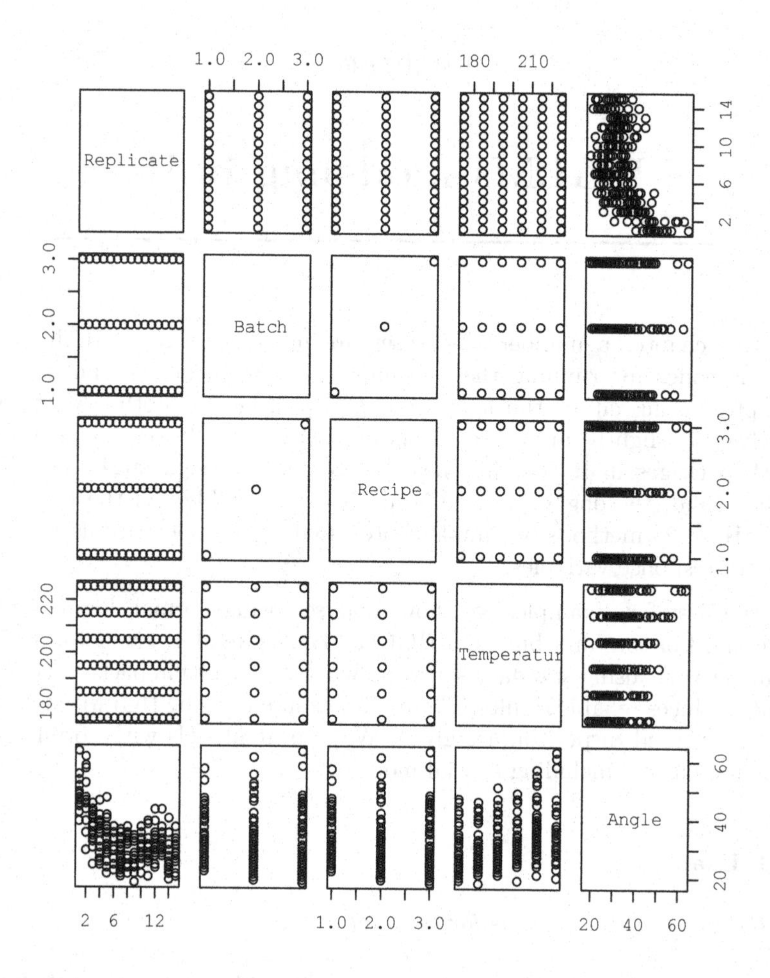


Figure 5.1 *Data included in the cake data with breaking angle as measure of interest.*

measurement presented here is the breaking angle (Figure 5.1). One half of a slab of cake is held fixed, while the other half is pivoted about the middle until breakage occurs. The angle through which the moving half has revolved is read on a circular scale. Since breakage is gradual, the reading tends to have a subjective element.

Consider the following normal linear mixed model: for $i = 1, ..., 3$ recipes,

$j = 1, ..., 6$ temperatures and $k = 1, ..., 15$ replicates,

$$y_{ijk} = \mu + \gamma_i + \tau_j + (\gamma\tau)_{ij} + v_k + v_{ik} + e_{ijk}$$

where $\gamma_i$ are main effects for recipe, $\tau_j$ are main effects for temperature, $(\gamma\tau)_{ij}$ are recipe-temperature interaction, $v_k$ are random replicates, $v_{ik}$ are whole-plot error components and $e_{ijk}$, are white noise. We assume all error components are independent and identically distributed. Cochran and Cox (1957) treated $v_k$ as fixed, but because of the balanced design structure the analyses are identical.

The same model but with responses $\log y_{ijk}$ gives a better fit. Since the deviance for the log-linear model is lower, we conclude that it has the best fit of the two models. The two models can be compared using the deviance

$$-2p_v(h|y_{ijk}; \beta, \tau) = 1639.1$$

and

$$-2p_v(h|\log(y_{ijk}); \beta, \tau) + 2\sum \log(y_{ijk}) = 1617.2,$$

where the second term is the Jacobian for the data transformation. Results from the log-normal linear mixed model are as follows and model checking plots are in Figure 5.2. The recipe with largest breaking angle seems to be Recipe I baked at 215°C. Furthermore, the variance between replicates is small ($\exp(-3.51) = 0.03$) compared to the residual variance ($\exp(2.996) = 20$).

```
Distribution of Main Response :
                          "gaussian"
[1] "Estimates from the model(mu)"
Angle~Recipe*Temperature+(1|Replicate)+(1|Replicate:Recipe)
[1] "identity"
                         Estimate Std. Error t-value
(Intercept)               3.36789    0.06151 54.7505
Recipe2                  -0.09715    0.06212 -1.5639
Recipe3                  -0.04392    0.06058 -0.7250
Temperature185            0.06114    0.05332  1.1467
Temperature195            0.04789    0.05365  0.8926
Temperature205            0.13079    0.05166  2.5317
Temperature215            0.27854    0.04870  5.7192
Temperature225            0.17573    0.05069  3.4669
Recipe2:Temperature185    0.03814    0.07808  0.4885
Recipe3:Temperature185   -0.03516    0.07718 -0.4555
Recipe2:Temperature195    0.12876    0.07696  1.6731
Recipe3:Temperature195    0.06739    0.07579  0.8890
```

```
Recipe2:Temperature205  0.04892     0.07553  0.6477
Recipe3:Temperature205 -0.04232     0.07487 -0.5653
Recipe2:Temperature215 -0.01590     0.07223 -0.2202
Recipe3:Temperature215 -0.08822     0.07115 -1.2399
Recipe2:Temperature225  0.09613     0.07345  1.3088
Recipe3:Temperature225  0.07012     0.07172  0.9777
[1] "Estimates for logarithm of lambda=var(u_mu)"
[1] "gaussian" "gaussian"
                  Estimate Std. Error t-value
Replicate           -3.509     0.3933   -8.92
Replicate:Recipe    -5.523     0.3556  -15.53
[1] "Estimates from the model(phi)"
phi ~ 1
<environment: 0x0000000018820ee0>
[1] "log"
            Estimate Std. Error t-value
(Intercept)    2.996    0.09465   31.65
[1] "== Likelihood Function Values and Condition AIC =="
                                       [,1]
-2 log(likelihood)             :  1632.1657
-2 log(restricted likelihood) :  1714.1769
cAIC                           :  1621.8752

Scaled deviance: 223.2533 on 223.2533 degrees of freedom
```

### *5.1.2 Gamma GLMM for cake data*

Here, the following gamma GLMM is considered, as it can be expected to give a better fit,

$$\log \mu_{ijk} = \mu + \gamma_i + \tau_j + (\gamma\tau)_{ij} + v_k + v_{ik}.$$

This model has a deviance $-2p_v(h|y_{ijk}; \beta, \theta) = 1616.1$ and therefore it is slightly better than the log-normal linear mixed model under the conditional AIC rule. Results from this model and model checking plots are as follows and in Figure 5.3. One can note that the histogram in the model checking plots shows a less skewed distribution for the studentized residuals. Furthermore, the normal probability plot has improved compared to the previous log-linear model.

```
Distribution of Main Response :
                              "gamma"
```

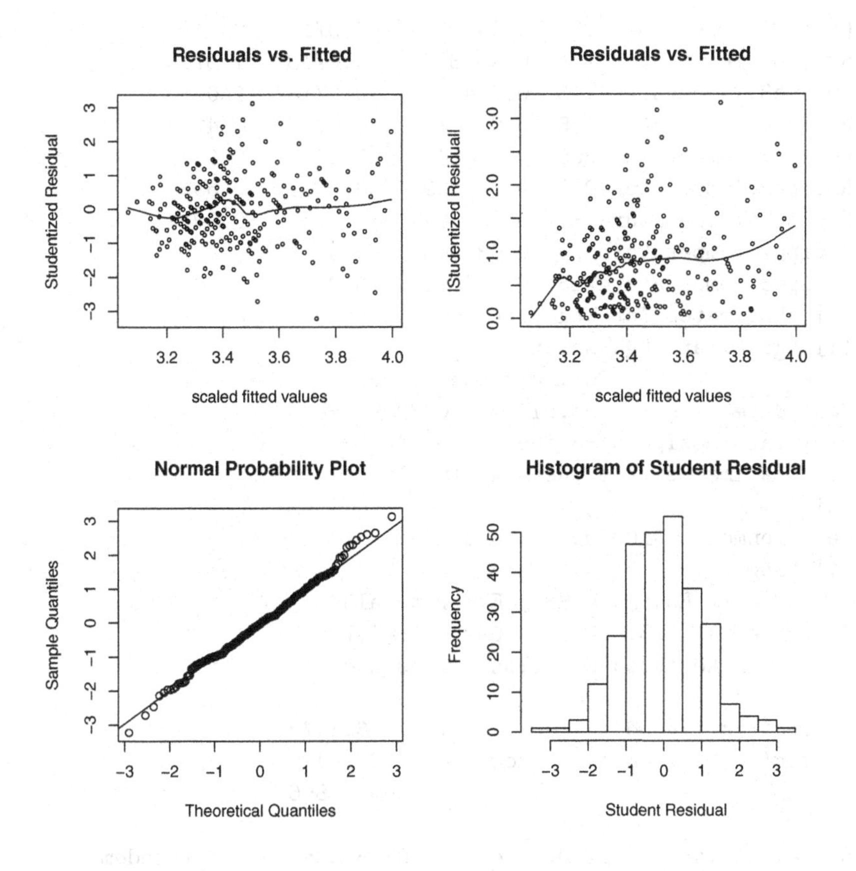


Figure 5.2 *Residual plot for the cake data under a linear mixed model.*

```
[1] "Estimates from the model(mu)"
Angle~Recipe*Temperature+(1|Replicate)+(1|Replicate:Recipe)
[1] "log"
                          Estimate Std. Error t-value
(Intercept)               3.353710    0.05965 56.2235
Recipe2                  -0.078756    0.05609 -1.4042
Recipe3                  -0.054051    0.05609 -0.9637
Temperature185            0.093259    0.05046  1.8480
Temperature195            0.061417    0.05046  1.2170
Temperature205            0.149099    0.05046  2.9546
Temperature215            0.287415    0.05046  5.6955
Temperature225            0.192808    0.05046  3.8207
Recipe2:Temperature185   -0.009348    0.07137 -0.1310
```

```
Recipe3:Temperature185 -0.051220    0.07137 -0.7177
Recipe2:Temperature195  0.097807    0.07137  1.3705
Recipe3:Temperature195  0.074481    0.07137  1.0436
Recipe2:Temperature205  0.028579    0.07137  0.4004
Recipe3:Temperature205 -0.038842    0.07137 -0.5443
Recipe2:Temperature215 -0.049499    0.07137 -0.6936
Recipe3:Temperature215 -0.064864    0.07137 -0.9089
Recipe2:Temperature225  0.081865    0.07137  1.1471
Recipe3:Temperature225  0.057664    0.07137  0.8080
[1] "Estimates for logarithm of lambda=var(u_mu)"
[1] "gaussian" "gaussian"
                  Estimate Std. Error t-value
Replicate           -3.514     0.3939  -8.922
Replicate:Recipe    -5.405     0.3426 -15.776
[1] "Estimates from the model(phi)"
phi ~ 1
<environment: 0x000000001c09d700>
[1] "log"
            Estimate Std. Error t-value
(Intercept)   -3.958    0.09472  -41.79
[1] "== Likelihood Function Values and Condition AIC =="
                                   [,1]
-2 log(likelihood)            :  1616.1173
-2 log(restricted likelihood) :  1697.8489
cAIC                          :  1604.3866

Scaled deviance: 222.0713 on 222.0713 degrees of freedom
```

### *5.1.3 Negative binomial model via Poisson HGLM for fabric data*

Consider fabric data in Bissell (1972). Response is the number of faults in a bolt of fabric of length $l$. By taking a look at the raw data (Figure 5.4), we can see that there seems to be a relationship between the number of faults and the length of the fabric. How is the data modeled appropriately?

If we fit the Poisson model

$$\log \mu = \alpha + x\beta,$$

where $x = \log l$ we have a deviance of 64.5 with 30 d.f., clearly indicating over-dispersion. However, it may be caused by the assumed Poisson regression model being incorrect. Azzalini et al. (1989) and Firth et al. (1991) introduced non-parametric tests for the goodness of fit of the

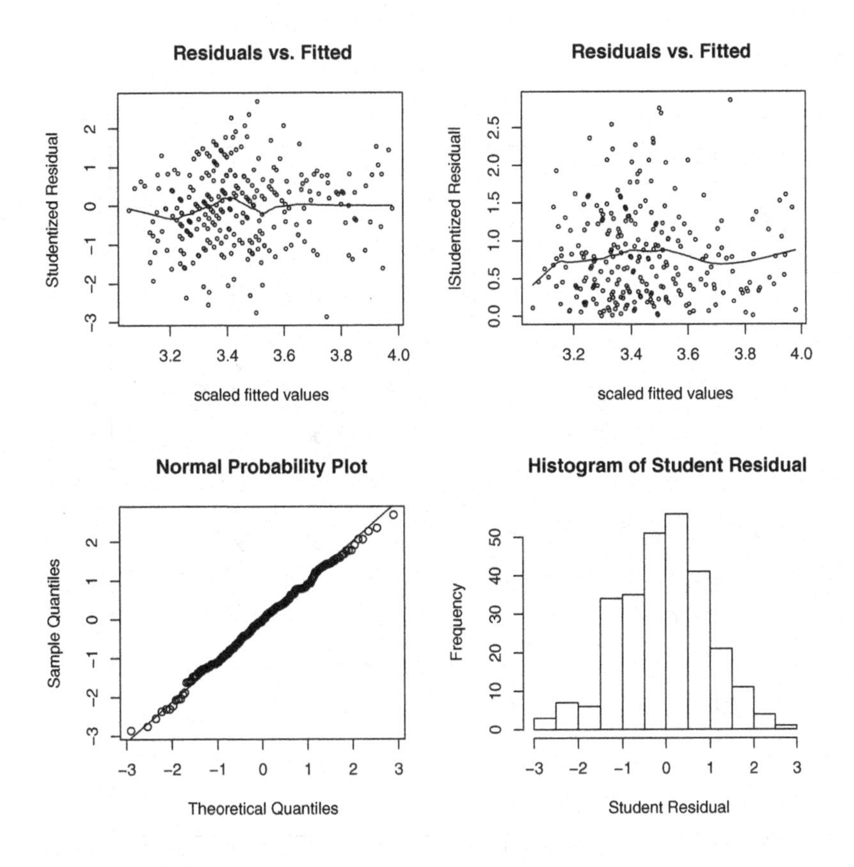


Figure 5.3 *Residual plot for the cake data under a gamma GLMM.*

Poisson regression model, and found that the over-dispersion is necessary for this dataset. For the analysis of such over-dispersed count data an exact likelihood approach is available. Bissell (1972) proposed the use of the negative binomial model, which can be fitted via a Poisson HGLM

$$\mu_c = E(y|u) = \exp(\alpha + x\beta)u,$$

$$\text{var}(y|u) = \mu_c,$$

where $u$ follows the gamma distribution with $E(u) = 1$ and $\text{var}(u) = \lambda$. This is a Poisson-gamma HGLM with saturated random effects. This leads to the model with an extra Poisson variation

$$\mu = E(y) = \exp(\alpha + x\beta)E(u) = \exp(\alpha + x\beta),$$

$$\text{var}(y) = E\{\text{var}(y|u)\} + \text{var}\{E(y|u)\} = \mu + \lambda\mu^2 > \mu.$$

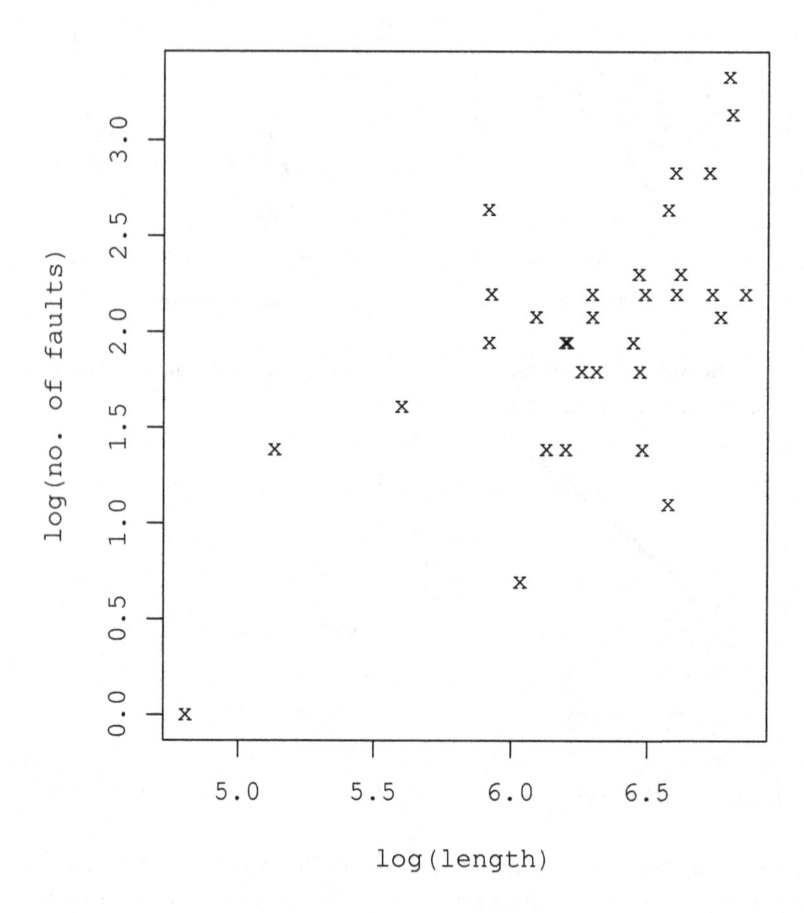


Figure 5.4 *Number of faults in the fabric dataset vs. the fabric length (on logarithmic scales).*

Hence, the model has the same marginal properties as a negative binomial model. From the following results, scaled deviance is 11.4 with degrees of freedom 12.8. With the negative-binomial model, we find no lack of fit (since the scaled deviance is approximately equal to the degrees of freedom) and no outliers in the normal probability plot (bottom left plot in Figure 5.5). Model checking plots are in Figure 5.5.

```
Distribution of Main Response :
```

```
                        "poisson"
[1] "Estimates from the model(mu)"
y ~ x + (1 | rf)
[1] "log"
            Estimate Std. Error t-value
(Intercept)  -3.7799     1.4431  -2.619
  x           0.9424     0.2257   4.176
[1] "Estimates for logarithm of lambda=var(u_mu)"
[1] "gamma"
   Estimate Std. Error t-value
rf   -2.076     0.3626  -5.726
[1] "== Likelihood Function Values and Condition AIC =="
                                          [,1]
-2 log(likelihood)             :   175.75612
-2 log(restricted likelihood) :   179.91955
cAIC                           :   172.76822

Scaled deviance: 14.33785 on 14.43390 degrees of freedom
```

### *5.1.4 Negative binomial model for train accident data continued*

Consider again British train accidents data presented in Chapter 2, where we investigated whether there was a trend in the number of accidents between trains and road vehicles (y) over the years (covariate x = number of years since 1975). Fitting the data assuming a Poisson GLM, there exist two outliers which give marginally significant lack of fit. We fit a negative binomial model via a Poisson-gamma HGLM with saturated random effects for full response, number of train accidents.

$$\log(\boldsymbol{\mu}) = \log(\boldsymbol{t}) + \alpha + \beta \boldsymbol{x} + \log(\boldsymbol{u}),$$

where $\boldsymbol{u}$ follows a gamma distribution with $E(u) = 1$ and $\text{var}(u) = \lambda$.

From the following results, scaled deviance is 11.4 with degrees of freedom 15.6 (p-value=0.72). Thus, with the negative-binomial model, there is no sign of lack of fit. From the normal probability plot in Figure 5.6, we see that the outlier corresponding to 1976 disappeared and that to 1986 reduced in size.

```
Distribution of Main Response :
                        "poisson"
[1] "Estimates from the model(mu)"
y ~ x + (1 | id)
```

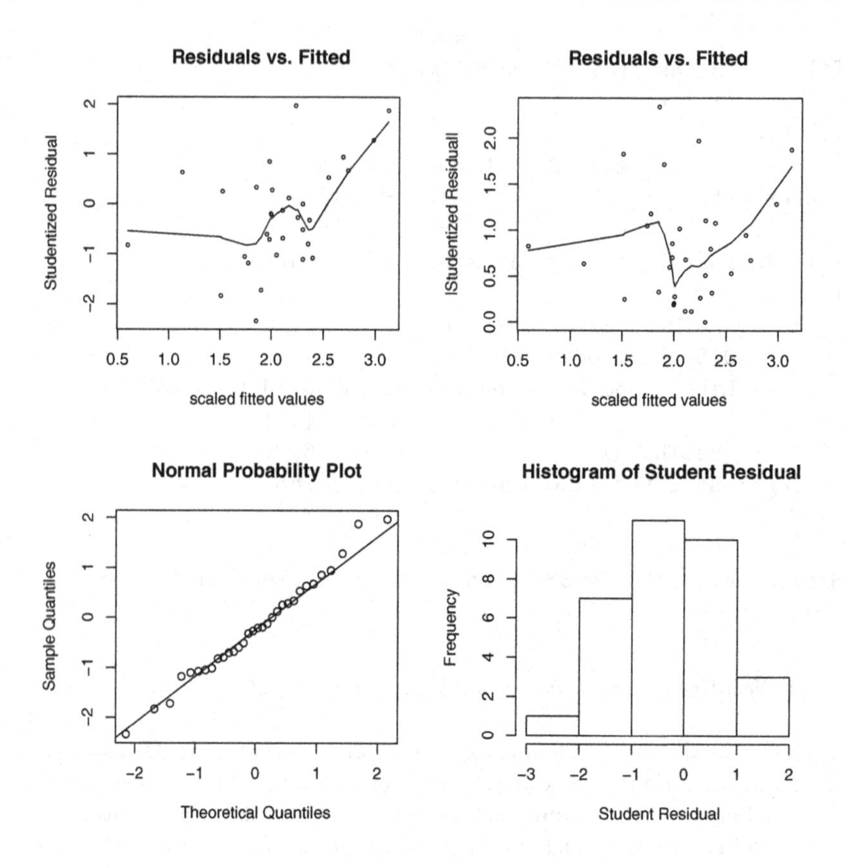


Figure 5.5 *Residual plot for the fabric data under a Poisson HGLM.*

```
[1] "log"
              Estimate Std. Error t-value
(Intercept) -4.13382    0.22286 -18.549
x           -0.03632    0.01451  -2.503
[1] "Estimates for logarithm of lambda=var(u_mu)"
[1] "gamma"
   Estimate Std. Error t-value
id   -1.753     0.4239  -4.137
[1] "== Likelihood Function Values and Condition AIC =="
                                        [,1]
-2 log(likelihood)              :  127.96315
-2 log(restricted likelihood) :  136.99910
cAIC                            :  129.85364
```

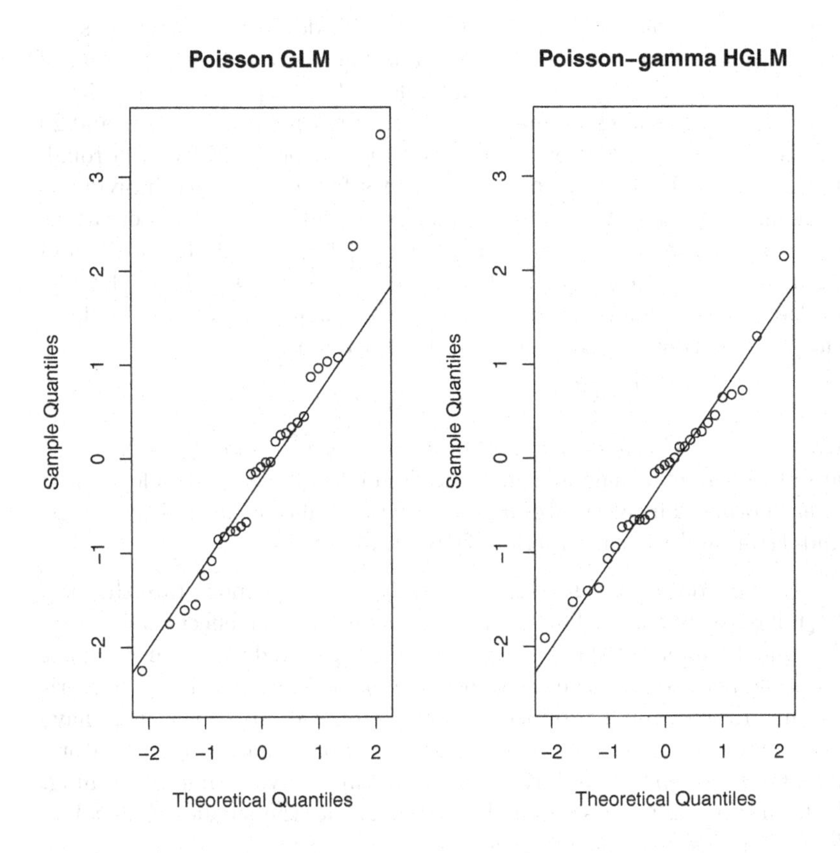


Figure 5.6 *Normal-probability plot for the train accident data under Poisson GLM and Poisson-gamma HGLM.*

```
Scaled deviance: 11.35076 on 15.58182 degrees of freedom
```

In the output, `id` is an observation identifier with $n_i = 1$ in (3.1) so that the number of levels in $u$ is equal to the total number of observations. There is still a negative trend ($\hat{\beta} = -0.03632$) but the standard error has increased from 0.0108 for the Poisson GLM to 0.0145. In general, neglecting over-dispersion gives too small standard errors.

### *5.1.5 Binomial GLMM for salamander data*

McCullagh and Nelder (1989) presented dataset on salamander mating.

Three experiments were conducted: two were done with the same salamanders in the summer and autumn and another one in the autumn of the same year using different salamanders. The response variable is binary, indicating success of mating. In each experiment, 20 females and 20 males from two populations called whiteside, denoted by W, and rough butt, denoted by R, were paired six times for mating with individuals from their own and the other population, resulting in 120 observations in each experiment. Covariates are, Trtf=0, 1 for female R and W and Trtm=0, 1 for male R and W. For $i, j = 1, ..., 20$ and $k = 1, 2, 3$, let $y_{ijk}$ be the outcome for the mating of the $i$th female with the $j$th male in the $k$th experiment. The model can be written as

$$\log\{p_{ijk}/(1-p_{ijk})\} = \boldsymbol{x}_{ijk}^t \boldsymbol{\beta} + v_{ik}^f + v_{jk}^m , \tag{5.1}$$

where $p_{ijk} = P(y_{ijk} = 1 | v_{ik}^f, v_{jk}^m)$, $v_{ik}^f \sim \mathrm{N}(0, \sigma_f^2)$, and $v_{jk}^m \sim \mathrm{N}(0, \sigma_m^2)$ are independent female and male random effects, assumed independent of each other. The covariates $\boldsymbol{x}_{ijk}$ comprise an intercept, indicators Trtf and Trtm, and their interaction Trtf·Trtm.

This binary data set has been extensively used to demonstrate that the PQL has severe biases. This model has crossed random effects, and therefore a high dimensional integral needs to be computed to obtain the exact marginal likelihood. Since the dimension of the integral is large, numerical integration cannot produce accurate computations. Thus, there have been many methods developed to obtain approximate ML estimators. EM, Gibbs sampling, MCMC type algorithms are very time consuming. Noh and Lee (2007a) showed that HL(1,2) has the smallest bias while HL(1,1) is fast with results as follows.

```
## Results for HL(1,1)
Distribution of Main Response :
                           "binomial"
[1] "Estimates from the model(mu)"
Mate~TypeM+TypeF+TypeM*TypeF+(1|Male)+(1|Female)
[1] "logit"
                Estimate Std. Error t-value
(Intercept)       1.0429     0.3928   2.655
TypeMW           -0.7288     0.4535  -1.607
TypeFW           -3.0037     0.5211  -5.764
TypeMW:TypeFW     3.7110     0.5567   6.666
[1] "Estimates for logarithm of lambda=var(u_mu)"
[1] "gaussian" "gaussian"
        Estimate Std. Error t-value
Male      0.1948     0.2798  0.6962
Female    0.3043     0.2725  1.1169
```

```
[1] "== Likelihood Function Values and Condition AIC =="
                                      [,1]
-2 log(likelihood)            :  418.7125
-2 log(restricted likelihood) :  419.0264
cAIC                          :  401.8995

Scaled deviance: 272.6058 on 295.3528 degrees of freedom

## Results for HL(1,2)
Distribution of Main Response :
                        "binomial"
[1] "Estimates from the model(mu)"
Mate~TypeM+TypeF+TypeM*TypeF+(1|Male)+(1|Female)
[1] "logit"
              Estimate Std. Error t-value
(Intercept)     1.0196     0.3964   2.572
TypeMW         -0.7239     0.4569  -1.584
TypeFW         -2.9727     0.5251  -5.661
TypeMW:TypeFW   3.6644     0.5591   6.554
[1] "Estimates for logarithm of lambda=var(u_mu)"
[1] "gaussian" "gaussian"
       Estimate Std. Error t-value
Male     0.2153     0.2786  0.7727
Female   0.3234     0.2715  1.1912
[1] "== Likelihood Function Values and Condition AIC =="
                                      [,1]
-2 log(likelihood)            :  418.7776
-2 log(restricted likelihood) :  419.0257
cAIC                          :  401.7016

Scaled Deviance:  271.2692 on 294.7835 degrees of freedom
```

### *5.1.6 Linear mixed model for integrated-circuit data*

In quality control experiments there is a need to model both the mean and the dispersion. Thus, Nelder and Lee (1991) proposed to use joint GLMs for the mean and dispersions. Suppose that a factory has 20 machines producing some product with repeated measurements on each machine. For such an example, it is reasonable to add a random machine effect. We wish to minimize the variance between machines and variance of products within machines. This can be modeled as an HGLM with structured dispersion. Between machine variation is modeled by the random effect variance and within machine variation is modeled by the dispersion (residual variance).

An experiment on integrated circuits was reported by Phadke et al. (1983). The width of lines made by a photoresist-nanoline tool were measured in five different locations on silicon wafers, measurements being taken before and after an etching process being treated separately. Here, the pre-etching data are analyzed. The eight experimental factors (A-H) were arranged in an $L_{18}$ orthogonal array and produced 33 measurements at each of five locations, giving a total of 165 observations. There were no whole-plot (i.e., between-wafer) factors. Wolfinger and Tobias (1998) developed a structured dispersion analysis for a normal HGLM, having wafers as random effects: Let $q$ be the index for wafers and $r$ for observations within wafers. The mean can be modeled as

$$y_{ijkop,qr} = \beta_0 + a_i + b_j + c_k + g_o + h_p + v_q + e_{qr},$$

where $v_q \sim \mathrm{N}(0, \lambda)$, $e_{qr} \sim \mathrm{N}(0, \phi)$, and $\lambda$ and $\phi$ represent the between-wafer and within-wafer variances respectively, which can be affected by the experimental factors (A-H). The dispersion and random effect variance can be modeled as

$$\log \phi_{imno} = \gamma_0^w + a_i^w + e_m^w + f_n^w + g_o^w$$

and

$$\log \lambda_m = \gamma_0^b + e_m^b.$$

where the superscripts $w$ and $b$ refer to within- and between-wafer variances. Results for parameter estimators are as follows and model checking plots are in Figure 5.7.

The between-wafer variance is $\exp(-4.778) = 0.008$ for the factor class E1; $\exp(-4.778 - 1.299) = 0.002$ and $\exp(-4.778 + 1.489) = 0.037$ for E2 and E3, respectively. The within-wafer (residual) variance varies for the different factors A, E, F, and G. For instance, the within-wafer variance for an observation from the set of factor classes A2, E1, F1, and G1 is $\exp(-4.711 - 0.862) = 0.004$.

```
Distribution of Main Response :
                        "gaussian"
[1] "Estimates from the model(mu)"
Width ~ A + B + C + G + H + (1 | Wafer)
[1] "identity"
              Estimate Std. Error    t-value
(Intercept)   2.452702    0.04932   49.72959
A2            0.377821    0.04639    8.14494
B2           -0.567583    0.04111  -13.80802
```

```
C2             0.387664    0.04349   8.91408
C3             0.521416    0.05229   9.97211
G2            -0.176381    0.05095  -3.46176
G3            -0.393040    0.04537  -8.66211
H2            -0.003333    0.04724  -0.07055
H3             0.306740    0.05129   5.98033
[1] "Estimates for logarithm of lambda=var(u_mu)"
[1] "gaussian"
            Estimate Std. Error t-value
(Intercept)   -4.778     0.6662  -7.172
E2            -1.299     1.1530  -1.127
E3             1.489     0.8625   1.726
[1] "Estimates from the model(phi)"
phi ~ A + E + F + G
[1] "log"
            Estimate Std. Error   t-value
(Intercept) -4.71051     0.3293 -14.30494
A2          -0.86224     0.2436  -3.53958
E2          -0.01587     0.3045  -0.05211
E3           0.67716     0.2923   2.31661
F2           0.69656     0.2998   2.32370
F3           1.04301     0.2990   3.48779
G2          -0.14500     0.2902  -0.49964
G3          -0.65146     0.2992  -2.17701
[1] "== Likelihood Function Values and Condition AIC =="
                                    [,1]
-2 log(likelihood)            :  -232.1114
-2 log(restricted likelihood) :  -188.7905
cAIC                          :  -255.0044

Scaled deviance: 137.0051 on 137.0051 degrees of freedom
```

### *5.1.7 Gamma HGLM for semiconductor data*

This example is taken from Myers et al. (2002). It involves a designed experiment in a semiconductor plant. Six factors are employed, and it is of interest to study the curvature or camber of the substrate devices produced in the plant. There is a lamination process, and the camber measurement is made four times on each device produced. The goal is to model the camber taken in $10^{-4}$ in./in. as a function of the design variables. Each design variable is taken at two levels and the design is a $2^{6-2}$ fractional factorial. The camber measurement is known to be

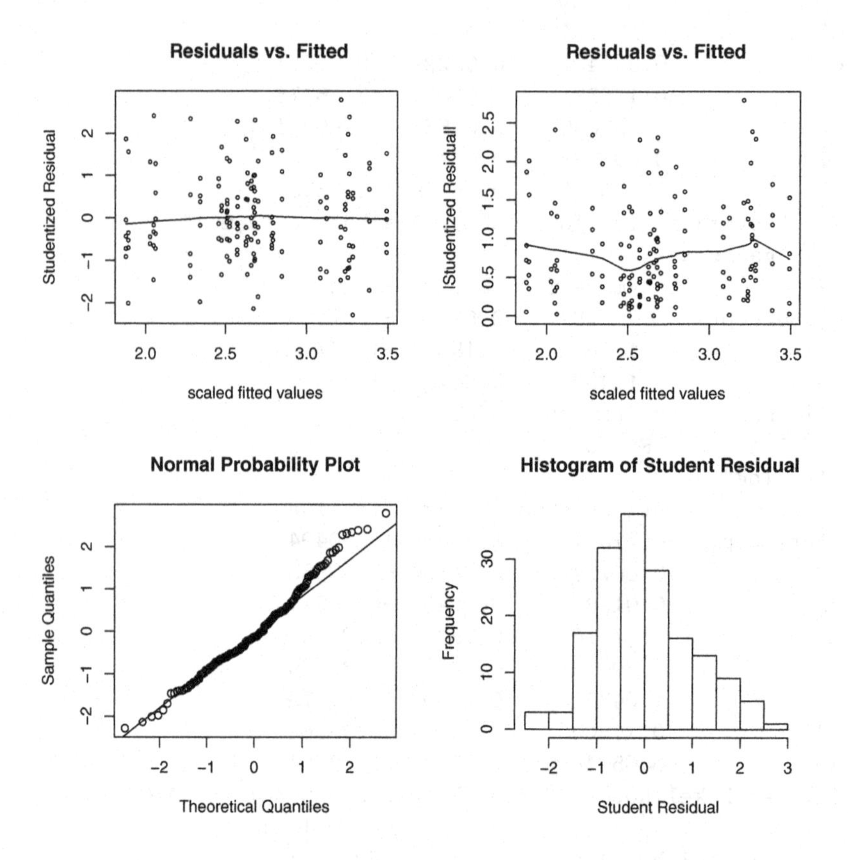


Figure 5.7 *Residual plot for the integrated-circuit data under a normal HGLM.*

nonnormal. Because the measurements were taken on the same device they are correlated. Myers et al. (2002) considered a gamma response model with a log link. They used a GEE approach assuming a working correlation to be AR(1). Because there are only four measurements on each device the compound symmetric and AR(1) correlation may not be very distinguishable. We considered a gamma HGLM by adding a random effect for the device in the mean model.

$$\log \mu = \beta_0 + x_1\beta_1 + x_3\beta_3 + x_5\beta_5 + x_6\beta_6 + v,$$

$$\log \phi = \gamma_0 + x_2\gamma_2 + x_3\gamma_3,$$

where $\mu = E(y|v)$ and the dispersion parameter $\lambda$ (variance of random effect) is for the between-variance $\text{var}(E(y|v))$, while the dispersion parameter $\phi$ is for the within-variance in a gamma HGLM with $\text{var}(y|v) = \phi\mu^2$.

Results for parameter estimators are as follows and model checking plots are in Figure 5.8.

The estimated variance of the random effect $v$ is $\exp(-3.014) = 0.049$. Furthermore, there are significant differences in the dispersion parameter $\phi$ within factors $x_2$ and $x_3$.

```
Distribution of Main Response :
                              "gamma"
[1] "Estimates from the model(mu)"
y ~ x1 + x3 + x5 + x6 + (1 | Device)
[1] "log"
             Estimate Std. Error t-value
(Intercept)  -4.7114     0.06676 -70.568
x1            0.2089     0.06623   3.154
x3            0.3281     0.06676   4.915
x5           -0.1739     0.06623  -2.626
x6           -0.3573     0.06618  -5.399
[1] "Estimates for logarithm of lambda=var(u_mu)"
[1] "gaussian"
       Estimate Std. Error t-value
Device   -3.014     0.5149  -5.854
[1] "Estimates from the model(phi)"
phi ~ x2 + x3
[1] "log"
             Estimate Std. Error t-value
(Intercept)  -2.6102      0.1954 -13.358
x2           -0.6730      0.1953  -3.446
x3           -0.4915      0.1952  -2.518
[1] "== Likelihood Function Values and Condition AIC =="
                                         [,1]
-2 log(likelihood)             :   -573.8579
-2 log(restricted likelihood)  :   -555.9127
cAIC                           :   -578.1991

Scaled deviance: 51.4562 on 51.4562 degrees of freedom
```

### *5.1.8 Binomial HGLM with structured dispersion for respiratory data*

This example is from a clinical trial comparing two treatments for a respiratory illness (Strokes et al., 1995). In each of two medical centers, eligible patients were randomly assigned to active treatment (=1) or placebo (=0). During treatment, respiratory status (poor=0, good=1)

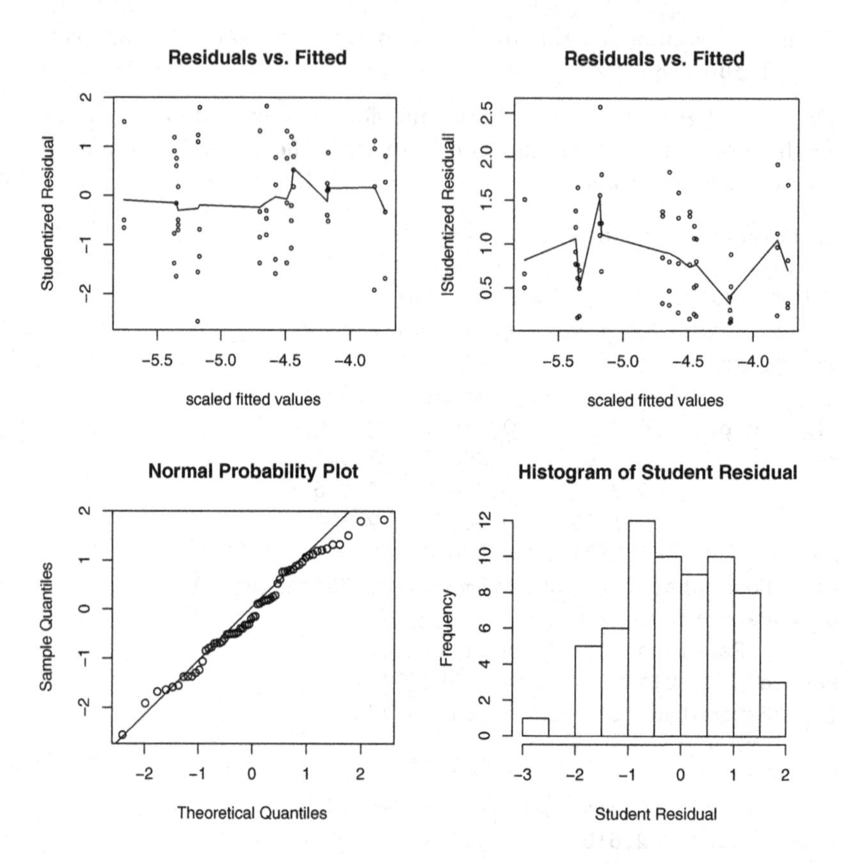


Figure 5.8 *Residual plot for the semiconductor data under a gamma HGLM.*

was determined at four visits. Potential explanatory variables were center, sex (male=1, female=0), and baseline respiratory status (poor=0, good=1), as well as age (in years) at the time of study entry. There were 111 patients (54 active, 57 placebo) with no missing data for responses or covariates. We consider the following HGLM: for $i = 1, \cdots, 111$ and $j = 1, \cdots, 4$

$$\log\left(\frac{p_{ij}}{1-p_{ij}}\right) = \beta_0^{(\mu)} + \beta_1^{(\mu)} trt_i + \beta_2^{(\mu)} msex_i + \beta_3^{(\mu)} age_i$$
$$+\beta_4^{(\mu)} center_i + \beta_5^{(\mu)} base_i + \beta_6^{(\mu)} y_{i(j-1)} + v_i, \tag{5.2}$$

where $p_{ij} = P(y_{ij} = 1|v_i, y_{i(j-1)})$, $v_i \sim \mathrm{N}(0, \lambda_i)$ and we take the baseline value for the ith subject as $y_{i(0)}$. The GLM with additional covariates $y_{i(j-1)}$ was introduced as a transition model by Diggle et al. (1994), as an alternative to a random-effect model. Furthermore, the random effects have a structured dispersion

$$\log \lambda_i = \beta_0^{(\lambda)} + \beta_1^{(\lambda)} age_i. \tag{5.3}$$

From the mean model treatment and baseline effects are significant. If the patient had a good baseline status with active treatment, (s)he will have more chance of having a good status in the next visit. From the dispersion model, heteroscedasticity increases with age implying that older people are more unstable in respiratory illness. Model-checking plots for $\lambda$ are in Figure 5.9. Note that the model checking plots give valuable information here even though the response values are binary.

```
Distribution of Main Response :
                      "binomial"
[1] "Estimates from the model(mu)"
y ~ trt + msex + age + center + base + past + (1 | patient)
[1] "logit"
            Estimate Std. Error t-value
(Intercept) -1.11053    1.03307 -1.0750
trt          1.25595    0.41480  3.0279
msex        -0.26118    0.59717 -0.4374
age         -0.03503    0.01854 -1.8890
center       0.68157    0.41922  1.6258
base         1.82060    0.44629  4.0794
past         0.57539    0.30431  1.8908
[1] "Estimates for logarithm of lambda=var(u_mu)"
[1] "gaussian"
            Estimate Std. Error t-value
(Intercept) -0.68345    0.73711 -0.9272
age          0.04726    0.02021  2.3387
[1] "== Likelihood Function Values and Condition AIC =="
                                  [,1]
-2 log(likelihood)           :  431.0405
-2 log(restricted likelihood) :  438.3807
cAIC                         :  422.4576

Scaled deviance: 298.0320 on 381.7868 degrees of freedom
```

### *5.1.9 Random slope model for orthodontic growth data*

The orthodontic growth data (Pinheiro and Bates, 2000) contain the growth measurements of 27 children for 11 girls and 16 boys from age 8

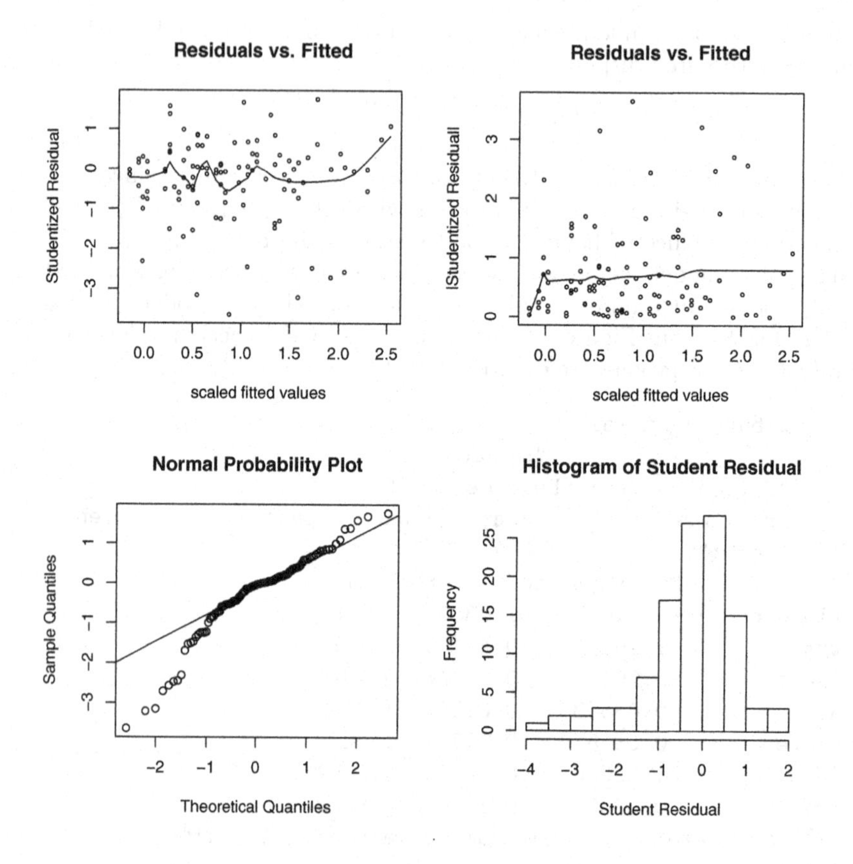


Figure 5.9 *Residual plot of $\lambda$ for the respiratory data.*

until age 14. Every two years, the distance (mm) between the pituitary and the pterygomaxillary fissure was recorded at ages 8, 10, 12, and 14 using x-ray images of the skull. Pinheiro and Bates (2000) considered the correlated random intercept and slope model similar to

$$y_{ij} = \beta_1 F_i + \beta_2 F_i A_{ij} + \beta_3 M_i + \beta_4 M_i A_{ij} + v_{1i} + A_{ij} v_{2i} + \epsilon_{ij} \quad (5.4)$$

where $y_{ij}$ is distance of the $i$th subject at the $j$th age $A_{ij}$, $F_i$ is 1 (or 0) for Female (or Male), $M_i = 1 - F_i$ and $\epsilon_{ij} \sim \mathrm{N}(0, \phi_{ij})$. Because the sex and age interaction is significant we consider different random intercept and slope models for gender; $\beta_1$ and $\beta_2$ ($\beta_3$ and $\beta_4$) are the overall mean intercept and slope for female (male), respectively. The random intercept $v_{1i}$ and random slope $v_{2i}$ are assumed to be bivariate normal distribution with mean 0, variances $\mathrm{var}(v_{1i} = \lambda_1)$ and $\mathrm{var}(v_{2i} = \lambda_2)$, and their correlation $\mathrm{Corr}(v_{1i}, v_{2i}) = \rho$.

For this HGLM, both hglm and dhglm give the same results as for instance lmer and the SAS procedure Proc Mixed. Model checking plots in Figure 5.10 identify three outliers. Two outliers are from the second and third observation of the ninth boy and one outlier is the first observation of the thirteenth boy. The fitting results are reported in Table 5.1 for data both with and without the outliers showing the sensitivity of the results. We will see in a later section how a DHGLM can be used for more robust analysis.

Table 5.1 *Estimates (SE) for the orthodontic growth data*

| | With full data | Without three outliers |
|---|---|---|
| $\beta_1$ | 17.37(1.23) | 17.37(0.90) |
| $\beta_2$ | 0.48(0.10) | 0.48(0.08) |
| $\beta_3$ | 16.34(1.02) | 17.39(0.76) |
| $\beta_4$ | 0.78(0.09) | 0.69(0.06) |
| $\log(\lambda_1)$ | 1.76(0.74) | 1.24(0.65) |
| $\log(\lambda_2)$ | −3.43(0.57) | −3.93(0.52) |
| $\rho$ | −0.67(0.24) | −0.39(0.17) |
| $\log(\phi)$ | 0.54(0.26) | −0.14(0.17) |

### *5.1.10 Spatial model for Scottish lip cancer data*

Clayton and Kaldor (1987) analyzed observed ($y_i$) and expected numbers ($n_i$) of lip cancer cases in the 56 administrative areas of Scotland with a view to produce a map that would display regional variation in cancer incidence and yet avoid the presentation of unstable rates for the smaller areas. The expected numbers had been calculated allowing for the different age distributions in the areas by using a fixed-effects multiplicative model; these were regarded for the purpose of analysis as constants (i.e., offset) based on an external set of standard rates. Presumably the spatial aggregation is due in large part to the effects of environmental risk factors. Data were available on the percentage of the work force in each area employed in agriculture, fishing, or forestry ($x_i$). This covariate exhibits spatial aggregation paralleling that for lip cancer itself. Because all three occupations involve outdoor work, exposure to sunlight, the principal known risk factor for lip cancer, might be the explanation. For analysis Breslow and Clayton (1993) considered the following Poisson

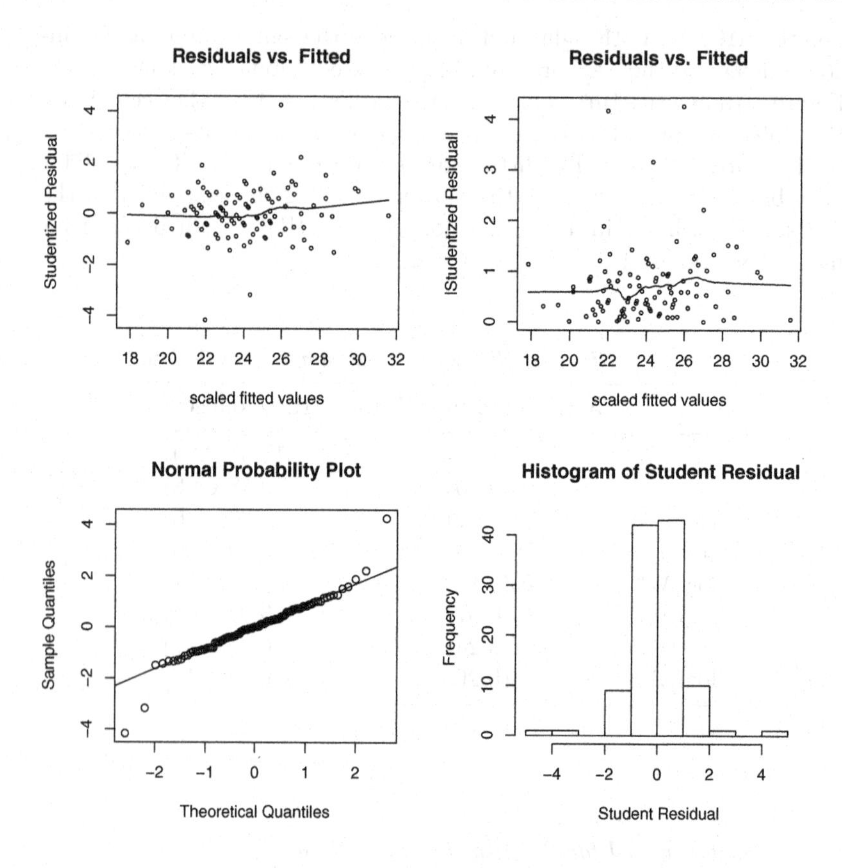


Figure 5.10 *Residual plot for orthodontic growth data including the three potential outliers.*

HGLM with the log link

$$
\begin{aligned}
y_i|v_i &\sim Poisson(\mu_i) \\
\eta_i &= \log\mu_i = \log n_i + \beta_0 + \beta_1 x_i/10 + v_i
\end{aligned}
$$

where $v_i$ represented unobserved area-specific log-relative risks. They tried models,

M1: $v_i \sim \mathrm{N}(0, \lambda)$ and

M2: $v_i \sim$ intrinsic autoregressive model (IAR).

M3: $v_i \sim$ MRF in which $[\mathrm{var}(v)]^{-1} = (I - \rho M)/\lambda$, where $M$ is the incidence matrix for neighbours. Thus, we consider three HGLMs with independent random effects (M1), IAR random effects (M2), and MRF random effects (M3). The latter two models assume spatially correlated random effects for spatial correlation.

Table 5.2 *Results from the three models: M1 (independent random effects), M2 (IAR correlated random effects) and M3 (MRF correlated random effects)*

| Model | Model part | Parameter | estimate | s.e. | t |
|---|---|---|---|---|---|
| M1 | Mean | (Intercept) | 1.909 | 0.175 | 10.90 |
| | | $x$ | 0.115 | 0.159 | 0.73 |
| | Random effect | $\log(\lambda)$ | $-1.036$ | 0.198 | |
| M2 | Mean | (Intercept) | $-0.177$ | 0.117 | $-1.51$ |
| | | $x$ | 0.357 | 0.122 | 2.92 |
| | Random effect | $\log(\lambda)$ | 0.525 | 0.180 | |
| M3 | Mean | (Intercept) | 0.271 | 0.207 | 1.31 |
| | | $x$ | 0.376 | 0.121 | 3.10 |
| | Random effect | $\log(\lambda)$ | 0.153 | 0.052 | |
| | | $\rho$ | 0.174 | 0.002 | |

Lee and Nelder (2001b) chose the model M3 as best. The MRF model with $\rho = 0$ is the M1 model. From Table 5.2, MRF with $\hat{\rho} = 0.174$ provides a suitable model. We found that the main difference between M1 and M3 is the prediction for county 49, which has the highest predicted value because it has the largest $n_i$. This gives the very large leverage value (or hat value) of 0.92, for example, under M3. For observed value of 28, M2 predicts 27.4, while M3 gives 29.8. It is the leverage which exaggerates the differences in predictions. Though model checking plots are useful, our eyes could be misled, so that objective criteria based upon the likelihood are also required in the model selection.

### *5.1.11 Matern model for loaloa data*

Rousset et al. (2016) presented the loaloa dataset which describes prevalence of infection by the nematode Loa loa in North Cameroon, 1991-2001. The study investigated the relationship between altitude, vegetation indices, and prevalence of the parasite. Let $y_i$ denote the number of infected individuals at $i$th location out of $n_i$ individuals ($i = 1, \cdots, 197$). Six covariates are considered:

- elev1 ($x_1$) : altitude, in m

- elev2 ($x_2$), elev3 ($x_3$), elev4 ($x_4$) : additional altitude variables derived from the previous one, provided for convenience, respectively, positive values of altitude-650, positive values of altitude-1000, and positive values of altitude-1300
- maxNDVI1 ($x_5$): a copy of maxNDVI modified as maxNDVI1 (maxNDVI1= 0.8 if maxNDVI > 0.8, maxNDVI1=maxNDVI otherwise) where maxNDVI is maximum normalized-difference vegetation index (NDVI) from repeated satellite scans
- seNDVI ($x_6$): standard error of NDVI

We consider binomial GLMM with the logit link

$$y_i|v_i \sim Binomial(n_i, p_i)$$
$$\log\left(\frac{p_i}{1-p_i}\right) = \beta_0 + \beta_1 x_1 + \beta_2 x_2 + \beta_3 x_3 + \beta_4 x_4 + \beta_5 x_5 + \beta_6 x_6 + v_i$$

where the random effect $v_i$ is for the $i$th location. Rousset et al. (2016) fitted HGLMs

M1: $v_i$ ~independent $\mathrm{N}(0, \lambda)$ and

M2: $v_i \sim$ normal distribution with variance $\lambda$ and Matern correation for two locations which is represented by

$$(1 - Nugget)\frac{(\rho d)^{\nu} K_{\nu}(\rho d)}{2^{\nu-1}\Gamma(\nu)},$$

where $Nugget$ is a parameter describing a discontinuous decrease in correlation at zero distance, $\rho$ is scaling parameter, $\nu$ is smoothness parameter, $K_{\nu}$ is the bessel K function of order $\nu$ and $d$ is distance computed by longitudes and latitudes for two locations. Following Rousset et al. (2016), in the dhglm package we let $Nugget = 0$, whereas $\rho$ and $\nu$ are estimated.

The M2 model has $cAIC = 1106.5$ which is less than $cAIC = 1127.0$ for M1, so that M2 model is selected between the two models. This means that the model with the Matern correlation fits better than the model with the independent random effects. The h-likelihood estimators from the dhglm package are equivalent to those from the spaMM package of Rousset et al. (2016). From Table 5.3, we see that $x_2$ does not show significance in M2, while it is misleadingly significant in M1 with a negative estimate.

### 5.1.12 State space model for gas consumption data

Durbin and Koopman (2000) analyzed the lagged quarterly demand for

Table 5.3 *Results for the loaloa data: M1 (independent random effects) and M2 (Matern correlated random effects)*

| Model | Model part | Parameter | estimate | s.e. | t |
|---|---|---|---|---|---|
| M1 | Mean | (Intercept) | −15.115 | 1.973 | −7.661 |
| | | $x_1$ | 0.00280 | 0.000526 | 5.334 |
| | | $x_2$ | −0.00395 | 0.00112 | −3.523 |
| | | $x_3$ | −0.00854 | 0.00226 | −3.784 |
| | | $x_4$ | 0.00873 | 0.00299 | 2.917 |
| | | $x_5$ | 14.9216 | 2.997 | 4.979 |
| | | $x_6$ | 1.862 | 4.919 | 0.377 |
| | Random effect | $\log(\lambda)$ | −0.411 | 0.116 | |
| | cAIC | | 1127 | | |
| M2 | Mean | (Intercept) | −10.330 | 2.977 | −3.471 |
| | | $x_1$ | −0.0000194 | 0.000620 | −0.0313 |
| | | $x_2$ | 0.000800 | 0.00148 | 0.541 |
| | | $x_3$ | −0.0117 | 0.00256 | −4.555 |
| | | $x_4$ | 0.0109 | 0.00318 | 3.434 |
| | | $x_5$ | 11.000 | 2.738 | 4.018 |
| | | $x_6$ | −3.028 | 4.424 | −0.685 |
| | Random effect | $\log(\lambda)$ | 1.826 | 0.128 | |
| | | $\rho$ | 0.0135 | | |
| | | $\nu$ | 0.243 | | |
| | cAIC | | 1106 | | |

gas in the UK from 1960 to 1986. They considered a local linear-trend model with quarterly seasonals which can be represented as a normal HGLM

$$y_t = \alpha + f_t + s_t + q_t + e_t$$

where $f_t = \sum_{j=1}^{t} r_j$ and $s_t = \sum_{j=1}^{t}(t-j+1)p_j$ are random effects for the local linear trend, the quarterly seasonals $q_t$ with $\sum_{j=0}^{3} q_{t-j} = w_t$, and $r_t \sim \mathrm{N}(0, \lambda_r)$, $p_t \sim \mathrm{N}(0, \lambda_p)$, $w_t \sim \mathrm{N}(0, \lambda_w)$, $e_t \sim \mathrm{N}(0, \phi_t)$. This model involves four independent random components $(r_t, p_t, w_t, e_t)$ of full size. Lee, Nelder, and Pawitan (2017) added a linear trend $\beta t$ and found that the random walk $f_t$ is not necessary. Thus, they considered a model

$$y_t = \alpha + \beta t + s_t + q_t + e_t. \tag{5.5}$$

Results for parameter estimators are as follows and model checking plots are in Figure 5.11. In the R output below, `st` $= \log(\widehat{\lambda}_p) = -5.368$ for $s_t$ and `qt` $= \log(\widehat{\lambda}_w) = -11.864$ for $q_t$, respectively. The residual plot displays apparent outliers, caused by a disruption in the gas supply in the third and fourth quarters of 1970.

```
Distribution of Main Response :
                       "gaussian"
[1] "Estimates from the model(mu)"
y ~ time + (1 | st) + (1 | qt)
[1] "identity"
             Estimate Std. Error   t-value
(Intercept) 4.807e+00   0.032301 148.81066
time        8.381e-05   0.007217   0.01161
[1] "Estimates for logarithm of lambda=var(u_mu)"
[1] "gaussian" "gaussian"
     Estimate Std. Error t-value
st    -5.368     0.1805  -29.74
qt   -11.864     0.5113  -23.20
[1] "Estimates from the model(phi)"
phi ~ 1
<environment: 0x0000000008472730>
[1] "log"
            Estimate Std. Error t-value
(Intercept)   -6.056     0.2325  -26.05
[1] "== Likelihood Function Values and Condition AIC =="
                                        [,1]
-2 log(likelihood)             :  -159.27062
-2 log(restricted likelihood) :  -145.50763
cAIC                           :  -298.60811

Scaled deviance: 31.01180 on 31.01181 degrees of freedom
```

Lee, Nelder, and Pawitan (2017) proposed to delete the random quarterly seasonals and add further fixed effects to model the 1970 disruption and seasonal effects

$$\begin{aligned} y_t &= \alpha + t\beta + \alpha_i + t\beta_i + \delta_1(t = 43) + \delta_2(t = 44) \\ &+ \gamma_1 \sin(2\pi t/104) + \gamma_2 \cos(2\pi t/104) + s_t + e_t, \end{aligned} \tag{5.6}$$

where $i = 1, ..., 4$ represents quarters, and $\delta_1$ and $\delta_2$ are explanatory variables for the third and fourth quarters of 1970. Lee, Nelder, and Pawitan (2017) further found extra dispersion in the third and fourth quarters, which led to a structured dispersion model

$$\log \phi_t = \varphi + \psi_i$$

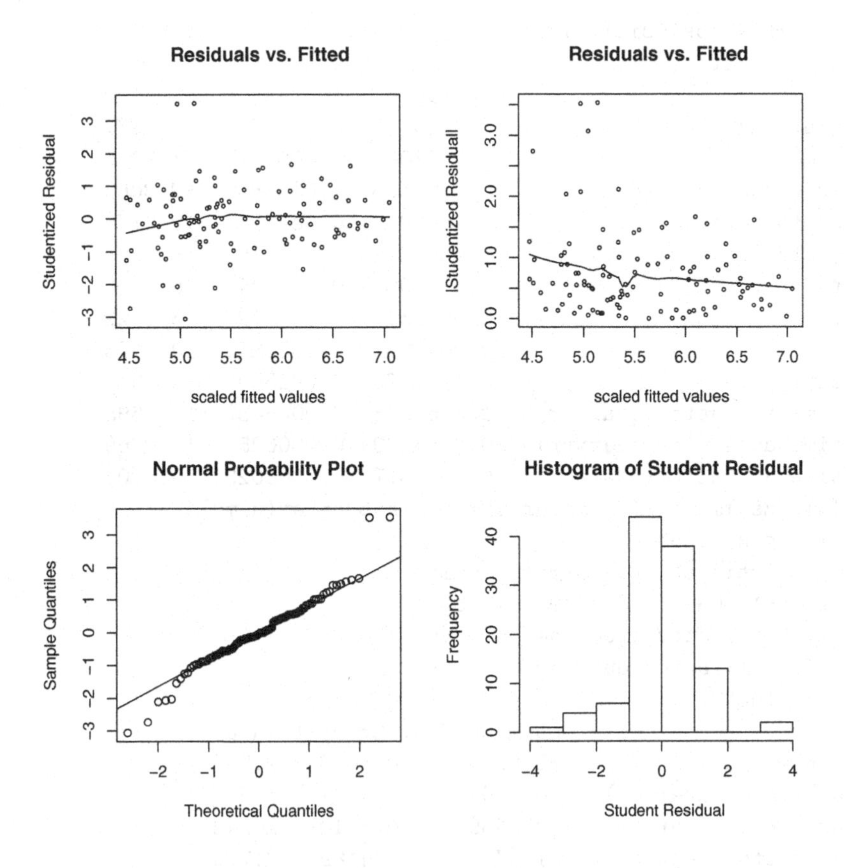


Figure 5.11 *Residual plot for gas consumption data under the model (5.5).*

where $\psi_i$ are the quarterly main effects. Results for parameter estimators are as follows and model checking plots are in Figure 5.12. The dispersion model clearly shows that the heterogeneity increases with quarters. Residual plot shows that most of the outliers have vanished. So the heterogeneity can be explained by adding covariates for the dispersion. This model has $cAIC = -228.1$ which is less than $cAIC = -298.6$ for model (5.5). Thus, both model-checking plots and cAICs clearly indicate that model (5.6) is the best among models considered for this dataset up to now.

```
Distribution of Main Response :
                       "gaussian"
[1] "Estimates from the model(mu)"
y ~ time+as.factor(quarter)+time*as.factor(quarter)+t43
```

```
    +t44+cos1+sin1+(1|time)
[1] "identity"
                          Estimate Std. Error  t-value
(Intercept)              5.0794620  0.1197486  42.4177
time                     0.0157092  0.0088362   1.7778
as.factor(quarter)2     -0.0945533  0.0336858  -2.8069
as.factor(quarter)3     -0.4892068  0.0395240 -12.3775
as.factor(quarter)4     -0.3597921  0.0514351  -6.9951
t43                      0.4715085  0.0891348   5.2898
t44                     -0.3907317  0.1212623  -3.2222
cos1                    -0.0610581  0.1035216  -0.5898
sin1                    -0.1431079  0.0915608  -1.5630
time:as.factor(quarter)2 -0.0060673  0.0005432 -11.1695
time:as.factor(quarter)3 -0.0094092  0.0006257 -15.0369
time:as.factor(quarter)4  0.0004577  0.0008025   0.5704
[1] "Estimates for logarithm of lambda=var(u_mu)"
[1] "gaussian"
     Estimate Std. Error t-value
st    -12.09     0.6979  -17.33
[1] "Estimates from the model(phi)"
phi ~ as.factor(quarter)
[1] "log"
                     Estimate Std. Error t-value
(Intercept)           -5.8645     0.3002 -19.539
as.factor(quarter)2    0.5402     0.4191   1.289
as.factor(quarter)3    0.9439     0.4210   2.242
as.factor(quarter)4    1.5791     0.4189   3.770
[1] "== Likelihood Function Values and Condition AIC =="
                                      [,1]
-2 log(likelihood)             :  -230.41079
-2 log(restricted likelihood)  :  -148.96252
cAIC                           :  -228.12313

Scaled deviance: 91.95559 on 91.95559 degrees of freedom
```

### *5.1.13 Additive non-parametric regression model for prestige data*

We consider an additive non-parametric regression model with the prestige data from the R-package car (Fox et al., 2016). Three variables are used in this example with 102 observations: income ($x_1$: average income of incumbents, dollars, in 1971), education ($x_2$: average education of occupational incumbents, years, in 1971) and prestige ($y$: Pineo-Porter

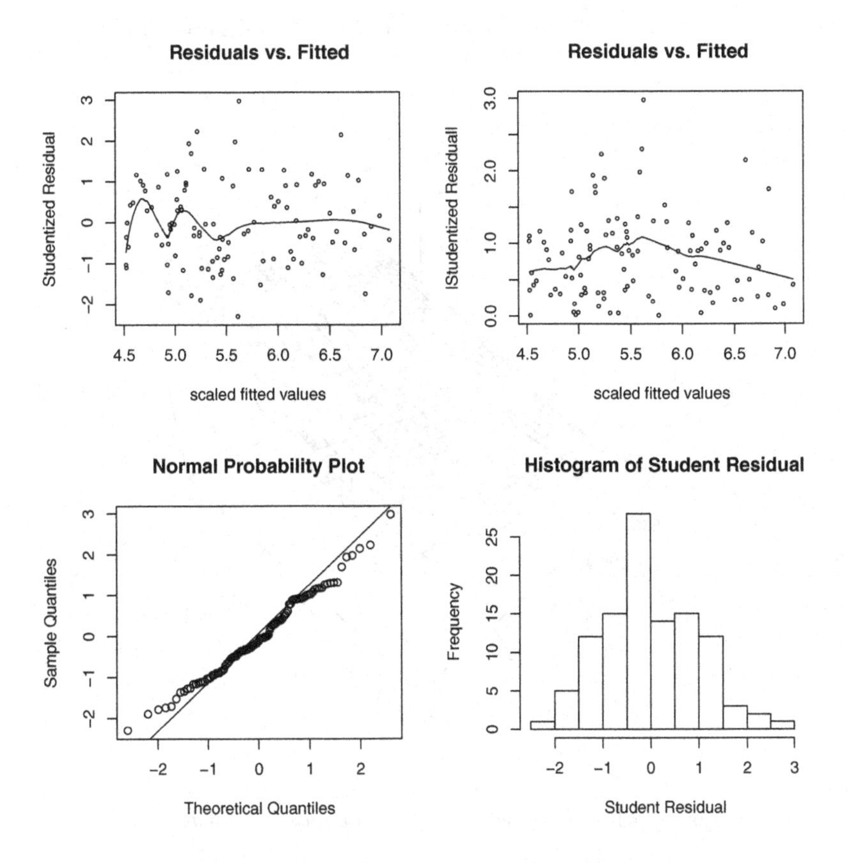


Figure 5.12 *Residual plot for gas consumption data under the model (5.6).*

prestige score for occupation, from a social survey conducted in the mid-1960s). The regression of prestige on income and education is illustrated here. Consider the following additive non-parametric regression model:

$$y_i = f_1(x_{1i}) + f_2(x_{2i}) + e_i,$$

where $f_1(\cdot)$ and $f_2(\cdot)$ are unknown functions and $e_i \sim \mathrm{N}(0, \sigma^2)$. Suppose that cubic smoothing splines are used to fit these unknown functions $f_1(\cdot)$ and $f_2(\cdot)$, which are characterized by singular precision matrices, $P_1$ and $P_2$, respectively (see Section 9.4 of Lee, Nelder, and Pawitan (2017)). This additive model can be fitted by using an HGLM,

$$y_i = \mathbf{x}_i^T \beta + v_{1i} + v_{2i} + e_i$$

where $\mathbf{x}_i^T = (1, x_{1i}, x_{2i})$ and $v_1 \sim \mathrm{N}(0, P_1^+)$ and $v_2 \sim \mathrm{N}(0, P_2^+)$ are random effects with $P^+$ being the Moore-Penrose inverse of $P$.

The fitted surface for the additive model is given in Figure 5.13.

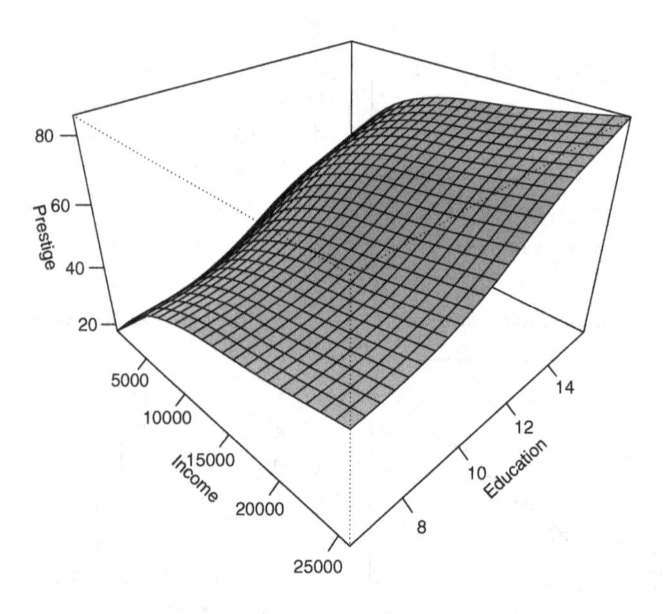


Figure 5.13 *The fitted surface for the additive model.*

The regression surface $\hat{f}_1(x_{1i}) + \hat{f}_2(x_{2i})$ from the additive model is in Figure 5.14, which shows that prestige increases with income and education.

## 5.2 R code

*The cake data (LMM)*

```
data(cake,package="mdhglm")
cake$Recipe<-as.factor(cake$Recipe)
cake$Temperature<-as.factor(cake$Temperature)
cake$Replicate<-as.factor(cake$Replicate)
model_mu<-DHGLMMODELING(Model="mean",Link="identity",
         LinPred=Angle~Recipe*Temperature
         +(1|Replicate)+(1|Replicate:Recipe),
```

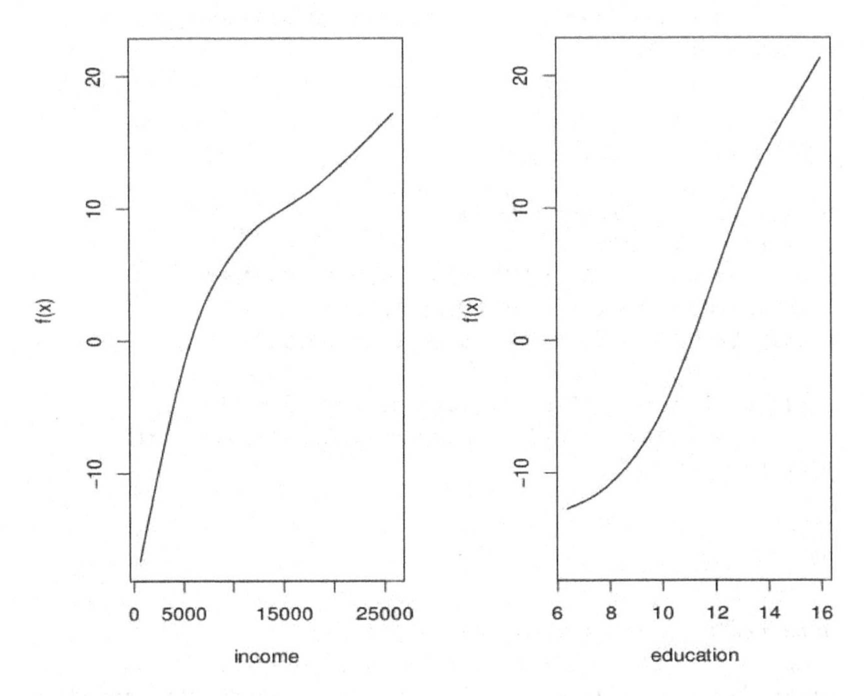


Figure 5.14 *The partial-regression functions for the additive model. The left is for income and the right is for education.*

```
    RandDist=c("gaussian","gaussian"))
model_phi<-DHGLMMODELING(Model="dispersion")

fit1<-dhglmfit(RespDist="gaussian",DataMain=cake,
        MeanModel=model_mu,DispersionModel=model_phi)
plotdhglm(fit1)
```

*The cake data (gamma GLMM)*

```
model_mu<-DHGLMMODELING(Model="mean",Link="log",
        LinPred=Angle~Recipe*Temperature
        +(1|Replicate)+(1|Replicate:Recipe),
    RandDist=c("gaussian","gaussian"))
model_phi<-DHGLMMODELING(Model="dispersion")

fit2<-dhglmfit(RespDist="gamma",DataMain=cake,
```

```
        MeanModel=model_mu,DispersionModel=model_phi)
plotdhglm(fit2)
```

*Fabric data*

```
data(fabric, package="mdhglm")
fabric$x<-log(fabric$x)
model_mu<-DHGLMMODELING(Model="mean",Link="log",
    LinPred=y~x+(1|rf),RandDist="gamma")
model_phi<-DHGLMMODELING(Model="dispersion")

fit1<-dhglmfit(RespDist="poisson",DataMain=fabric,
        MeanModel=model_mu,DispersionModel=model_phi)
plotdhglm(fit1)
```

*Train accident data*

```
data(train,package="mdhglm")
model_mu<-DHGLMMODELING(Model="mean",Link="log",
LinPred=y~x+(1|id), Offset=log(train$t), RandDist="gamma")
model_phi<-DHGLMMODELING(Model="dispersion")

fit<-dhglmfit(RespDist="poisson",DataMain=train,
    MeanModel=model_mu,DispersionModel=model_phi)
qqnorm(fit[[2]],main="Poisson-gamma HGLM")
qqline(fit[[2]])
```

*Salamander data*

```
data(salamander, package="mdhglm")
model_mu<-DHGLMMODELING(Model="mean",Link="logit",
 LinPred=Mate~TypeM+TypeF+TypeM*TypeF+(1|Male)+(1|Female),
 RandDist=c("gaussian","gaussian"))
model_phi<-DHGLMMODELING(Model="dispersion")

fit11<-dhglmfit(RespDist="binomial",
 DataMain=salamander,mord=1,dord=1,
 MeanModel=model_mu,DispersionModel=model_phi)

fit12<-dhglmfit(RespDist="binomial",
```

```
 DataMain=salamander,mord=1,dord=2,
 MeanModel=model_mu,DispersionModel=model_phi)
```

*Integrated-circuit data*

```
data(circuit, package="mdhglm")
circuit$A<-as.factor(circuit$A)
circuit$B<-as.factor(circuit$B)
circuit$C<-as.factor(circuit$C)
circuit$D<-as.factor(circuit$D)
circuit$E<-as.factor(circuit$E)
circuit$F<-as.factor(circuit$F)
circuit$G<-as.factor(circuit$G)
circuit$H<-as.factor(circuit$H)
circuit$I<-as.factor(circuit$I)

model_mu<-DHGLMMODELING(Model="mean", Link="identity",
 LinPred=Width~A+B+C+G+H+(1|Wafer),RandDist="gaussian",
 LinkRandVariance="log",LinPredRandVariance=lambda~1+E)
model_phi<-DHGLMMODELING(Model="dispersion", Link="log",
 LinPred=phi~A+E+F+G)

circuit_model<-dhglmfit(RespDist="gaussian",DataMain=circuit,
 MeanModel=model_mu,DispersionModel=model_phi)
plotdhglm(circuit_model)
```

*Semiconductor data*

```
data(semicon, package="mdhglm")
model_mu<-DHGLMMODELING(Model="mean", Link="log",
 LinPred=y~x1+x3+x5+x6+(1|Device),RandDist="gaussian")
model_phi<-DHGLMMODELING(Model="dispersion", Link="log",
 LinPred=phi~x2+x3)

fit3<-dhglmfit(RespDist="gamma",DataMain=semicon,
 MeanModel=model_mu,DispersionModel=model_phi)
plotdhglm(fit3)
```

*Respiratory data*

```
data(respiratory, package="mdhglm")
respiratory$trt<-as.numeric(respiratory$treatment=="active")
respiratory$msex<-as.numeric(respiratory$sex=="male")
respiratory$base<-respiratory$baseline
model_mu1<-DHGLMMODELING(Model="mean", Link="logit",
 LinPred=y~trt+msex+age+center+base+past+(1|patient),
 RandDist="gaussian",LinkRandVariance="log",
 LinPredRandVariance=lambda~1+age)
model_phi1<-DHGLMMODELING(Model="dispersion",Link="log")

fit1<-dhglmfit(RespDist="binomial",DataMain=respiratory,
 MeanModel=model_mu1,DispersionModel=model_phi1)
plotdhglm(fit1,type="lambda")
```

*Orthodontic growth data*

```
data(Orthodont,package="nlme")
Orthodont$nsex <- as.numeric(Orthodont$Sex=="Male")
Orthodont$nsexage <- with(Orthodont, nsex*age)
model_mu<-DHGLMMODELING(Model="mean", Link="identity",
LinPred=distance~ nsex + age + nsexage +
(1|Subject)+(age|Subject),RandDist="gaussian")
model_phi<-DHGLMMODELING(Model="dispersion")

res_hglm<-dhglmfit(RespDist="gaussian",DataMain=Orthodont,
MeanModel=model_mu,DispersionModel=model_phi)
```

*Gas consumption data*

```
data(gas,package="mdhglm")
gas$t43<-ifelse(gas$time==43,1,0)
gas$t44<-ifelse(gas$time==44,1,0)

ft<-matrix(0,108,108)
st<-matrix(0,108,108)
wt<-matrix(0,108,108)
for (i in 1:108) {
   for (j in 1:i) {
       ft[i,j]<-1
```

```
    }
}
for (i in 1:108) {
    for (j in 1:i) {
       st[i,j]<-(i-j+1)
    }
}
for (i in 1:108) {
    if (i==1) wt[i,1]<-1
    if (i==2) {
          wt[i,1]<-1
          wt[i,2]<-1
    }
    if (i==3) {
          wt[i,1]<-1
          wt[i,2]<-1
          wt[i,3]<-1
    }
    if (i>3) {
       for (j in 1:4) {
          ii<-i-j+1
          wt[i,ii]<-1
       }
   }
}

qt<-solve(wt)
gas$cos1<-cos(6.2831853070*gas$time/104)
gas$sin1<-sin(6.2831853070*gas$time/104)

model_mu<-DHGLMMODELING(Model="mean", Link="identity",
LinPred=y~time+(1|time)+(1|time),
RandDist=c("gaussian","gaussian"),
LMatrix=list(st,qt))

model_phi<-DHGLMMODELING(Model="dispersion")

res_hglm1<-dhglmfit(RespDist="gaussian",DataMain=gas,
MeanModel=model_mu,DispersionModel=model_phi)

plothglm(res_hglm1)
```

```
model_mu<-DHGLMMODELING(Model="mean", Link="identity",
LinPred=y~time+as.factor(quarter)+
time*as.factor(quarter)+t43+t44+cos1+sin1
+(1|time),RandDist="gaussian",
LMatrix=list(st))

model_phi<-DHGLMMODELING(Model="dispersion",Link="log",
LinPred=phi~as.factor(quarter))

res_hglm2<-dhglmfit(RespDist="gaussian",DataMain=gas,
MeanModel=model_mu,DispersionModel=model_phi)

plotdhglm(res_hglm2)
```

*Scottish lip cancer data*

*hglm code*

```
data(cancer)
logE <- log(E)
XX <- model.matrix(~ Paff)
#M3
cancerCAR <- hglm(X = XX, y = O, Z = diag(56),
         family = poisson(),
         rand.family = CAR(D = nbr),
         offset = logE, conv = 1e-8,
         maxit = 200, fix.disp = 1)
summary(cancerCAR)
```

*dhglm code*

```
data(cancer, package="hglm.data")
county <- factor(1:56)
lip <- data.frame(cbind(logE, O, Paff, county))
#M1
model_mu<-DHGLMMODELING(Model="mean", Link="log",
 LinPred=O~Paff+(1|county),
RandDist="gaussian",LinkRandVariance="log",
 LinPredRandVariance=lambda~1, Offset=logE)
```

```
model_phi<-DHGLMMODELING(Model="dispersion")

res_hglm<-dhglmfit(RespDist="poisson",DataMain=lip,
 MeanModel=model_mu,DispersionModel=model_phi)
plotdhglm(res_hglm)

#M2
res_hglm2<-HGLMREML.h(y~x+(1|country),Offset=lip$n,
 DataMain=lip,
 RespDist="Poisson",RespLink="log",RandDist="normal",
 mord=0,dord=1,spatial="IAR",Neighbor=adjacent)

#M3
res_hglm3<-HGLMREML.h(y~x+(1|country),Offset=lip$n,
 DataMain=lip,
 RespDist="Poisson",RespLink="log",RandDist="normal",
 mord=0,dord=1,spatial="MRF",Neighbor=adjacent)
```

*Loaloa data*

```
data(Loaloa)
Loaloa$y <- Loaloa$npos
Loaloa$n <- Loaloa$ntot
Loaloa$x1 <- Loaloa$elev1
Loaloa$x2 <- Loaloa$elev2
Loaloa$x3 <- Loaloa$elev3
Loaloa$x4 <- Loaloa$elev4
Loaloa$x5 <- Loaloa$maxNDVI1
Loaloa$x6 <- Loaloa$seNDVI
Loaloa$LOC <- c(1:197)

#M1
model_mu<-DHGLMMODELING(Model="mean", Link="logit",
LinPred=y~x1+x2+x3+x4+x5+x6+(1|LOC),
RandDist="gaussian")
model_phi<-DHGLMMODELING(Model="dispersion")
res_hglm<-dhglmfit(RespDist="binomial",
BinomialDen=Loaloa$n,DataMain=Loaloa,
MeanModel=model_mu,DispersionModel=model_phi)

#M2
```

```
model_mu<-DHGLMMODELING(Model="mean", Link="logit",
LinPred=y~x1+x2+x3+x4+x5+x6+(1|LOC),
RandDist="gaussian",spatial=Matern,
longitude=longitude,
latitude=latitude)
model_phi<-DHGLMMODELING(Model="dispersion")
res_hglm<-dhglmfit(RespDist="binomial",
BinomialDen=Loaloa$n,DataMain=Loaloa,
MeanModel=model_mu,DispersionModel=model_phi)
```

*Prestige data*

```
data(Prestige,package="car")
model_mu<-DHGLMMODELING(Model="mean", Link="identity",
LinPred=prestige~income+education,spline=c("cubic"))
model_phi<-DHGLMMODELING(Model="dispersion")

res<-dhglmfit(RespDist="gaussian",DataMain=Prestige,
MeanModel=model_mu,DispersionModel=model_phi)
```

## 5.3 Exercises

1. Test the significance of the interaction effect of Recipe×Temperature for the cake data using the gamma GLMM.

2. Compute 95% confidence intervals for the two variance components in the salamander dataset using the binomial GLMM with Male and Female as random effects.

3. Elston et al. (2001) studied the number of ticks on red grouse chicks. The reference and data (grouseticks) are available in the lme4 package. Run the following Poisson GLMM in lme4 with YEAR and HEIGHT (i.e., altitude) as fixed effects and BROOD, INDEX and LOCATION as random effects:

```
library(lme4)
data(grouseticks)
lin_pred <- TICKS~YEAR+HEIGHT+
            (1|BROOD)+(1|INDEX)+(1|LOCATION)
```

```
full_mod1 <- glmer(lin_pred, family="poisson",
           data=grouseticks))
```

There are convergence problems with this model and therefore it may be useful to study the hat values to check whether they are close to 1. Use the hglm package to check the hat values and explain why there are more hat values than observations. Why can hat values close to 1 cause convergence problems?

```
library(hglm)
full_mod2 <- hglm2(lin_pred, family=poisson(),
             data=grouseticks)
full_mod2$hv
```

4. Crowder (1978) analyzed the proportion of seeds that germinated on each of 21 plates arranged according to a 2 by 2 factorial layout by seed ($x_1$) and type of root extract ($x_2$). The data are shown in the table below, where $y_i$ and $n_i$ are the number of germinated and the total number of seeds on the $i$-th plate, $i = 1, ..., 21$. These data are also analyzed by, for example, Breslow and Clayton (1993), and can be downloaded at http://www.math.ntnu.no/~hrue/r-inla.org/examples/Winbugs-data

/Seeds.txt or directly in R as data(seeds, package="mdhglm").

| Plate | y | n | seed ($x_1$) | root ($x_2$) |
|---|---|---|---|---|
| 1 | 10 | 39 | 0 | 0 |
| 2 | 23 | 62 | 0 | 0 |
| 3 | 23 | 81 | 0 | 0 |
| | | | ... | |
| 6 | 5 | 6 | 0 | 1 |
| 7 | 53 | 74 | 0 | 1 |
| | | | ... | |
| 12 | 8 | 16 | 1 | 0 |
| 13 | 10 | 30 | 1 | 0 |
| | | | ... | |
| 20 | 32 | 51 | 1 | 1 |
| 21 | 3 | 7 | 1 | 1 |

Fit a random effects logistic model, allowing for over-dispersion. If $p_i$ is the probability of germination on the $i$-th plate:

$$\begin{aligned} y_i|v_i &\sim Bin(n_i, p_i) \\ \log\left(\frac{p_i}{1-p_i}\right) &= \beta_0 + \beta_1 x_{i1} + \beta_2 x_{i2} + \beta_{12}(x_{i1} \cdot x_{i2}) + v_i \\ v_i &\sim \mathrm{N}(0, \lambda). \end{aligned}$$

What is the estimated effect of root extract type? Is it significant? Compute a 95% CI for the random effect variance.

5. Breslow (1984) analyzed some mutagenicity assay data (shown below) on salmonella in which three plates have been processed at each dose $i$ of quinoline and the number of revertant colonies of TA98 Salmonella measured. The data are available at http://www.math.ntnu.no/~hrue/r-inla.org/examples/Winbugs-data/Salm.txt or data(salm) in the mdhglm package.

| $i$ | dose ($\mu g$) | No. of colonies | | |
|---|---|---|---|---|
| 1 | 0 | 15 | 21 | 29 |
| 2 | 10 | 16 | 18 | 21 |
| 3 | 33 | 16 | 26 | 33 |
| 4 | 100 | 27 | 41 | 69 |
| 5 | 333 | 33 | 38 | 41 |
| 6 | 1000 | 20 | 27 | 42 |

Fit a random effects Poisson model allowing for over-dispersion. Let $y_{ij}$ be the number of colonies in the $j$-th plate of the $i$-th dose ($i = 1, \cdots, 6,\ j = 1, 2, 3$):

$$\begin{aligned} y_{ij}|v_i &\sim Possion(\mu_i) \\ \log(\mu_i) &= \beta_0 + \beta_1 \log(dose_i + 10) + \beta_2 dose_i + v_i \\ v_i &\sim \mathrm{N}(0, \lambda). \end{aligned}$$

Are your estimates similar to those using the INLA package in R? Discuss similarities and differences between the HGLM and INLA approaches. [The INLA package can be installed using the R command `install.packages ("INLA", repos="http://www.math.ntnu.no/inla/R/stable")`, and the R code to run the INLA example is available at

http://www.r-inla.org/examples/volume-1/code-for-salm-example.]

6. The blackcap dataset in the Rousset et al. (2016) is extracted from a

study of genetic polymorphisms potentially associated to migration behavior in the blackcap (Sylvia atricapilla). Across different populations in Europe and Africa, the average migration behavior was found to correlate with average allele size (dependent on the number of repeats of a small DNA motif) at the locus ADCYAP1, encoding a neuropeptide. The data frame includes 14 observations on the following variables. For response variable "migStatus" with covariate "means", fit normal HGLM with Matern correlation by using distance between two locations.

```
data(blackcap,package="spaMM")
# latitude : latitude, indeed.
# longitude : longitude, indeed.
# migStatus : migration status as determined by Mueller
  et al,
#  from 0 (resident populations) to 2.5
#  (long-distance migratory populations)
# means : Mean allele sizes in each population
# pos : Numerical index for the populations
```

CHAPTER 6

# Double HGLMs - using the **dhglm** package

The likelihood principle means that likelihood methods can give an efficient way of data analysis if the model is right. Thus, it is important to check the model to verify the analysis of data. However, it may be difficult to check all the model assumptions. In binary data, for instance, it is often difficult to identify the distribution of random effects. Lee, Nelder, and Pawitan (2017) showed that the normal assumption in binary GLMMs can give serious biases if the normal assumption on random effects is not right. Thus, it could lead to a wrong analysis if the likelihood method is too model sensitive and model assumptions are difficult to check. Furthermore, likelihood inferences for some models are known to be sensitive to outliers of data contaminations. If the data size is small we may check the data carefully to identify outliers, but for big-sized data it may be difficult to identify outliers or contaminated data. Even though we are able to identify them, it is difficult to decide whether to throw them away (possibly causing information loss) or to keep them (possibly causing distorted analysis if the method is sensitive). A widely cited weakness of the likelihood method is that it is not robust against model distributional assumptions or presence of outliers or data contamination. Thus, it is desirable to develop models, whose likelihood inferences will be robust against such violations. This is possible by assuming a model covering a broad class of distributions.

In this chapter, we present Double HGLM (DHGLM) which is a very general class of models for a single response variable. HGLMs have component models for parameters $(\mu, \phi, \lambda)$, where only the model for $\mu$ allows random effects. DHGLMs allow random effects for all component models for $(\mu, \phi, \lambda)$:

1. Random effects for $\mu$ provide covariance modeling and extra-Poisson or extra-binomial variation.

2. Random effects for $\phi$ provide a heavy tailed distribution for residual variance to give a robust analysis against outliers or data contaminations.

3. Random effects for $\lambda$ provide heavy tailed distributions for random effects to give a robust analysis against misspecifications of distributional assumptions on random effects. Furthermore, some distributional assumptions provide variable selection as we shall study in Chapter 10.

The GLM notation shows how the models can be easily extended for further modeling of the dispersion parameters. Various DHGLMs will be presented for analysis of data together with the R code for fitting these models using the dhglm package. We also show that many models can be unified and extended by using DHGLMs. For example, they include such models in the field of finance as autoregressive conditional heteroscedasticity (ARCH; Engle (1982)) and generalized ARCH (GARCH; Bollerslev (1986)). An other application of DHGLMs is variable selection for high-dimensional models, having more parameters than observations in Chapter 10.

In summary, DHGLMs allow random effects in variance components and error variance, so that various heavy-tailed distributions can be accommodated for various components of the single model.

## 6.1 Model description for DHGLMs

In DHGLMs there are many design matrices and parameters to be introduced, which requires some new notation. Here, superscripts are used to indicate which part of the model the design matrix, or parameter, belongs to. Consider for instance the linear mixed model that we now write as

$$\boldsymbol{y} = \mathbf{X}^{(\mu)}\boldsymbol{\beta}^{(\mu)} + \mathbf{Z}^{(\mu)}\boldsymbol{v}^{(\mu)} + \boldsymbol{e}$$

with random effects $v_i^{(\mu)} \sim \mathrm{N}(0, \lambda)$ and residuals $e_i \sim \mathrm{N}(0, \phi_i)$. Here $\mathbf{X}^{(\mu)}$ and $\mathbf{Z}^{(\mu)}$ are design matrices and $\beta^{(\mu)}$ and $\boldsymbol{v}^{(\mu)}$ are fixed and random effects, respectively, for the component model for $\mu$.

Up to now, we have modeled the mean and the dispersion parameters, $\mu$ and $\phi$, simultaneously by adding random effects for the corresponding linear predictors. Now we will also model the dispersion parameter for the random effects $\lambda$ together with $\mu$ and $\phi$, which will give a wide range of models. To keep track of the various possibilities, some logical classification of the models is required. So, to explain the model structure of DHGLMs, we represent a DHGLM as

$$\{\text{model}(\mu), \text{model}(\phi)\},$$

where model($\mu$) is the model for the mean and model($\phi$) is the model for the dispersion. Using this notation we can summarize model developments as follows:

i) The original GLM by Nelder and Wedderburn (1972) can be represented by {GLM($\mu$), constant}:

$$\boldsymbol{\eta}^{(\mu)} = g^{(\mu)}(\mu) = \mathbf{X}^{(\mu)}\boldsymbol{\beta}^{(\mu)}$$

with a constant $\phi$.

ii) The joint GLM by Nelder and Lee (1991) is represented by {GLM($\mu$), GLM($\phi$)}:

$$\begin{aligned}\boldsymbol{\eta}^{(\mu)} &= g^{(\mu)}(\mu) = \mathbf{X}^{(\mu)}\boldsymbol{\beta}^{(\mu)}\\ \boldsymbol{\eta}^{(\phi)} &= g^{(\phi)}(\phi) = \mathbf{X}^{(\phi)}\boldsymbol{\beta}^{(\phi)}\end{aligned}$$

iii) The HGLM by Lee and Nelder (1996) is represented by {HGLM($\mu$), constant}:

$$\boldsymbol{\eta}^{(\mu)} = g^{(\mu)}(\mu) = \mathbf{X}^{(\mu)}\boldsymbol{\beta}^{(\mu)} + \mathbf{Z}^{(\mu)}\boldsymbol{v}^{(\mu)}$$

with constant of $\phi$ and $\lambda$.

iv) HGLM with structured dispersion (Lee and Nelder, 2001a) is represented by {HGLM($\mu$), GLM($\phi$)}.

$$\begin{aligned}\boldsymbol{\eta}^{(\mu)} &= g^{(\mu)}(\mu) = \mathbf{X}^{(\mu)}\boldsymbol{\beta}^{(\mu)} + \mathbf{Z}^{(\mu)}\boldsymbol{v}^{(\mu)}\\ \boldsymbol{\eta}^{(\lambda)} &= g^{(\lambda)}(\lambda) = \mathbf{X}^{(\lambda)}\boldsymbol{\beta}^{(\lambda)}\\ \boldsymbol{\eta}^{(\phi)} &= g^{(\phi)}(\phi) = \mathbf{X}^{(\phi)}\boldsymbol{\beta}^{(\phi)}.\end{aligned}$$

Here we allow structured dispersions for $\phi$ and $\lambda$.

v) Lee and Nelder (2006) introduced DHGLMs, represented by {HGLM ($\mu$), HGLM($\phi$)}:

$$\begin{aligned}\boldsymbol{\eta}^{(\mu)} &= g^{(\mu)}(\mu) = \mathbf{X}^{(\mu)}\boldsymbol{\beta}^{(\mu)} + \mathbf{Z}^{(\mu)}\boldsymbol{v}^{(\mu)}\\ \boldsymbol{\eta}^{(\lambda)} &= g^{(\lambda)}(\lambda) = \mathbf{X}^{(\lambda)}\boldsymbol{\beta}^{(\lambda)}\\ \boldsymbol{\eta}^{(\phi)} &= g^{(\phi)}(\phi) = \mathbf{X}^{(\phi)}\boldsymbol{\beta}^{(\phi)} + \mathbf{Z}^{(\phi)}\boldsymbol{v}^{(\phi)}\\ \boldsymbol{\eta}^{(\alpha)} &= g^{(\alpha)}(\alpha) = \mathbf{X}^{(\alpha)}\boldsymbol{\beta}^{(\alpha)},\end{aligned}$$

Here we allow structured dispersion for $\lambda$ and random-effect model for $\phi$.

vi) Noh et al. (2005) introduce the model {DHGLM($\mu$), GLM($\phi$)}:

$$\begin{aligned}
\boldsymbol{\eta}^{(\mu)} &= g^{(\mu)}(\mu) = \mathbf{X}^{(\mu)}\boldsymbol{\beta}^{(\mu)} + \mathbf{Z}^{(\mu)}\boldsymbol{v}^{(\mu)} \\
\boldsymbol{\eta}^{(\lambda)} &= g^{(\lambda)}(\lambda) = \mathbf{X}^{(\lambda)}\boldsymbol{\beta}^{(\lambda)} + \mathbf{Z}^{(\lambda)}\boldsymbol{v}^{(\lambda)} \\
\boldsymbol{\eta}^{(\tau)} &= g^{(\tau)}(\tau) = \mathbf{X}^{(\tau)}\boldsymbol{\beta}^{(\tau)} \\
\boldsymbol{\eta}^{(\phi)} &= g^{(\phi)}(\phi) = \mathbf{X}^{(\phi)}\boldsymbol{\beta}^{(\phi)}
\end{aligned}$$

Here we allow structured dispersion for $\phi$ and random-effect model for $\lambda$

vii) Noh and Lee (2017) introduced the model, {DHGLM($\mu$), HGLM($\phi$)}:

$$\begin{aligned}
\boldsymbol{\eta}^{(\mu)} &= g^{(\mu)}(\mu) = \mathbf{X}^{(\mu)}\boldsymbol{\beta}^{(\mu)} + \mathbf{Z}^{(\mu)}\boldsymbol{v}^{(\mu)} \\
\boldsymbol{\eta}^{(\lambda)} &= g^{(\lambda)}(\lambda) = \mathbf{X}^{(\lambda)}\boldsymbol{\beta}^{(\lambda)} + \mathbf{Z}^{(\lambda)}\boldsymbol{v}^{(\lambda)} \\
\boldsymbol{\eta}^{(\tau)} &= g^{(\tau)}(\tau) = \mathbf{X}^{(\tau)}\boldsymbol{\beta}^{(\tau)} \\
\boldsymbol{\eta}^{(\phi)} &= g^{(\phi)}(\phi) = \mathbf{X}^{(\phi)}\boldsymbol{\beta}^{(\phi)} + \mathbf{Z}^{(\phi)}\boldsymbol{v}^{(\phi)} \\
\boldsymbol{\eta}^{(\alpha)} &= g^{(\alpha)}(\tau) = \mathbf{X}^{(\alpha)}\boldsymbol{\beta}^{(\alpha)}.
\end{aligned}$$

Here we allow random-effect models for both $\phi$ and $\lambda$. Lee and Nelder (2006) introduced random effects in the dispersion to allow heavy-tailed distribution for the distribution of $y|v$. Noh and Lee (2007b) showed that this gives robust analysis against outliers. By introducing random effects in the model for the random effect variances we can have robust analysis against misspecification of the distribution of the random effects (Noh et al., 2005). For more discussion see Chapter 11 of Lee, Nelder, and Pawitan (2017).

## 6.2 Examples

In this section, the code for running the dhglm package is shown together with examples. In Section 6.4, the notation used in the dhglm package is defined in detail. Implementation details are also found in Section 6.4.

### *6.2.1 Crack growth data continued*

Consider the crack growth data in Chapter 4 again. Using this dataset, we illustrate how to fit all sub-models of DHGLM models. Using dhglmfit, the model assuming a gamma response can be specified as follows:

```
> data(crack_growth)
> dhglmfit(RespDist="gamma",DataMain=crack_growth,
+ MeanModel=model_mu,DispersionModel=model_phi)
```

Now we show how to specify the mean (model_mu) and dispersion (model _phi) in the dhglm package for models ranging from a simple GLM to more advanced DHGLMs.

(i) GLM {GLM($\mu$), constant}:

$$\eta_{ij}^{(\mu)} = \log \mu_{ij} = \beta_0^{(\mu)} + \beta_1^{(\mu)} l_{ij-1}.$$

```
> model_mu<-DHGLMMODELING(Model="mean", Link="log",
+ LinPred=y~crack0)
> model_phi<-DHGLMMODELING(Model="dispersion")
```

This model can also be fitted by using the existing glm function as follows.

```
> glm(y~crack0,crack_growth,family=Gamma(link = "log"))
```

(ii) Joint GLM {GLM($\mu$), GLM($\phi$)}:

$$\begin{aligned} \eta_{ij}^{(\mu)} &= \log \mu_{ij} = \beta_0^{(\mu)} + \beta_1^{(\mu)} l_{ij-1}, \\ \eta_{ij}^{(\phi)} &= \log \phi_{ij} = \beta_0^{(\phi)} + \beta_1^{(\phi)} t_j. \end{aligned}$$

```
> model_mu<-DHGLMMODELING(Model="mean", Link="log",
+ LinPred=y~crack0)
> model_phi<-DHGLMMODELING(Model="dispersion", Link="log",
+ LinPred=phi~cycle)
```

This model can also be fitted by using the existing fitjoint function as follows.

```
> fitjoint("glm", y~crack0, d~cycle, data = crack_growth,
+ family.mean=Gamma(link="log"),
+ family.disp=Gamma(link="log"))
```

(iii) HGLM1 {HGLM($\mu$), constant}:

$$\eta_{ij}^{(\mu)} = \log \mu_{ij} = \beta_0^{(\mu)} + \beta_1^{(\mu)} l_{ij-1} + v_i^{(\mu)},$$

where $u_i^{(\mu)} = \exp(v_i^{(\mu)})$ follows inverse-gamma distribution with $E(u_i^{(\mu)}) = 1$ and $\text{var}(u_i^{(\mu)}) = \lambda_i = \lambda$ with $\eta_i^{(\lambda)} = \log \lambda = \beta_0^{(\lambda)}$.

```
> model_mu<-DHGLMMODELING(Model="mean", Link="log",
+ LinPred=y~crack0+(1|specimen), RandDist="inverse-gamma")
> model_phi<-DHGLMMODELING(Model="dispersion")
```

For fitting HGLM1, the existing glmer and glmmPQL functions can be used as follows.

```
> glmer(y~crack0+(1|specimen),family = Gamma(link="log"),
+ data = crack_growth,nAGQ = 20)
> glmmPQL(y ~ crack0, random = ~ 1 | specimen,
+ family = Gamma(link="log"), data = crack_growth)
```

(iv) HGLM2 {HGLM($\mu$), GLM($\phi$)}:

$$\eta_{ij}^{(\mu)} = \log \mu_{ij} = \beta_0^{(\mu)} + \beta_1^{(\mu)} l_{ij-1} + v_i^{(\mu)},$$
$$\eta_{ij}^{(\phi)} = \log \phi_{ij} = \beta_0^{(\phi)} + \beta_1^{(\phi)} t_j.$$

```
> model_mu<-DHGLMMODELING(Model="mean", Link="log",
+ LinPred=y~crack0+(1|specimen),RandDist="inverse-gamma")
> model_phi<-DHGLMMODELING(Model="dispersion", Link="log",
+ LinPred=phi~cycle)
```

The nlmixed procedure of SAS can fit HGLM2. However, there is no package available to fit DHGLMs.

v) DHGLM1 {DHGLM($\mu$), GLM($\phi$)}.

$$\eta_{ij}^{(\mu)} = \log \mu_{ij} = \beta_0^{(\mu)} + \beta_1^{(\mu)} l_{ij-1} + v_i^{(\mu)}$$
$$\eta_i^{(\lambda)} = \log(\lambda_i) = \beta_0^{(\lambda)} + v_i^{(\lambda)},$$
$$\eta_{ij}^{(\phi)} = \log \phi_{ij} = \beta_0^{(\phi)} + \beta_1^{(\phi)} t_j,$$

where $v_i^{(\lambda)} \sim \mathrm{N}(0, \tau)$.

```
> model_mu<-DHGLMMODELING(Model="mean", Link="log",
+ LinPred=y~crack0+(1|specimen),RandDist="inverse-gamma",
+ LinPredRandVariance=lambda~1+(1|specimen),
+ LinkRandVariance="log")
> model_phi<-DHGLMMODELING(Model="dispersion", Link="log",
+ LinPred=phi~cycle)
```

(vi) DHGLM2 {HGLM($\mu$), HGLM($\phi$)}:

$$\eta_{ij}^{(\mu)} = \log \mu_{ij} = \beta_0^{(\mu)} + \beta_1^{(\mu)} l_{ij-1} + v_i^{(\mu)}$$
$$\eta_{ij}^{(\phi)} = \log \phi_{ij} = \beta_0^{(\phi)} + \beta_1^{(\phi)} t_j + v_i^{(\phi)},$$

where $v_i^{(\phi)} \sim \mathrm{N}(0, \alpha)$.

```
> model_mu<-DHGLMMODELING(Model="mean", Link="log",
+ LinPred=y~crack0+(1|specimen),RandDist="inverse-gamma")
> model_phi<-DHGLMMODELING(Model="dispersion", Link="log",
+ LinPred=phi~cycle+(1|specimen),RandDist="gaussian")
```

vii) DHGLM3 {DHGLM($\mu$), HGLM($\phi$)}.

$$\begin{aligned}
\eta_{ij}^{(\mu)} &= \log \mu_{ij} = \beta_0^{(\mu)} + \beta_1^{(\mu)} l_{ij-1} + v_i^{(\mu)} \\
\eta_i^{(\lambda)} &= \log(\lambda_i) = \beta_0^{(\lambda)} + v_i^{(\lambda)}, \\
\eta_{ij}^{(\phi)} &= \log \phi_{ij} = \beta_0^{(\phi)} + \beta_1^{(\phi)} t_j + v_i^{(\phi)},
\end{aligned}$$

where $v_i^{(\phi)} \sim \mathrm{N}(0, \alpha)$ and $v_i^{(\lambda)} \sim \mathrm{N}(0, \tau)$.

```
> model_mu<-DHGLMMODELING(Model="mean", Link="log",
+ LinPred=y~crack0+(1|specimen),RandDist="inverse-gamma",
+ LinPredRandVariance=lambda~1+(1|specimen),
+ LinkRandVariance="log")
> model_phi<-DHGLMMODELING(Model="dispersion", Link="log",
+ LinPred=phi~cycle+(1|specimen),RandDist="gaussian")
```

In Table 6.1, we see that glmer and glmmPQL give unstable estimates for dispersion parameters, while dhglmfit and nlmixed in SAS produce similar parameter estimates, which are reliable (Lee, Nelder, and Pawitan, 2017). cAIC selects DHGLM2 as the best-fitting model. Although unreported, the likelihood-ratio test based on the restricted likelihoods selects the same model. The mean of increment in crack length ($\mu$) increases as the previous crack length increases and the corresponding dispersion ($\phi$) decreases as the number of cycles increases. This means that the residual variance changes as the number of cycles increases. We can also conclude that heteroscedasticity between metallic specimens exists significantly in the mean as well as in the dispersion.

By using the studentized deviance residuals, we can obtain model-checking plots of the model objects `res_crack_growth_hglm2` for HGLM2 and `res_crack_growth_dhglm` for DHGLM2, as in Figures 6.1 and 6.2, respectively. This is done as follows.

```
> plotdhglm(res_crack_growth_hglm2)
> plotdhglm(res_crack_growth_dhglm2)
```

From the figures, we see that most of the outliers in HGLM2, caused by abrupt changes among repeated measures, disappear when random effects are allowed in the model for the residual variance.

Table 6.1 *Estimates (SE) for the crack growth data*

| model | function | $\beta_0^{(\mu)}$ | $\beta_1^{(\mu)}$ | $\beta_0^{(\lambda)}$ | $\beta_0^{(\phi)}$ | $\beta_1^{(\phi)}$ | cAIC |
|---|---|---|---|---|---|---|---|
| GLM | dglmfit | −5.85(0.11) | 2.57(0.09) | | −2.71(0.10) | | −1418 |
| | glm | −5.85(0.11) | 2.57(0.09) | | | | |
| JGLM | dhglmfit | −5.94(0.10) | 2.63(0.09) | | −2.11(0.20) | −10.5(2.8) | −1432 |
| | fitjoint | −5.94(0.10) | 2.63(0.09) | | −2.11(0.20) | −10.5(2.8) | |
| | nlmixed | −5.93(0.11) | 2.63(0.09) | | −2.13(0.20) | −10.5(2.8) | |
| HGLM1 | dhglmfit | −5.65(0.09) | 2.38(0.07) | −3.37(0.37) | −3.40(0.12) | | −1548 |
| | glmer | −5.82(0.004) | 2.57(0.003) | 0.00002 | 0.00007 | | |
| | glmmPQL | −5.65(0.09) | 2.38(0.07) | −1.71 | −1.70 | | |
| | nlmixed | −5.65(0.09) | 2.38(0.07) | −3.43(0.34) | −3.33(0.09) | | |
| HGLM2 | dhglmfit | −5.69(0.09) | 2.41(0.07) | −3.47(0.37) | −2.72(0.25) | −11.5(3.06) | −1561 |
| | nlmixed | −5.68(0.09) | 2.41(0.07) | −3.51(0.34) | −2.72(0.20) | −10.7(2.90) | |
| DHGLM1[a] | dhglmfit | −5.71(0.09) | 2.41(0.07) | −3.48(0.26) | −2.72(0.20) | −10.6(2.85) | −1560 |
| DHGLM2[b] | dhglmfit | −5.62(0.07) | 2.36(0.05) | −3.41(0.36) | −3.01(0.26) | −11.5(2.74) | −1621 |
| DHGLM3[c] | dhglmfit | −5.65(0.08) | 2.36(0.05) | −3.46(0.25) | −3.00(0.20) | −10.2(2.24) | −1618 |

[a] : the estimate (SE) for $\log(\tau)$ is −3.2(9.8).
[b] : the estimate (SE) for $\log(\alpha_0)$ is −0.41(0.15).
[c] : the estimate (SE) for $\log(\tau)$ and $\log(\alpha_0)$ is −2.5(6.5) and −1.1(0.30), respectively.

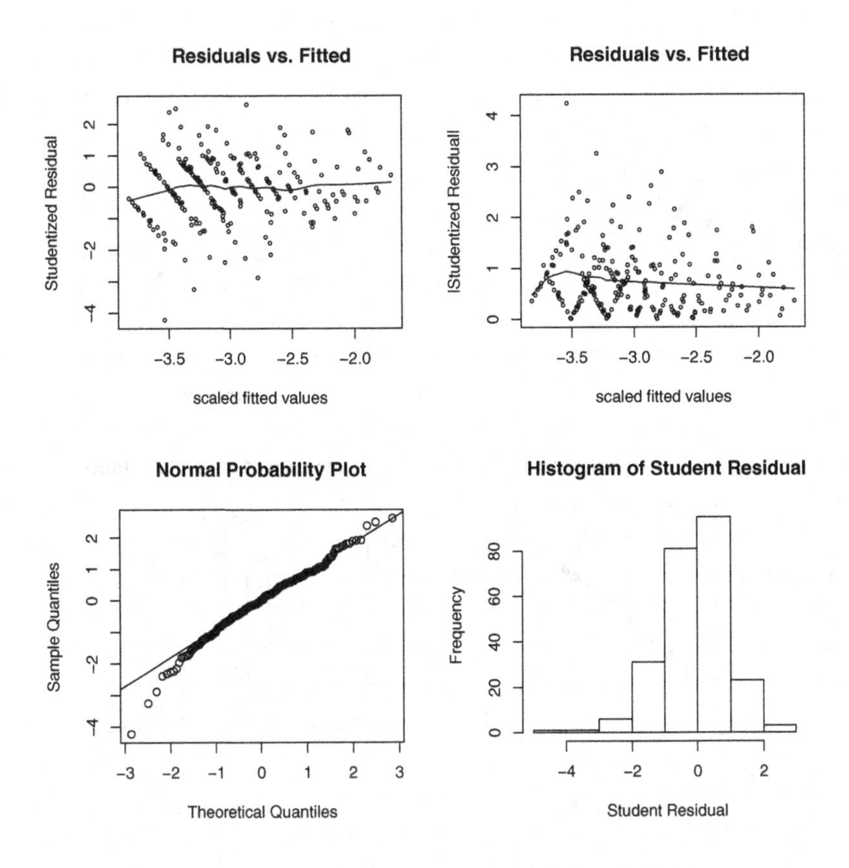


Figure 6.1 *Model checking plots for HGLM2 of the crack growth data.*

### *6.2.2 Gas consumption data continued*

Consider the gas consumption data introduced in Chapter 5, again. Residual plots for the model (5.6), shown in Figure 5.11, display apparent outliers. There was a disruption in the gas supply in the third and fourth quarters of 1970. Durbin and Koopman (2000) pointed out that this might lead to a distortion in the seasonal pattern when a Gaussian assumption is made for the error component $e_t$, so that they proposed to use heavy-tailed models, such as those with a t-distribution for the error component $e_t$.

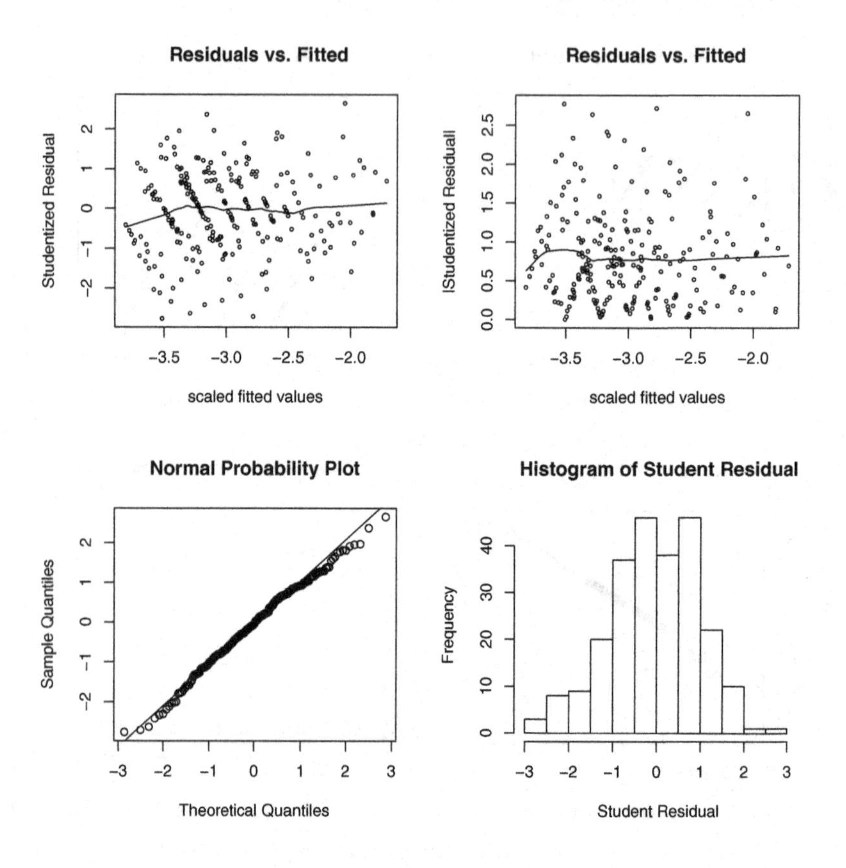


Figure 6.2 *Model checking plots for DHGLM2 of the crack growth data*

Consider the following DHGLM, allowing heavy-tailed distribution for $e_t$

$$\begin{aligned} y_t &= \alpha + t\beta + \alpha_i + t\beta_i + \delta_1(t = 43) + \delta_2(t = 44) \\ &+ \gamma_1 \sin(2\pi t/104) + \gamma_2 \cos(2\pi t/104) + s_t + e_t, \\ \log \phi_t &= \varphi + \psi_i + v_t^{(\phi)}, \end{aligned} \tag{6.1}$$

where $v_t^{(\phi)} \sim \text{N}(0, \alpha)$. In Table 6.2, we compare the results from the DHGLM and the HGLM. cAIC selects DHGLM as the best-fitting model. The likelihood-ratio test for $H_0 : \alpha = 0$, based on the restricted likelihood, rejects the null hypothesis because the deviance difference for HGLM and DHGLM is $18.8(= 167.8 - 149.0) > \chi^2_{2\delta}(1) = 2.71$ with a significant level $\delta = 0.05$. From Figures 5.12 and 6.3, we see that a big outlier in HGLM disappeared under the DHGLM (6.1). As Durbin and

Koopman (2000) pointed out, a disruption in the gas supply in the third and fourth quarters of 1970 may lead to a distortion in the seasonal pattern. From Table 6.2, we see that the effects of $\beta$, $\gamma_1$, and $\psi_2$ become significant in the DHGLM. Thus, the use of a heavy tailed distribution using DHGLM is helpful in drawing a strong conclusion.

Table 6.2 *Estimates from analyses of the gas consumption data under the HGLM and DHGLM*

| | HGLM (5.6) | | | DHGLM (6.1) | | |
|---|---|---|---|---|---|---|
| Coefficient | Estimate | SE | t-value | Estimate | SE | t-value |
| mean model | | | | | | |
| $\alpha$ | 5.0795 | 0.1197 | 42.42 | 5.0617 | 0.0903 | 56.07 |
| $\alpha_2$ | −0.0946 | 0.0337 | −2.81 | −0.0956 | 0.0309 | −3.09 |
| $\alpha_3$ | −0.4892 | 0.0395 | −12.38 | −0.4740 | 0.0373 | −12.70 |
| $\alpha_4$ | −0.3507 | 0.0514 | −7.00 | −0.3573 | 0.0414 | −8.63 |
| $\beta$ | 0.0157 | 0.0088 | 1.78 | 0.0163 | 0.0065 | 2.50 |
| $\beta_2$ | −0.0061 | 0.0005 | −11.20 | −0.0061 | 0.0005 | −12.22 |
| $\beta_3$ | −0.0094 | 0.0006 | −15.04 | −0.0098 | 0.0006 | −16.31 |
| $\beta_4$ | 0.0005 | 0.0008 | 0.57 | 0.0005 | 0.0006 | 0.71 |
| $\delta_1$ | 0.4715 | 0.0891 | 5.29 | 0.4722 | 0.0813 | 5.81 |
| $\delta_2$ | −0.3907 | 0.1213 | −3.22 | −0.3941 | 0.0925 | −4.26 |
| $\gamma_1$ | −0.1431 | 0.0916 | −1.56 | −0.1598 | 0.0658 | −2.43 |
| $\gamma_2$ | −0.0611 | 0.1035 | −0.59 | −0.0293 | 0.0759 | −0.39 |
| dispersion model | | | | | | |
| $\varphi$ | −5.864 | 0.300 | −19.54 | −6.263 | 0.237 | −26.45 |
| $\psi_2$ | 0.540 | 0.419 | 1.29 | 0.927 | 0.335 | 2.769 |
| $\psi_3$ | 0.944 | 0.421 | 2.24 | 1.227 | 0.335 | 3.664 |
| $\psi_4$ | 1.579 | 0.419 | 3.77 | 1.605 | 0.335 | 4.793 |
| $\log(\lambda_p)$ | −12.09 | 0.698 | −17.33 | −12.77 | 0.762 | −16.77 |
| $\log(\alpha)$ | | | | −1.669 | 4.158 | −0.401 |
| likelihood values and cAIC | | | | | | |
| −2 log (likelihood) | | −230.4 | | −249.2 | | |
| −2 log (restricted likelihood) | | −149.0 | | −167.8 | | |
| cAIC | | −228.1 | | −249.2 | | |

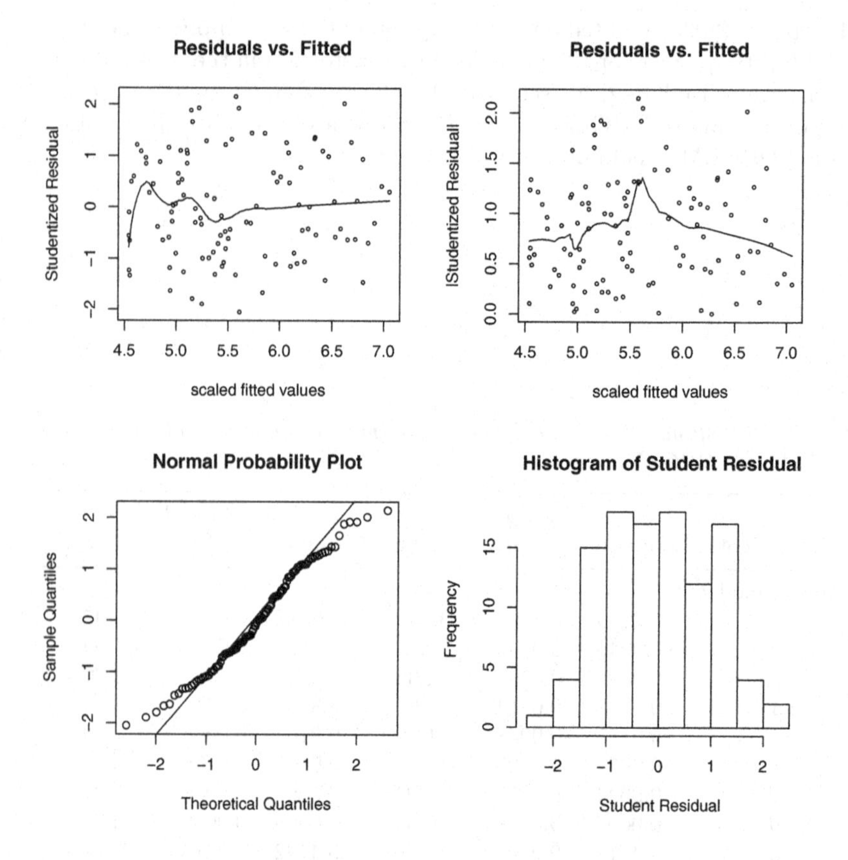


Figure 6.3 *Residual plot for gas consumption data under the model 6.1.*

### *6.2.3 Pound-dollar exchange-rate data*

Modeling of variance has been a special interest in two areas, quality improvement and finance. In the analysis of data from quality improvement experiments we aim to minimize variation among products. Thus, we need a structured dispersion model to minimize within variance among product and between batches (or wafer). We illustrate the analysis of data from quality improvement experiments by using injection molding data (Chapter 4) and semi-conductor data (Chapter 5). In finance, variance is often called volatility, which cannot be controlled. Thus, one wants to predict the volatility to reduce future risk. Various time-series models have been developed for volatilities, and we will examine a few of them using DHGLMs.

Consider daily observations for the weekday closing exchange rates for the U.K. Sterling/U.S. Dollar from 1/10/81 to 28/6/85. Following Harvey et al. (1994), we use the 936 mean-corrected returns as responses:

$$y_t = 100 * \{\log(r_t/r_{t-1}) - \sum \log(r_i/r_{i-1})/n\},$$

where $r_t$ denotes the exchange rate at time $t$.

Consider the model

$$y_t = \sqrt{\phi_t} z_t,$$

where $z_t$ is a standard normal random variable and $\phi_t$ is a volatility at time $t$. Engle (1982) introduced the ARCH model of order 1,

$$\phi_t = \beta_0^{(\phi)} + \beta_1^{(\phi)} y_{t-1}^2.$$

This is a joint GLM {GLM($\mu = 0$), GLM($\phi$)}, which can be fitted by specifying the identity link function for GLM($\phi$) and fixing the mean null using the `BetaFix` argument as follows.

```
> data(exch,package="mdhglm")
> model_mu<-DHGLMMODELING(Model="mean", Link="identity",
+ LinPred=yt~1)
> model_phi<-DHGLMMODELING(Model="dispersion",
+ Link="identity", LinPred=phi~yt12)
> res_arch<-dhglmfit(RespDist="gaussian",DataMain=exch,
+ MeanModel=model_mu, DispersionModel=model_phi, BetaFix=0)
```

In the coding, "yt" represents the response and "yt12" is $y_{t-1}^2$.

The ARCH(1) model was extended to the GARCH(1,1) model by Bollerslev (1986) as follows:

$$\phi_t = \beta_0^{(\phi)} + \beta_1^{(\phi)} y_{t-1}^2 + \gamma \phi_{t-1}.$$

The ARCH and GARCH models can be fitted by proc autoreg of SAS and garch() function in the tseries package (Trapletti et al., 2016). The GARCH(1,1) model can be fitted by specifying `corr="GARCH"` for the dispersion model as follows:

```
> model_mu<-DHGLMMODELING(Model="mean", Link="log",
+ LinPred=yt~1)
> model_phi<-DHGLMMODELING(Model="dispersion",
+ Link="identity", LinPred=phi~1+(1|date),
+ RandDist="gaussian",corr="GARCH")
> res_garch<-dhglmfit(RespDist="gaussian",DataMain=exch,
+ MeanModel=model_mu, DispersionModel=model_phi, BetaFix=0)
```

The model is implemented in the dhglm package by letting $\beta_0^{*(\phi)} = \beta_0^{(\phi)}/(1-\rho)$ with $\rho = \beta_1^{(\phi)} + \gamma$. Then

$$\begin{aligned}
v_t^{(\phi)} &= \phi_t - \beta_0^{*(\phi)} \\
&= \beta_0^{(\phi)} + \beta_1^{(\phi)} y_{t-1}^2 + \gamma\phi_{t-1} - \beta_0^{*(\phi)} \\
&= \beta_0^{(\phi)} + \beta_1^{(\phi)} y_{t-1}^2 + \rho(\phi_{t-1} - \beta_0^{*(\phi)}) - \beta_1^{(\phi)}\phi_{t-1} - (1-\rho)\beta_0^{*(\phi)} \\
&= \beta_0^{(\phi)} + \beta_1^{(\phi)} y_{t-1}^2 + \rho(\phi_{t-1} - \beta_0^{*(\phi)}) - \beta_1^{(\phi)}\phi_{t-1} - \beta_0^{(\phi)} \\
&= \rho(\phi_{t-1} - \beta_0^{*(\phi)}) + \beta_1^{(\phi)}(y_{t-1}^2 - \phi_{t-1}) \\
&= \rho v_{t-1}^{(\phi)} + r_t^{(\phi)},
\end{aligned}$$

where $r_t^{(\phi)} = \beta_1^{(\phi)}(y_{t-1}^2 - \phi_{t-1})$. Thus, the GARCH(1,1) can be written as a dispersion model with correlated random effects

$$\phi_t = \beta_0^{*(\phi)} + v_t^{(\phi)},$$

where $v_t^{(\phi)} = \rho v_{t-1}^{(\phi)} + r_t^{(\phi)}$.

Let $\eta_t^{(\phi)} = \log \phi_t$. To avoid negative volatility, we can consider the exponential GARCH (EGARCH), with a log link and

$$\eta_t^{(\phi)} = \beta_0^{(\phi)} + \beta_1^{(\phi)} y_{t-1}^2 + \gamma\eta_{t-1}^{(\phi)},$$

which is equivalent to

$$\eta_t^{(\phi)} = \beta_0^{*(\phi)} + v_t^{(\phi)}.$$

We can easily fit the EGARCH by taking log link for $\phi$.

Now if we take $r_t^{(\phi)} \sim N(0, \alpha)$, i.e., $v_t^{(\phi)} = \rho v_{t-1}^{(\phi)} + r_t^{(\phi)} \sim AR(1)$, we have the stochastic volatility (SV) model, originating from Harvey et al. (1994). Thus, this SV model can be fitted by specifying HGLM($\phi$) as follows.

```
> model_mu<-DHGLMMODELING(Model="mean", Link="log",
+ LinPred=yt~1)
> model_phi<-DHGLMMODELING(Model="dispersion", Link="log",
+ LinPred=phi~1+(1|date), RandDist="gaussian",corr="AR")
> res_sv<-dhglmfit(RespDist="gaussian",DataMain=exch,
+ MeanModel=model_mu, DispersionModel=model_phi, BetaFix=0)
```

If we take positive-valued responses $y_t^2$, all these finance models become mean models. For example, SV models become gamma HGLMs with temporal random effects satisfying

$$E(y_t^2|v_t^{(\phi)}) = \phi_t \quad \text{and} \quad \text{var}(y_t^2|v_t^{(\phi)}) = 2\phi_t^2,$$

Table 6.3 *Estimates (SE) for the exchange-rate data in the ARCH(1) and GARCH(1,1) models*

| model | function | $\beta_0^{(\phi)}$ | $\beta_1^{(\phi)}$ | $\gamma$ |
|---|---|---|---|---|
| ARCH(1) | dhglmfit | 0.430(0.020) | 0.183(0.036) | |
| | autoreg | 0.429(0.019) | 0.183(0.032) | |
| | garch | 0.394(0.013) | 0.250(0.033) | |
| GARCH(1,1) | dhglmfit | 0.0079(0.0044) | 0.063(0.015) | 0.917(0.018) |
| | autoreg | 0.0079(0.0043) | 0.062(0.013) | 0.923(0.020) |
| | garch | 0.0101(0.0039) | 0.096(0.014) | 0.887(0.019) |

which is equivalent to assuming $y_t^2|v_t^{(\phi)} \sim \phi_t\chi^2(1)$: see Section 11.6.4 of Lee, Nelder, and Pawitan (2017). Thus, the method for {HGLM($\mu$), constant} can be used directly to fit the SV model as follows.

```
> exch$yt2<-exch$yt^2
> model_mu<-DHGLMMODELING(Model="mean", Link="log",
+ LinPred=yt2~1+(1|date),
+ RandDist="gaussian",corr="AR")
> model_phi<-DHGLMMODELING(Model="dispersion",
+ Link="identity")
> res_sv1<-dhglmfit(RespDist="gamma",DataMain=exch,
+ MeanModel=model_mu, DispersionModel=model_phi,PhiFix=2)
```

We can set $\phi = 2$ by specifying `PhiFix=2`. The SAS procedure proc autoreg and the R function garch() give the MLEs for the ARCH and GARCH models. From Table 6.3 we can see that the two analyses by dhglmfit() and proc autoreg are similar, however the garch() function in the tseries packagegives somewhat different estimates.

SV models have attracted much attention as a way of allowing clustered volatility in asset returns. Despite their intuitive appeal, SV models have been used less frequently than the ARCH and GARCH models in empirical applications because there is no computationally fast and accurate estimation algorithm available. For the SV model, Harvey et al.'s (1994) estimator was improved by Shephard and Pitt (1997) using the MCMC method. Durbin and Koopman (2000) developed an importance-sampling method for both the ML and Bayesian procedures. The dhglm package gives parameter estimates without resorting to computationally intensive simulation methods such as the importance-sampling or Monte-Carlo methods. SEs are directly computed using the Hessian matrix.

For the data, SV model has $cAIC = 1807$ which is less than $cAIC = 2006$

for ARCH and $cAIC = 1863$ for GARCH models, so that SV model is the best one among alternative models. From Table 6.4, we can see that the dhglm package and Durbin and Koopman (2000) importance-sampling method give similar estimates and SE values.

Table 6.4 *Estimates (SE) for the exchange-rate data in the SV model*

| | Durbin and Koopman (2000) Estimate (SE) | dhglm Estimate (SE) |
|---|---|---|
| $\beta_0^{*(\phi)}$ | −0.911(0.206) | −0.894(0.202) |
| $\log(\alpha)$ | −3.514(0.434) | −3.506(0.416) |
| $\log(\rho/(1-\rho))$ | 3.588(0.501) | 3.496(0.491) |

### *6.2.4 Random slope model for orthodontic growth data continued*

Consider the orthodontic growth data in Chapter 5 again., where we considered the HGLM with random slope (5.4)

$$y_{ij} = \beta_1^{(\mu)} F_i + \beta_2^{(\mu)} F_i A_{ij} + \beta_3^{(\mu)} M_i + \beta_4^{(\mu)} M_i A_{ij} + v_{1i}^{(\mu)} + A_{ij} v_{2i}^{(\mu)} + \epsilon_{ij}$$

where $\epsilon_{ij} \sim \mathrm{N}(0, \phi_{ij})$ and the random intercept $v_{1i}^{(\mu)}$ and random slope $v_{2i}^{(\mu)}$ are assumed to be bivariate normal distributed with mean 0, variances $\mathrm{var}(v_{1i}^{(\mu)}) = \lambda_1$ and $\mathrm{var}(v_{2i}^{(\mu)}) = \lambda_2$, and correlation $\mathrm{Corr}(v_{1i}^{(\mu)}, v_{2i}^{(\mu)}) = \rho$. In Chapter 5, we detected three outliers (two outliers are from the second and third observation of the ninth boy and one outlier is the first observation of the thirteenth boy). Noh and Lee (2007b) showed that a robust analysis against such outliers can be obtained by adding random effects to the residual variances $\phi_{ij}$. Thus, we consider the following DHGLM:

$$\log(\phi_{ij}) = \beta_0^{(\phi)} + v_i^{(\phi)},$$

where $v_i^{(\phi)} \sim \mathrm{N}(0, \alpha)$. This model can be fitted as follows:

```
> data(Orthodont,package="nlme")
> Orthodont$nsex <- as.numeric(Orthodont$Sex=="Male")
> Orthodont$nsexage <- with(Orthodont, nsex*age)
> model_mu<-DHGLMMODELING(Model="mean", Link="identity",
+ LinPred=distance~ nsex + age + nsexage +
+ (1|Subject)+(age|Subject),RandDist="gaussian")
> model_phi<-DHGLMMODELING(Model="dispersion",
+ Link="log",
```

```
+ phi~1+(1|Subject),RandDist="gaussian")

> res_dhglm<-dhglmfit(RespDist="gaussian",
+ DataMain=Orthodont,
+ MeanModel=model_mu,DispersionModel=model_phi)

> plotdhglm(res_dhglm)
```

From Table 6.5, we see that the DHGLM has slightly larger standard errors for fixed-effect estimators of the mean model to reflect the presence of outliers. Among models we considered, cAIC selects DHGLM as the best-fitting model. Also, the likelihood-ratio test based on the RL confirms the choice: For testing $\alpha = 0$ (equivalent of the absence of random effects $v_i^{(\phi)}$), we use the critical value (from a 50:50 mixture of the $\chi^2(0)$ and $\chi^2(1)$ distributions) of $\chi^2_{2\delta}(1)$ for a size-$\delta$ test because $\alpha = 0$ is at the boundary of the parameter space (Self and Liang, 1987). Thus, the deviance difference for HGLM and DHGLM is $37.9(= 432.6 - 394.7) > \chi^2_{2\delta}(1) = 2.71$ with $\delta = 0.05$. Thus, the likelihood-ratio test selects the DHGLM. Model checking plots for the DHGLM in Figure 6.4 show that all large outliers (whose sizes are bigger than 4) disappear.

For comparison, we also fit the data removing three outliers. Table 6.5 shows that the intercept and age estimators for male, $\beta_3^{(\mu)}$ and $\beta_4^{(\mu)}$, from HGLMs are sensitive to outliers, while those from DHGLMs are not.

### *6.2.5 DHGLM for schizophrenic behavior data*

Rubin and Wu (1997) analyzed schizophrenic behavior data from an eye-tracking experiment with a visual target moving back and forth along a horizontal line on a screen. The outcome measurement is called the gain ratio, which is eye velocity divided by target velocity, and it is recorded repeatedly at the peak velocity of the target during eye-tracking under three conditions. The first condition is plain sine (PS), which means the target velocity is proportional to the sine of time and the color of the target is white. The second condition is color sine (CS), which means the target velocity is proportional to the sine of time, as for PS, but the colors keep changing from white to orange or blue. The third condition is triangular (TR), in which the target moves at a (Intercept) speed equal to the peak velocity of PS, back and forth, but the color is always white.

There are 43 non-schizophrenic subjects, 22 females and 21 males, and

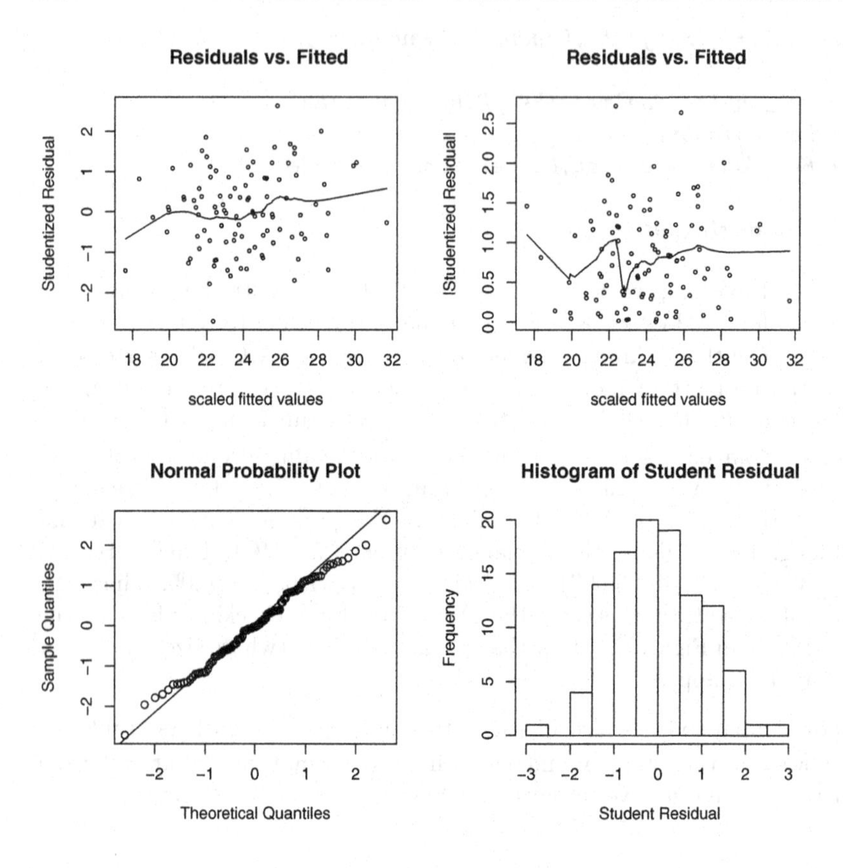


Figure 6.4 *Residual plot for orthodontic growth data under DHGLM.*

43 schizophrenic subjects, 13 females and 30 males. In the experiment, each subject is exposed to five trials, usually three PS, one CS, and one TR. During each trial, there are 11 cycles, and a gain ratio is recorded for each cycle. However, for some cycles, the gain ratios are missing because of eye blinks, so that there are, on average, 34 observations out of 55 cycles for each subject (see Table 6.6). In the next chapter, we will model the missing data mechanism, but here we assume (for simplicity) that the missing data are missing at random (MAR). MAR means that missingness only depends upon the observed data, so that under MAR assumption, we can perform the analysis using only observed data.

For observed responses $y_{ij}$, gain ratios for the $j$th measurement of the

Table 6.5 *Estimates (SE) for the orthodontic growth data*

| | With full data | | Without three outliers | |
|---|---|---|---|---|
| | HGLM | DHGLM | HGLM | DHGLM |
| $\beta_1^{(\mu)}$ | 17.37(1.23) | 17.51(1.31) | 17.37(0.90) | 17.65(0.92) |
| $\beta_2^{(\mu)}$ | 0.48(0.10) | 0.47(0.10) | 0.48(0.08) | 0.45(0.08) |
| $\beta_3^{(\mu)}$ | 16.34(1.02) | 17.33(1.07) | 17.39(0.76) | 17.44(0.77) |
| $\beta_4^{(\mu)}$ | 0.78(0.09) | 0.70(0.10) | 0.69(0.06) | 0.69(0.06) |
| $\log(\lambda_1)$ | 1.76(0.74) | 1.84(0.80) | 1.24(0.65) | 1.29(0.67) |
| $\log(\lambda_2)$ | −3.43(0.57) | −3.41(0.55) | −3.93(0.52) | −4.07(0.53) |
| $\rho$ | −0.67(0.24) | −0.50(0.21) | −0.39(0.17) | −0.37(0.16) |
| $\beta_0^{(\phi)}$ | 0.54(0.26) | 0.43(0.21) | −0.14(0.17) | −0.18(0.19) |
| $\log(\alpha)$ | | −1.34(0.72) | | −5.96(2.35) |
| −2 log (likelihood) | 429.7 | 388.9 | 365.4 | 371.9 |
| −2 log (restricted likelihood) | 432.6 | 394.7 | 373.1 | 378.9 |
| cAIC | 392.4 | 352.7 | 311.2 | 312.0 |

$i$th subject, consider the following HGLM:

$$\begin{aligned} y_{ij} = \quad & \beta_0^{(\mu)} + x_{1ij}\beta_1^{(\mu)} + x_{2ij}\beta_2^{(\mu)} + t_j\beta_3^{(\mu)} + sch_i\beta_4^{(\mu)} + sch_i \cdot x_{1ij}\beta_5^{(\mu)} \\ & + sch_i \cdot x_{2ij}\beta_6^{(\mu)} + v_i^{(\mu)} + e_{ij} \end{aligned}$$

where $v_i \sim \mathrm{N}(0, \lambda)$ is the subject random effect, $e_{ij} \sim \mathrm{N}(0, \phi)$ is a white noise, $sch_i = 1$ if a subject is schizophrenic and 0 otherwise; $t_j$ is the measurement time; $x_{1ij}$ is the effect of PS vs. CS; $x_{2ij}$ is the effect of TR vs. the average of CS and PS. We find that schizophrenic patients have a larger variance, which can be modeled as follows

$$\log(\phi_i) = \beta_0^{(\phi)} + sch_i\beta_1^{(\phi)}.$$

This HGLM can be fitted as follows.

```
> data(sch, package="mdhglm")
> model_mu<-DHGLMMODELING(Model="mean", Link="identity",
+ LinPred=y~x1+x2+time+sch+sch*x1+sch*x2+(1|subject),
+ RandDist="gaussian")
> model_phi<-DHGLMMODELING(Model="dispersion",Link="log",
+ LinPred=phi~sch)
> reshglm<-dhglmfit(RespDist="gaussian",DataMain=sch,
```

Table 6.6 *Repeated measures of three schizophrenics having abrupt changes*

| ID | trt | 1 | 2 | 3 | 4 | 5 | 6 | 7 | 8 | 9 | 10 | 11 | |
|---|---|---|---|---|---|---|---|---|---|---|---|---|---|
| 25 | PS | .916 | .831 | .880 | .908 | .951 | .939 | .898 | .909 | .939 | .896 | .826 | |
| | PS | .887 | .900 | .938 | .793 | .794 | .935 | .917 | .882 | .635 | .849 | .810 | |
| | CS | .836 | .944 | .889 | .909 | .863 | .838 | .844 | .784 | ★ | ★ | ★ | |
| | PS | .739 | **.401**[a] | .787 | .753 | .853 | .731 | .862 | .882 | .835 | .862 | .883 | |
| 129 | CS | .893 | .702 | .902 | ★ | ★ | .777 | ★ | ★ | ★ | ★ | ★ | |
| | PS | ★ | ★ | ★ | .849 | .774 | ★ | ★ | ★ | ★ | ★ | **.209**[a] | |
| 207 | PS | ★ | ★ | .862 | .983 | ★ | ★ | ★ | .822 | .853 | ★ | .827 | |
| | CS | .881 | .815 | .886 | **.519**[a] | ★ | .657 | ★ | .879 | ★ | ★ | .881 | |
| | CS | .782 | ★ | ★ | ★ | | .840 | ★ | .837 | ★ | ★ | .797 | ★ |

★ indicate missing; [a] abrupt change

```
+ MeanModel=model_mu,DispersionModel=model_phi)
Distribution of Main Response :
                      "gaussian"
[1] "Estimates from the model(mu)"
y ~ x1+x2+time+sch+sch*x1+sch*x2+(1|subject)
[1] "identity"
             Estimate Std. Error  t-value
(Intercept)  0.811250   0.013744  59.0240
x1           0.006425   0.004525   1.4197
x2          -0.121455   0.004930 -24.6351
time        -0.002430   0.000436  -5.5740
sch         -0.036096   0.019525  -1.8487
x1:sch      -0.028978   0.007009  -4.1343
x2:sch      -0.007240   0.007847  -0.9226
[1] "Estimates for logarithm of lambda=var(u_mu)"
[1] "gaussian"
        Estimate Std. Error t-value
subject   -4.839     0.1565  -30.91
[1] "Estimates from the model(phi)"
phi ~ sch
[1] "log"
            Estimate Std. Error  t-value
(Intercept)   -5.320    0.03667 -145.082
sch            0.251    0.05337    4.703
[1] "== Likelihood Function Values and Condition AIC =="
                                     [,1]
-2 log(likelihood)             :  -6550.164
-2 log(restricted likelihood) :  -6488.259
cAIC                           :  -6780.296

Scaled deviance: 2817.382 on 2817.382 degrees of freedom
```

Psychological theory suggests a model in which schizophrenics suffer from an attention deficit on some trials, as well as general motor reflex retardation; both aspects lead to relatively slower responses for schizophrenics, with motor retardation affecting all trials and attentional deficiency only some. Also, psychologists have known for a long time about large variations in within-schizophrenic performance on almost any task (Silverman, 1967). Thus, abrupt changes among repeated responses may be peculiar to schizophrenics and such volatility may differ for each patient. Such heteroscedasticity among schizophrenics cannot be modeled by the fixed effect model above, but can be modeled by a DHGLM, introducing a random effect in the dispersion

$$\log(\phi_i) = \beta_0^{(\phi)} + sch_i\beta_1^{(\phi)} + sch_i v_i^{(\phi)} \tag{6.2}$$

where $v_i^{(\phi)} \sim \mathrm{N}(0, \alpha)$ are random effects in dispersion. Given the random effects $(v_i^{(\mu)}, v_i^{(\phi)})$, the repeated measurements are independent, and $v_i^{(\mu)}$ and $v_i^{(\phi)}$ are independent. Thus the ith subject has a dispersion $\phi_i = \exp(\beta_0^{(\phi)} + sch_i\beta_1^{(\phi)} + sch_i v_i^{(\phi)})$ if he or she is schizophrenic, and $\phi_i = \exp(\beta_0^{(\phi)})$ otherwise.

This DHGLM can be fitted as follows.

```
> model_mu<-DHGLMMODELING(Model="mean", Link="identity",
+ LinPred=y~x1+x2+time+sch+sch*x1+sch*x2+(1|subject),
+ RandDist="gaussian")
> model_phi<-DHGLMMODELING(Model="dispersion",Link="log",
+ LinPred=phi~sch+(sch|subject),RandDist="gaussian")
> resdhglm<-dhglmfit(RespDist="gaussian",DataMain=sch,
+ MeanModel=model_mu,DispersionModel=model_phi)
Distribution of Main Response :
                        "gaussian"
[1] "Estimates from the model(mu)"
y~x1+x2+time+sch+sch*x1+sch*x2+(1|subject)
[1] "identity"
             Estimate Std. Error  t-value
(Intercept)  0.812285  0.0136394  59.5541
x1           0.003068  0.0038326   0.8004
x2          -0.117130  0.0042559 -27.5218
time        -0.002267  0.0003695  -6.1368
sch         -0.037156  0.0193825  -1.9170
x1:sch      -0.019516  0.0058599  -3.3305
x2:sch      -0.008516  0.0065643  -1.2973
[1] "Estimates for logarithm of lambda=var(u_mu)"
[1] "gaussian"
        Estimate Std. Error t-value
subject   -4.853     0.1566  -30.98
[1] "Estimates from the model(phi)"
phi ~ sch + (sch | subject)
[1] "log"
            Estimate Std. Error t-value
(Intercept)  -5.4609     0.0873 -62.553
sch           0.2976     0.1246   2.389
[1] "Estimates for logarithm of tau=var(u_phi)"
        Estimate Std. Error t-value
subject   -1.123    0.08297  -13.53
```

```
[1] "== Likelihood Function Values and Condition AIC =="
                                   [,1]
-2 log(likelihood)            :  -7037.247
-2 log(restricted likelihood) :  -6973.585
cAIC                          :  -7280.780

Scaled deviance: 2948.361 on 2817.492 degrees of freedom
```

cAIC shows that DHGLM (cAIC=-7280) has a better fit than HGLM (cAIC=-6780). By using the studentized deviance residuals, we can obtain model-checking plots of the model objects `resschhglm` for HGLM and `reschdhglm` for DHGLM, as in Figures 6.5 and 6.6, respectively. This is done as follows.

```
> plotdhglm(resschhglm)
> plotdhglm(resschdhglm)
```

From the figures, we see that most of the outliers in HGLM, caused by abrupt changes among repeated measures, disappear when random effects are allowed in the model for the residual variance.

### 6.2.6 *Respiratory data continued*

With binary data it is difficult to identify the distribution of random effects. Noh et al. (2005) noted that the use of a heavy-tailed distribution for random effects, by allowing random effects for $\lambda$, removes sensitivity of the parameter estimation to the choice of random-effect distribution. For binary data, they showed that GLMM estimators can give serious biases if the true distribution is not normal. Furthermore, for ascertained samples, these biases become more serious. Thus, in binary data, we recommend the use of DHGLM.

Consider model (5.2) in Chapter 5 but with a heavy-tailed distribution for random effects, by allowing random effects in the variance for random effects:

$$
\begin{aligned}
\log\left(\frac{p_{ij}}{1-p_{ij}}\right) &= \beta_0^{(\mu)} + \beta_1^{(\mu)} trt_i + \beta_2^{(\mu)} msex_i + \beta_3^{(\mu)} age_i \\
&+ \beta_4^{(\mu)} center_i + \beta_5^{(\mu)} base_i + \beta_6^{(\mu)} y_{i(j-1)} + v_i^{(\mu)}, \\
\log \lambda_i &= \beta_0^{(\lambda)} + age_i \beta_1^{(\lambda)} + v_i^{(\lambda)}
\end{aligned}
$$

where $v_i^{(\mu)} \sim \mathrm{N}(0, \lambda_i)$ and $v_i^{(\lambda)} \sim \mathrm{N}(0, \tau)$. This DHGLM can be fitted as follows.

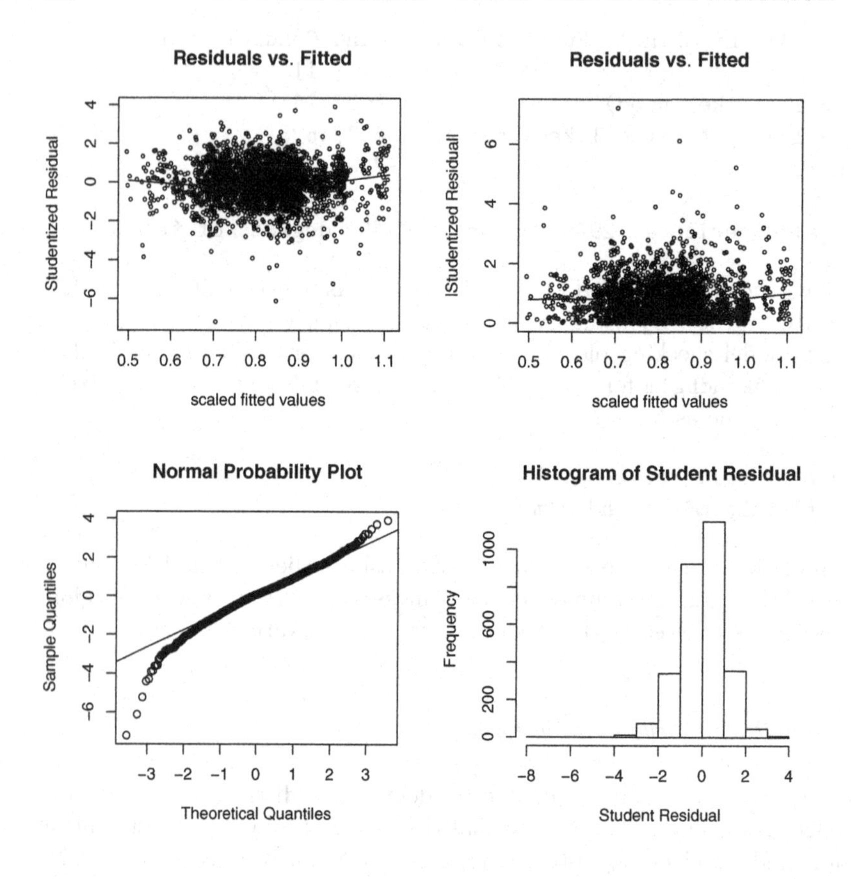


Figure 6.5 *Model checking plots for HGLM of the schizophrenic behavior data.*

```
> data(respiratory,package="mdhglm")
> respiratory$trt<-
+ as.numeric(respiratory$treatment=="active")
> respiratory$msex<-as.numeric(respiratory$sex=="male")
> model_mu2<-DHGLMMODELING(Model="mean", Link="logit",
+ LinPred=y~trt+msex+age+center+base+past+(1|patient),
+ RandDist="gaussian",LinkRandVariance="log",
+ LinPredRandVariance=lambda~1+age+(1|patient),
+ RandDistRandVariance="gaussian")
> model_phi2<-DHGLMMODELING(Model="dispersion",Link="log")

> fit2<-dhglmfit(RespDist="binomial",DataMain=respiratory,
+ MeanModel=model_mu2,DispersionModel=model_phi2)
```

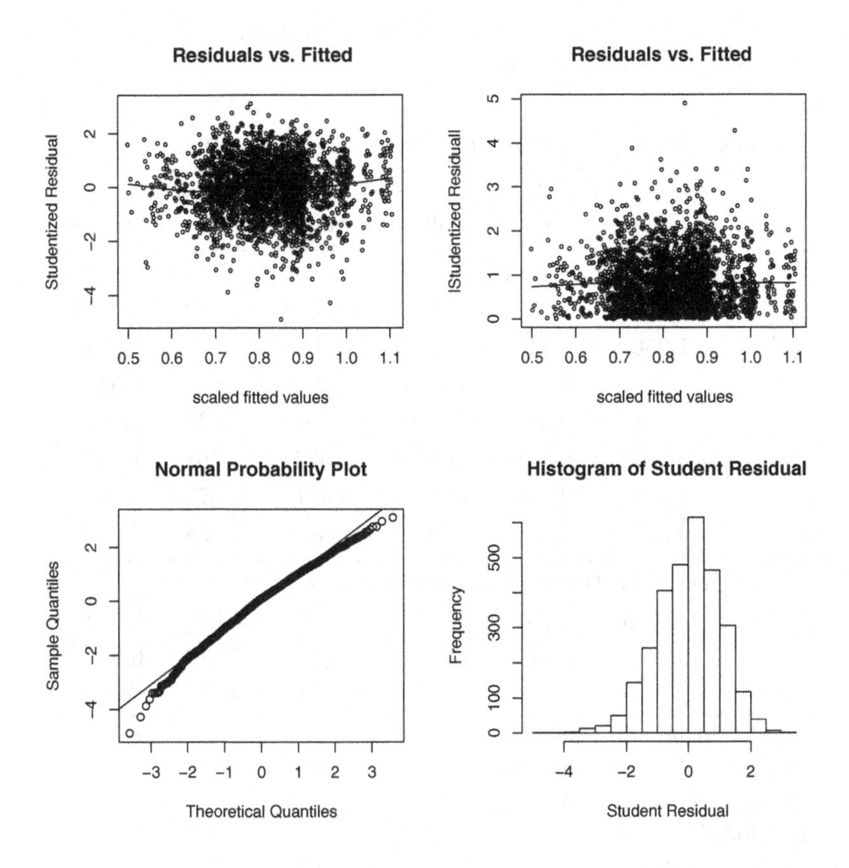


Figure 6.6 *Model checking plots for DHGLM of the schizophrenic behavior data.*

```
> plotdhglm(fit2,type="lambda")
```

In Table 6.7, the cAIC from the HGLM is 431.0 and for the DHGLM it is 413.6. The likelihood-ratio test for $H_0 : \tau = 0$, based on the restricted likelihood, rejects the null hypothesis because the deviance difference for HGLM and DHGLM is $3.3(= 438.4 - 435.1) > \chi^2_{2\delta}(1) = 2.71$ with a significant level $\delta = 0.05$. From Figures 5.9 and 6.7 for model $\lambda$, we see that large outliers, and an unpleasant pattern in the normal probability plot under the HGLM, disappear under the DHGLM. Thus, the DHGLM is preferred. Furthermore, in this example there are apparent differences between parameter estimates. In this case, we should report the results from the DHGLM because a distributional assumption of random effects is hard to identify with the binary data.

Table 6.7 *HGLM and DHGLM results for the respiratory data*

| | HGLM | | | DHGLM | | |
|---|---|---|---|---|---|---|
| | Estimate | SE | t-value | Estimate | SE | t-value |
| | model for the mean | | | | | |
| $\beta_0^{(\mu)}$ | −1.111 | 1.033 | −1.075 | −0.290 | 1.683 | −0.172 |
| $\beta_1^{(\mu)}$ | 1.256 | 0.415 | 3.028 | 1.601 | 0.640 | 2.503 |
| $\beta_2^{(\mu)}$ | −0.261 | 0.597 | 0.437 | −0.541 | 1.030 | 0.525 |
| $\beta_3^{(\mu)}$ | −0.035 | 0.019 | −1.889 | −0.060 | 0.032 | −1.873 |
| $\beta_4^{(\mu)}$ | 0.682 | 0.419 | 1.626 | 0.672 | 0.654 | 1.027 |
| $\beta_5^{(\mu)}$ | 1.821 | 0.446 | 4.079 | 2.411 | 0.672 | 3.586 |
| $\beta_6^{(\mu)}$ | 0.575 | 0.304 | 1.891 | −0.051 | 0.338 | −0.152 |
| | model for the random effect variance | | | | | |
| $\beta_0^{(\lambda)}$ | −0.683 | 0.737 | −0.927 | 0.015 | 0.345 | 0.042 |
| $\beta_1^{(\lambda)}$ | 0.047 | 0.020 | 2.339 | 0.067 | 0.010 | 6.976 |
| $\log(\tau)$ | | | | −1.246 | 3.020 | −0.413 |
| | likelihood values and cAIC | | | | | |
| −2 log (likelihood) | 431.0 | | | 428.2 | | |
| −2 log (restricted likelihood) | 438.4 | | | 435.1 | | |
| cAIC | 422.5 | | | 413.6 | | |

### *6.2.7 Salamander data continued*

Consider the HGLM (5.1) for salamander data again. For this binary data set, we fit a DHGLM model:

$$\log\{p_{ijk}/(1-p_{ijk})\} = x_{ijk}^t\beta^{(\mu)} + v_{fik}^{(\mu)} + v_{mjk}^{(\mu)},$$
$$\log(\lambda_{fik}) = \beta_{f0}^{(\lambda)} + b_{fik}^{(\lambda)} \text{ and } \log(\lambda_{mik}) = \beta_{m0}^{(\lambda)} + b_{mik}^{(\lambda)},$$

where $v_{fik}^{(\mu)} \sim \mathrm{N}(0, \lambda_{fik})$, $v_{mjk}^{(\mu)} \sim \mathrm{N}(0, \lambda_{mjk})$, $b_{fik}^{(\lambda)} \sim \mathrm{N}(0, \tau_f)$, and $b_{mik}^{(\lambda)} \sim \mathrm{N}(0, \tau_m)$. In this model, we consider two random effects $v_{fik}^{(\mu)}$ and $v_{mjk}^{(\mu)}$, which have heavy tailed distributions. This DHGLM can be fitted as follows.

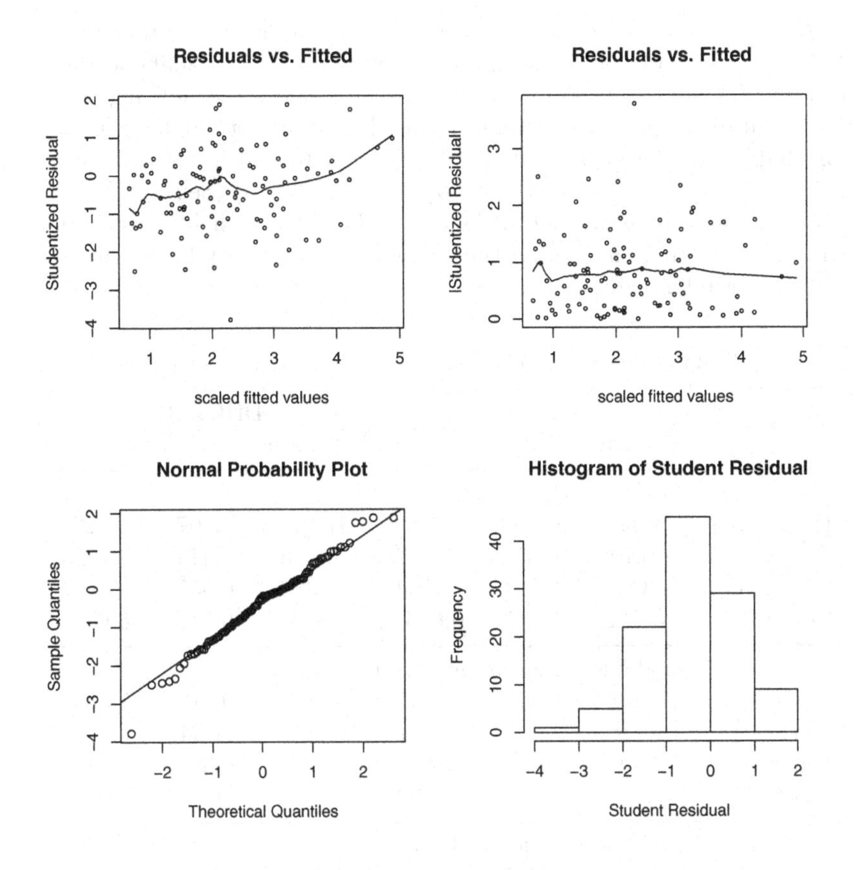


Figure 6.7 *Residual plot of $\lambda$ for the respiratory data.*

```
> data(salamander, package="mdhglm")
> model_mu<-DHGLMMODELING(Model="mean",Link="logit",
+ LinPred=Mate~TypeM+TypeF+TypeM*TypeF+(1|FeMale)+(1|Male),
+ RandDist=c("gaussian","gaussian"),
+ LinkRandVariance="log",
+ LinPredRandVariance=lambda~1+(1|FeMale)+(1|Male),
+ RandDistRandVariance=c("gaussian","gaussian"))
> model_phi<-DHGLMMODELING(Model="dispersion")

> fit11<-dhglmfit(RespDist="binomial",DataMain=salamander,
+ MeanModel=model_mu,DispersionModel=model_phi)
```

In Table 6.8, the cAIC from the HGLM is 401.9, while that from the DHGLM is 401.5. In this example, the cAIC difference is less than

1, so that there would be no advantage to use the heavy-tailed distribution, compared with the normal distribution. The likelihood-ratio test for $H_0 : \tau_f = 0$ and $\tau_m = 0$ does not reject the null hypothesis because the deviance difference based on the restricted likelihood for HGLM and DHGLM is $2.4(= 419.0 - 416.6)$ which has p-value of $0.243 = 0.5 \times P(\chi^2(1) > 2.4) + 0.25 \times P(\chi^2(2) > 2.4)$ (Self and Liang, 1987). Estimates between HGLM and DHGLM are slightly different, which also strongly indicating the adequacy of normality for the distribution of random effects.

Table 6.8 *HGLM and DHGLM results for the salamander data*

| | HGLM | | | DHGLM | | |
|---|---|---|---|---|---|---|
| | Estimate | SE | t-value | Estimate | SE | t-value |
| | model for the mean | | | | | |
| (Intercept) | 1.043 | 0.393 | 2.655 | 1.107 | 0.407 | 2.720 |
| TypeMW | −0.729 | 0.454 | −1.607 | −0.736 | 0.474 | −1.553 |
| TypeFW | −3.004 | 0.521 | −5.764 | −3.086 | 0.557 | −5.540 |
| MW:FW | 3.711 | 0.557 | 6.666 | 3.823 | 0.573 | 6.672 |
| | model for the random effect variance | | | | | |
| (Female) | 0.304 | 0.273 | 1.117 | 0.288 | 0.276 | 1.043 |
| $\log(\tau_f)$ | | | −0.791 | 0.511 | −1.549 | |
| (Male) | 0.195 | 0.280 | 0.696 | 0.161 | 0.301 | 0.535 |
| $\log(\tau_m)$ | | | −1.432 | 1.413 | −1.013 | |
| | likelihood values and cAIC | | | | | |
| −2 log (likelihood) | | 418.7 | | 415.4 | | |
| −2 log (restricted likelihood) | | 419.0 | | 416.6 | | |
| cAIC | | 401.9 | | 401.5 | | |

### *6.2.8 Bacteria data continued*

Consider the bacteria data data in Chapter 4 again. We fit a DHGLM model:

$$\begin{aligned}
\log(p_{ij}/(1-p_{ij})) &= \beta_0^{(\mu)} + \beta_1^{(\mu)} I(i = drug) + \beta_2^{(\mu)} I(i = drug+) + v_i^{(\mu)} \\
\log(\lambda_i) &= \beta_0^{(\lambda)} + v_i^{(\lambda)}
\end{aligned}$$

where $v_i^{(\mu)} \sim \text{N}(0, \lambda_i)$ and $v_i^{(\lambda)} \sim \text{N}(0, \tau)$. This DHGLM can be fitted as follows.

```
> data(bacteria, package="mdhglm")
> model_mu<-DHGLMMODELING(Model="mean", Link="logit",
+ LinPred=y1 ~ trt  + (1 | ID),RandDist="gaussian",
+ LinkRandVariance="log",
+ LinPredRandVariance=lambda~1+(1|ID),
+ RandDistRandVariance=c("gaussian"))
> model_phi<-DHGLMMODELING(Model="dispersion")

> fit<-dhglmfit(RespDist="binomial",DataMain=bacteria,
+ MeanModel=model_mu,DispersionModel=model_phi)
```

In Table 6.9, the cAIC from the HGLM is 205.0, while that from the DHGLM is 204.5. In this example the cAIC difference is less than 1, so that there would be no advantage to use the heavy-tailed distribution. The likelihood-ratio test for $H_0 : \tau = 0$ does not reject the null hypothesis because the deviance difference based on the restricted likelihood for HGLM and DHGLM is $0.8(= 205.9 - 205.1) < \chi^2_{2\delta}(1) = 2.71$ with a significant level $\delta = 0.05$. Estimates between HGLM and DHGLM are only slightly different, which also strongly indicates the adequacy of normality for the distribution of random effects.

### 6.2.9 *Epilepsy data continued*

Consider the epilepsy data in Section 4.1.3 again. We found that the distributional assumption for random effects is suspicious. Thus, we may fit the following Poisson DHGLMs:

$$
\begin{aligned}
\log(\mu_{ij}) &= \beta_0^{(\mu)} + x_{B_i}\beta_B^{(\mu)} + x_{T_i}\beta_T^{(\mu)} + x_{A_i}\beta_A^{(\mu)} + x_{V_j}\,\beta_V^{(\mu)} \\
&+ x_{B_iT_i}\beta_{BT}^{(\mu)} + v_i^{(\mu)} + v_{ij}^{(\mu)}, \\
\log(\lambda_{1i}) &= \beta_0^{(\lambda_1)} + v_i^{(\lambda_1)} \text{ and } \log(\lambda_{2ij}) = \beta_0^{(\lambda_2)} + v_{ij}^{(\lambda_2)}
\end{aligned}
$$

i) Poisson-normal DHGLM: $v_i^{(\mu)} \sim \text{N}(0, \lambda_{1i})$, $v_i^{(\lambda_1)} \sim \text{N}(0, \tau_1)$ and $v_{ij}^{(\mu)} = 0$, $v_{ij}^{(\lambda_2)} = 0$,

ii) Poisson-normal-gamma DHGLM: $v_i^{(\mu)} \sim \text{N}(0, \lambda_{1i})$, $v_i^{(\lambda_1)} = 0$ and $v_{ij}^{(\mu)} \sim G(\lambda_{2ij})$, $v_{ij}^{(\lambda_2)} \sim \text{N}(0, \tau_2)$,

iii) Poisson-gamma-gamma DHGLM1: $v_i^{(\mu)} \sim G(\lambda_{1i})$, $v_i^{(\lambda_1)} = 0$ and $v_{ij}^{(\mu)} \sim G(\lambda_{2ij})$, $v_{ij}^{(\lambda_2)} \sim \text{N}(0, \tau_2)$,

Table 6.9 *HGLM and DHGLM results for the bacteria data*

| | HGLM | | | DHGLM | | |
|---|---|---|---|---|---|---|
| | Estimate | SE | t-value | Estimate | SE | t-value |
| | model for the mean | | | | | |
| (Intercept) | 2.412 | 0.441 | 5.473 | 2.257 | 0.385 | 5.861 |
| trt(drug) | −1.256 | 0.622 | −2.021 | −1.216 | 0.534 | −2.275 |
| trt(drug+) | −0.752 | 0.637 | −1.182 | −0.733 | 0.550 | −1.333 |
| | model for the random effect variance | | | | | |
| (Intercept) | 0.290 | 0.333 | 0.870 | −0.398 | 0.149 | −2.672 |
| log($\tau$) | | | −2.182 | 3.704 | −0.589 | |
| | likelihood values and cAIC | | | | | |
| −2 log (likelihood) | | 206.5 | | 204.8 | | |
| −2 log (restricted likelihood) | | 205.9 | | 205.1 | | |
| cAIC | | 205.0 | | 204.5 | | |

iv) Poisson-gamma-gamma DHGLM2: $v_i^{(\mu)} \sim G(\lambda_{1i})$, $v_i^{(\lambda_1)} \sim \mathrm{N}(0, \tau_1)$ and $v_{ij}^{(\mu)} \sim G(\lambda_{2ij}), v_{ij}^{(\lambda_2)} \sim \mathrm{N}(0, \tau_2)$,

v) quasi Poisson-normal DHGLM: $v_i^{(\mu)} \sim \mathrm{N}(0, \lambda_{1i})$, $v_i^{(\lambda_1)} \sim \mathrm{N}(0, \tau_1)$ and $\mathrm{var}(y_{ij}|v_i^{(\mu)}, v_{ij}^{(\mu)}) = \phi\mu_{ij}$.

The Poisson-gamma-gamma DHGLM1 of the above models can be fitted as follows.

```
> data(epilepsy, package="dhglm")
> model_mu<-DHGLMMODELING(Model="mean", Link="log",
+ LinPred=y~B+T+A+B:T+V+(1|patient)+(1|id),
+ RandDist=c("gamma","gamma"),
+ LinPredRandVariance=lambda~1+(1|id),
+ LinkRandVariance="log",
+ RandDistRandVariance="gaussian")
> model_phi<-DHGLMMODELING(Model="dispersion")

> res2<-dhglmfit(RespDist="poisson",DataMain=epilepsy,
+ MeanModel=model_mu,DispersionModel=model_phi)
```

The likelihood-ratio test for $H_0 : \tau_2 = 0$, based on the restricted likelihood, rejects the null hypothesis because the deviance difference for negative binomial-gamma HGLM and Poisson-gamma-gamma DHGLM1

Table 6.10 *cAIC and rAIC from Poisson DHGLMs for the epilepsy data*

| model | cAIC | rAIC |
|---|---|---|
| NB-gamma HGLM | 1,163.9 | 1,274.8 |
| Poisson-normal DHGLM | 1,270.5 | 1,349.1 |
| Poisson-normal-gamma DHGLM | 1,183.0 | 1,282.7 |
| Poisson-gamma-gamma DHGLM1 | 1,144.2 | 1,244.4 |
| Poisson-gamma-gamma DHGLM2 | 1,146.1 | 1,246.4 |
| quasi Poisson-normal DHGLM | 1,217.2 | 1,319.6 |

is $32.4(=1,270.8-1,238.4) > \chi^2_{2\delta}(1) = 2.71$ with a significant level $\delta = 0.05$. However, the likelihood-ratio test for $H_0 : \tau_1 = 0$, does not reject the null hypothesis because the deviance difference based on the restricted likelihood for Poisson-gamma-gamma DHGLM1 and Poisson-gamma-gamma DHGLM2 is $0.0(=1,238.4-1,238.4)$. Thus, the likelihood-ratio test selects the Poisson-gamma-gamma DHGLM1.

In Table 6.10, both the cAIC and rAIC select the Poisson-gamma-gamma DHGLM1 as the final model. Parameter estimates from the negative binomial-gamma HGLM in Section 4.1 and the Poisson-gamma-gamma DHGLM1 are shown as Table 6.11. There are some differences between parameter estimates and we would report the results from the DHGLM because the analysis is robust against misspecifications of distribution of random effects.

Model-checking plots based on residuals are shown in Figure 6.8. From the model checking plots for $v_i$ ($\lambda_1$) and $v_{ij}$ ($\lambda_2$) in Figure 6.9, we see that most outliers in the negative-binomial HGLMs in Figure 3.2, are disappeared under the DHGLM. Thus, the DHGLM is preferred.

### *6.2.10 Joint cubic splines for stroke patient data*

Approximately 30% of hospitalized patients due to acute ischemic stroke are placed in the risk of early neurologic deterioration (END) at their hospital stay. The patient's risk to END can be monitored by following their blood pressure (BP). We used dataset `stroke` in the R package mdhglm which has systolic BP (SBP) with time in hours after arriving at the emergency room for two stroke patients (one is END; the other is non-END).

For detection of changes of SBP with respect to time, we use cubic splines (Silverman, 1967; Green and Silverman, 1994) not only for the

Table 6.11 *HGLM and DHGLM results for the epilepsy data*

| | NB-gamma | | | Poisson-gamma-gamma | | |
|---|---|---|---|---|---|---|
| | HGLM | | | DHGLM1 | | |
| | Estimate | SE | t-value | Estimate | SE | t-value |
| | model for the mean | | | | | |
| (Intercept) | −1.304 | 1.342 | −0.972 | −1.341 | 0.972 | −1.380 |
| B | 0.900 | 0.143 | 6.290 | 0.892 | 0.103 | 8.643 |
| T | −1.831 | 0.429 | −1.938 | −0.810 | 0.330 | −2.456 |
| A | 0.508 | 0.399 | 1.273 | 0.510 | 0.288 | 1.770 |
| V | −0.028 | 0.017 | −1.684 | −0.026 | 0.016 | −1.616 |
| B:T | 0.325 | 0.219 | 1.480 | 0.321 | 0.165 | 1.943 |
| | model for the random effect variance | | | | | |
| | for patient | | | | | |
| (Intercept) | −1.290 | 0.222 | −5.824 | −1.472 | 0.155 | −9.506 |
| | for id | | | | | |
| (Intercept) | −1.989 | 0.161 | −12.336 | −1.735 | 0.077 | −22.415 |
| $\log(\tau_2)$ | | | | −2.182 | 3.704 | −0.589 |
| | likelihood values and cAIC | | | | | |
| −2 log (likelihood) | | 1,255.6 | | 1,220.1 | | |
| −2 log (restricted likelihood) | | 1,270.8 | | 1,238.4 | | |
| cAIC | | 1,163.9 | | 1,144.2 | | |

mean changes but also for variance changes, using the joint cubic splines model (Lee and Nelder, 2006):

$$y_t = f_m(t) + e_t \text{ and } \log \phi_t = f_d(t),$$

where $y_t$ is SBP measurement at time $t$, $e_t \sim \mathrm{N}(0, \phi_t)$ and $f_m(t)$ and $f_d(t)$ are unknown functions of the mean and variance, respectively. For joint fitting of the mean $\mu_t$ and variance $\phi_t$, we use the DHGLM

$$\mu_t = \beta_0^{(\mu)} + \beta_1^{(\mu)} t + v_t^{(\mu)} \text{ and } \log \phi_t = \beta_0^{(\phi)} + \beta_1^{(\phi)} t + v_t^{(\phi)},$$

where $v_t^{(\mu)}$ $[v_t^{(\phi)}]$ is the random component with mean 0 and a singular precision matrix $P/\lambda^{(\mu)}$ $[P/\lambda^{(\phi)}]$. To fit joint splines, we need joint linear trends as fixed effects: for more detailed explanation see Chapter 9 of Lee, Nelder, and Pawitan (2017).

This model can be fitted by specifying the option `spline="cubic"` in the DHGLMMODELING function. From Figure 6.10, we have found that

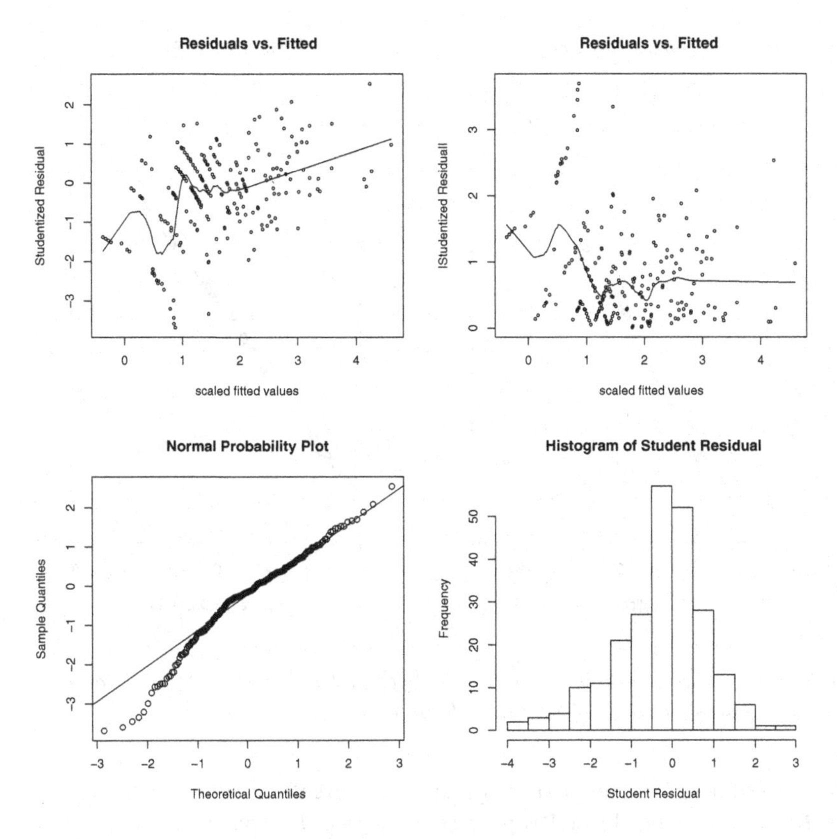


Figure 6.8 *Model-checking plot for the Poisson-gamma-gamma DHGLM1 on the epilepsy data.*

mean patterns for END and non-END patients are similar, so that it has been very difficult to predict the potential END patients. However, it can be noticed from the plot that the END patient has higher variance in SBP than non-END patient. Thus, the variance of the SBP is used as a covariate for predicting an END event, which greatly prevents the occurrence of END patients in the emergency room in Korea.

```
> data(stroke,package="mdhglm")
> # y1 : SBP for END patient
> # y2 : SBP for non-END patient
> model_mu<-DHGLMMODELING(Model="mean", Link="identity",
+ LinPred=y1~1+(1|time),RandDist="gaussian",spline="cubic")
> model_phi<-DHGLMMODELING(Model="dispersion",Link="log",
+ LinPred=y1~1+(1|time),RandDist="gaussian",spline="cubic")
```

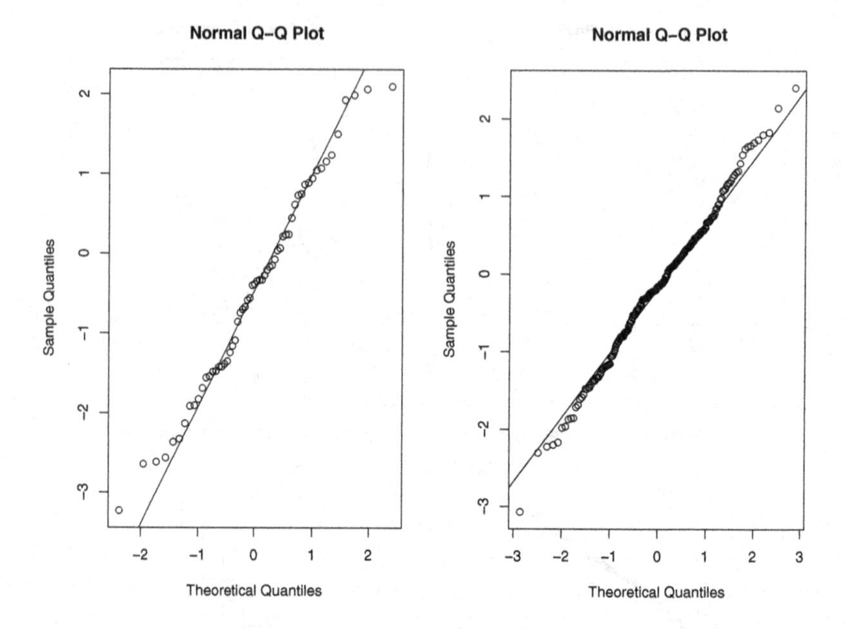


Figure 6.9 *Model-checking plot for $v_i$ (left) and $v_{ij}$ (right) for the Poisson-gamma-gamma DHGLM1 on the epilepsy data.*

```
> res<-dhglmfit(RespDist="gaussian",DataMain=sch,
+ MeanModel=model_mu,DispersionModel=model_phi)
```

## 6.3 An extension of linear mixed models via DHGLM

We consider a DHGLM of Example 2 in Chapter 1 again and see how it can be presented as an interconnected GLM. For simplicity assume that the random effects at each level of the model are i.i.d.. The IWLS algorithm gives fast computations using GLM estimation (see Figure 6.11).

Consider the DHGLM introduced in Chapter 1 (1.4) with applications in animal breeding. Using our new notation with superscripts to denote sub-models, the model is written as

$$\boldsymbol{y} = \mathbf{X}^{(\mu)}\boldsymbol{\beta}^{(\mu)} + \mathbf{Z}^{(\mu)}\boldsymbol{v}^{(\mu)} + \boldsymbol{e} \tag{6.3}$$

$$\boldsymbol{e} \sim \mathrm{N}(0, exp(\boldsymbol{X}^{(\phi)}\boldsymbol{\beta}^{(\phi)} + \boldsymbol{Z}^{(\phi)}\boldsymbol{v}^{(\phi)})). \tag{6.4}$$

In this chapter we start by looking at models having independent random

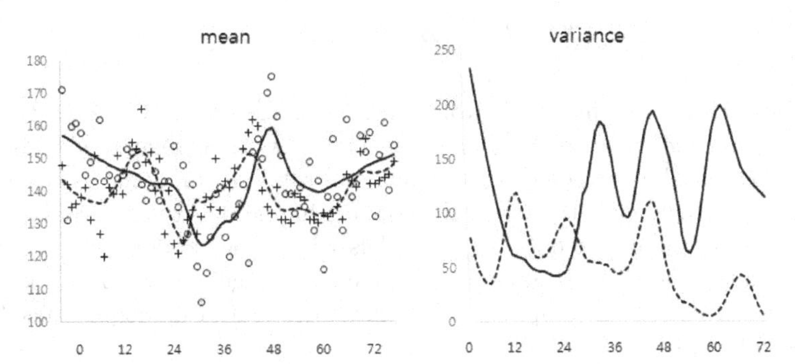


Figure 6.10 *Joint cubic splines for SBP of two stroke patients (END: solid line, non-END: dashed-line). In the mean model, circle point [plus point] stands for SBP measurements of END (non-END) patient.*

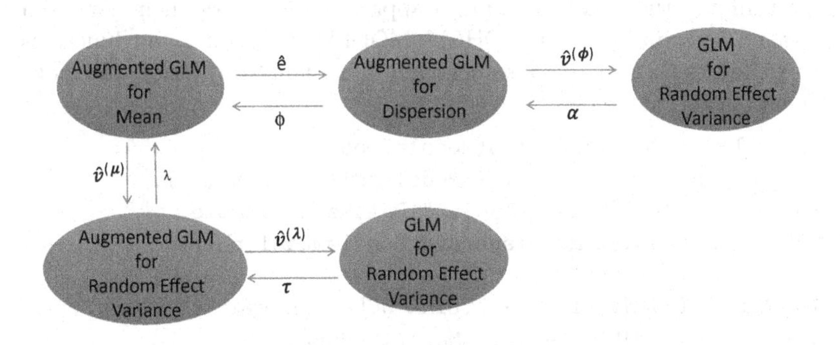


Figure 6.11 *Interconnected GLMs for fitting DHGLMs.*

effects, such that with $\boldsymbol{v}^{(\mu)} \sim \mathrm{N}(0, \lambda\mathbf{I})$, $\boldsymbol{v}^{(\phi)} \sim \mathrm{N}(0, \alpha\mathbf{I})$ and $cor(\boldsymbol{v}^{(\mu)}, \boldsymbol{v}^{(\phi)}) = 0$.

Here we also specify the h-likelihood to show that it is rather easy to specify even though the model is rather advanced.

$$
\begin{aligned}
h =& \log(f(\boldsymbol{y}|\boldsymbol{v}^{(\mu)}, \boldsymbol{v}^{(\phi)})) + \log(f(\boldsymbol{v}^{(\mu)})) + \log(f(\boldsymbol{v}^{(\phi)})) = \\
& -\tfrac{1}{2}\log(|\boldsymbol{V}|) \\
& -\tfrac{1}{2}(\boldsymbol{y} - (\boldsymbol{X}^{(\mu)}\boldsymbol{\beta}^{(\mu)} + \boldsymbol{Z}^{(\mu)}\boldsymbol{v}^{(\mu)}))^T\boldsymbol{V}^{-1}(\boldsymbol{y} - (\boldsymbol{X}^{(\mu)}\boldsymbol{\beta}^{(\mu)} + \boldsymbol{Z}^{(\mu)}\boldsymbol{v}^{(\mu)})) \\
& -\tfrac{m}{2}\log(\lambda) - \tfrac{1}{2\lambda^2}(\boldsymbol{v}^{(\mu)})^T\boldsymbol{v}^{(\mu)} \\
& -\tfrac{m}{2}\log(\alpha) - \tfrac{1}{2\alpha^2}(\boldsymbol{v}^{(\phi)})^T\boldsymbol{v}^{(\phi)}
\end{aligned}
$$

where $\boldsymbol{V} = \text{diag}(\exp(\boldsymbol{X}^{(\phi)}\boldsymbol{\beta}^{(\phi)} + \boldsymbol{Z}^{(\phi)}\boldsymbol{v}^{(\phi)}))$.

In the example of Chapter 1, Rönnegård et al. (2010) and Felleki et al. (2012) considered an animal breeding model with correlated $\boldsymbol{v}^{(\mu)}$ and $\boldsymbol{v}^{(\phi)}$. We could allow correlations among all random components $\boldsymbol{e}$, $\boldsymbol{v}^{(\mu)}$ and $\boldsymbol{v}^{(\phi)}$, which leads to many other interesting models to explore. However, in this book, we only cover models where the random effects are independent between components. Current packages allow correlations within each random component. For example, spatial modeling considered in Chapter 5.

## 6.4 Implementation in the dhglm package

The models for $\mu$ and $\phi$ are defined by the DHGLMMODELING function, by specifying link functions, linear predictors, and distributions of random effects. The dhglmfit function with DHGLMMODELING objects specifies the distribution of $y|u^{(\mu,\lambda,\phi)}$ and the method of fitting etc., for example, the order of Laplace approximation. Argument lists and detailed descriptions for the DHGLMMODELING and dhglmfit functions are in Tables 6.12 and 6.13, respectively. The structure of these functions that we use is as follows:

```
> model_mu<-DHGLMMODELING(Model="mean", ...)
> model_phi<-DHGLMMODELING(Model="dispersion", ...)
> dhglmfit(RespDist="gaussian",DataMain=example,
+ MeanModel=model_mu,DispersionModel=model_phi)
```

```
DHGLMMODELING(Model="mean",Link=NULL,LinPred="constant",
RandDist=NULL,Offset=NULL,LMaxtrix=NULL,
LinkRandVariance=NULL,LinPredRandVariance=NULL,
RandDistRandVaraince="log", LinkRandVariance2=NULL,
LinPredRandVariance2=NULL, spline=NULL,
temporalspatial=NULL, Neighbor=NULL)
```

```
DHGLMMODELING(Model="dispersion",Link=NULL,
LinPred="constant",RandDist=NULL,Offset=NULL,
LMaxtrix=NULL,LinkRandVariance=NULL,
LinPredRandVariance=NULL,spline=NULL,
temporalspatial=NULL, Neighbor=NULL)
```

```
dhglmfit(RespDist="gaussian",BinomialDen=NULL, DataMain,
MeanModel,DispersionModel=NULL,PhiFix=NULL,LamFix=NULL,
mord=1,dord=1,Maxiter=200,convergence=1e-06)
```

Table 6.12 *Description specifying arguments of DHGLMMODELING function*

<table>
<tr><th>Arguments</th><th>Details</th></tr>
<tr><td colspan="2">Specifying the model for $\mu$</td></tr>
<tr><td>`Model`</td><td>Model = "mean"</td></tr>
<tr><td>`Link`</td><td>$r^{(\mu)}(\cdot)$ : "identity", "logit", "cloglog", "log", or "inverse"</td></tr>
<tr><td>`LinkRandVariance`</td><td>$r^{(\lambda)}(\cdot)$ : "log", or "inverse"</td></tr>
<tr><td>`LinkRandVariance2`</td><td>$r^{(\alpha)}(\cdot)$ : "log", or "inverse"</td></tr>
<tr><td>`LinPred`</td><td>$\eta^{(\mu)}$ : `LinPred=y~x1+x2+(1|id1)+(1|id2)`<br>y : main response, x1 and x2 : fixed covariates, id1 and id2 : random terms</td></tr>
<tr><td>`LinPredRandVariance`</td><td>$\eta^{(\lambda)}$ : `c(lambda~xx1+(1|id11),lambda~xx2+(1|id12))`<br>xx1 and xx2 are fixed covariates and id11 and id12 are random terms in $\eta^{(\lambda)}$</td></tr>
<tr><td>`LinPredRandVariance2`</td><td>$\eta^{(\alpha)}$ : `c(alpha~xxx1,alpha~xxx2)`; xxx1 and xxx2 are fixed covariates in $\eta^{(\alpha)}$</td></tr>
<tr><td>`RandDist`</td><td>distributions of $u^{(\mu)}$ : "gaussian", "beta", "gamma", or "inverse-gamma"<br>different distributions can be allowed, e.g., c("gaussian", "gamma")</td></tr>
<tr><td>`RandDistRandVariance`</td><td>distributions of $u^{(\lambda)}$ : "gaussian", "beta", "gamma", or "inverse-gamma"</td></tr>
<tr><td>`Offset`</td><td>specifying a known component offset to be included in $\eta^{(\mu)}$</td></tr>
<tr><td colspan="2">Specifying the model for $\phi$</td></tr>
<tr><td>`Model`</td><td>Model = "dispersion"</td></tr>
<tr><td>`Link`</td><td>$r^{(\phi)}(\cdot)$ : "log", or "inverse"</td></tr>
<tr><td>`LinkRandVariance`</td><td>$r^{(\tau)}(\cdot)$ : "log", or "inverse"</td></tr>
<tr><td>`LinPred`</td><td>$\eta^{(\phi)}$ : `LinPred=phi~x1+x2+(1|id1)+(1|id2)`</td></tr>
<tr><td>`LinPredRandVariance`</td><td>$\eta^{(\tau)}$ : `c(tau~xx1,tau~xx2)`</td></tr>
<tr><td>`RandDist`</td><td>distributions of $u^{(\phi)}$ : "gaussian", "beta", "gamma", or "inverse-gamma"</td></tr>
<tr><td>`Offset`</td><td>specifying a known component offset to be included in $\eta^{(\phi)}$</td></tr>
</table>

Table 6.13 *Arguments of the* ***dhglmfit*** *function*

| Arguments | Details |
|---|---|
| `RespDist` | distributions of $y\|u^{(\mu,\lambda,\phi)}$ : "gaussian", "binomial", "poisson", or "gamma" |
| `BinomialDen` | the denominator when `RespDist` = "binomial"<br>`NLUU` gives the denominator as one |
| `DataMain` | the data frame to be used (non-optional) |
| `MeanModel` | the DHGLMMODLING object having the option `Model` = "mean" |
| `DispersionModel` | the DHGLMMODLING object having the option `Model` = "dispersion" |
| `PhiFix` | $\phi$ is to be estimated using `PhiFix = NULL` or maintained constant<br>when fixed at a value specifying `PhiFix` |
| `LamFix` | $\lambda$ is to be estimated using `LamFix = NULL` or maintained constant<br>when fixed at a value specifying `LamFix` |
| `mord` | the order of Laplace approximation to the marginal likelihood for fitting<br>mean parameters; choice is either 0 or 1 |
| `dord` | the order of Laplace approximation to the adjusted profile likelihood for fitting<br>dispersion parameters; choice is either 1 or 2 |
| `MaxIter` | maximum number of iterations for estimating all parameters |
| `convergence` | the criterion for convergence is the sum of absolute differences<br>between the previous and current iterations for the values of<br>all the estimated parameters and estimated random effects |
| `Dmethod` | method of fitting dispersion model<br>"deviance" or "Pearson" |

The function dhglmfit returns the dhglm object. We refer to the help file of this package for documentation pertaining to all elements in the dhglm object. In the help file possibilities of fitting models having random effects with various correlation structures are also described. The function plotdhglm creates model-checking plots by using studentized deviance residuals in the dhglm object.

## 6.5 Exercises

1. Consider the epileptic seizure count data again. Lee, Nelder, and Pawitan (2017) showed that over-dispersion is necessary to give an adequate fit to the data. Using residual plots they showed their final model to be better than other models they considered. However, those plots still showed apparent outliers. Thus, fit the following Poisson DHGLM (Lee, Nelder, and Pawitan (2017); Section 11.5.3):

$$\begin{aligned}\log(\boldsymbol{\mu}) &= \boldsymbol{\beta}_0 + x_B\boldsymbol{\beta}_B + x_T\boldsymbol{\beta}_T + x_A\boldsymbol{\beta}_A + x_{BT}\boldsymbol{\beta}_{BT} + \boldsymbol{Zv} \\ \log(\boldsymbol{\phi}) &= x_B\gamma_B + \boldsymbol{Zw}\end{aligned}$$

where $\boldsymbol{v} \sim \mathrm{N}(0, \lambda\boldsymbol{I})$ and $\boldsymbol{w} \sim \mathrm{N}(0, \alpha\boldsymbol{I})$. Assess the model using model-checking plots.

2. Consider the seeds data in Exercise 1 of Chapter 5 again. Fit a binomial DHGLM allowing a heavy-tailed distribution for the random effect as follows:

$$\begin{aligned}y_i|v_i &\sim Bin(n_i, p_i) \\ \log\left(\frac{p_i}{1-p_i}\right) &= \beta_0 + \beta_1 x_{i1} + \beta_2 x_{i2} + \beta_{12}(x_{i1}\cdot x_{i2}) + v_i \\ v_i &\sim \mathrm{N}(0, \lambda_i) \\ \log(\lambda_i) &= \log(\lambda_0) + b_i \\ b_i &\sim \mathrm{N}(0, \alpha).\end{aligned}$$

Check the significance for presence of random effects $b_i$ and assess the model-checking plots.

3. Consider the sleep study data in Exercise 3 of Chapter 4 again. Execute model checking procedure for the linear mixed model with random intercept and slope, as

```
data(sleepstudy,package="nlme")
model_mu<-DHGLMMODELING(Model="mean", Link="identity",
```

```
LinPred=Reaction~ Days +
(1|Subject)+(Days|Subject),
RandDist="gaussian")
model_phi<-DHGLMMODELING(Model="dispersion")
res_hglm<-dhglmfit(RespDist="gaussian",DataMain=sleepstudy,
MeanModel=model_mu,DispersionModel=model_phi)
```

Fit a DHGLM allowing a heavy-tailed distribution for the residual variance. Compare model-checking plots for two models.

4. Consider the mutagenicity assay data in Exercise 2 of Chapter 5 again. Fit a Poisson DHGLM allowing a heavy-tailed distribution for the random effect as follows:

$$\begin{aligned}
y_{ij}|v_i &\sim Possion(\mu_i) \\
\log(\mu_i) &= \beta_0 + \beta_1 \log(dose_i + 10) + \beta_2 dose_i + v_i \\
v_i &\sim \mathrm{N}(0, \lambda_i) \\
\log(\lambda_i) &= \log(\lambda_0) + b_i \\
b_i &\sim \mathrm{N}(0, \alpha).
\end{aligned}$$

Check the significance for presence of random effects $b_i$ and assess the model-checking plots.

CHAPTER 7

# Fitting multivariate HGLMs

As a most general model for a single response, we presented a DHGLM, which has a great room for further generalization by including more general correlation patterns among random effects. In this chapter we introduce multivariate models for various types of responses including continuous, proportion, counts, events, etc. We show that general multivariate models can be generated by connecting DHGLMs for various responses with correlated random effects.

The R package mdhglm (Lee et al., 2016b) was developed for fitting multivariate HGLMs, as well as multivariate DHGLMs. Correlation between random components is essential in the definition of joint models, where correlations among multivariate responses are modeled via correlated random effects. Using h-likelihood, a bivariate binary-normal model was described in Yun and Lee (2004). Another early multivariate model in the h-likelihood framework can be found in Ha et al. (2003), combining a time to event response and a longitudinal Gaussian response, where their correlation is modeled by a shared random effect, described in Chapter 9. H-likelihood and related fitting algorithm are in Chapter 9.

Several longitudinal models can be linked together to compose a multivariate hierarchical generalized linear model (Fieuws and Verbeke, 2006). Such models are jointly estimated upon the introduction of correlations between the random effects of several longitudinal processes.

The main function in the R package mdhglm is `jointfit()`. For instance, the following code fits a bivariate HGLM with Gaussian and binomial responses.

```
> jm1<-DHGLMMODELING(Link="identity", LinPred=y1~x1+(1|id1),
+          RandDist="gaussian")
> jm2<-DHGLMMODELING(Link="logit", LinPred=y2~x2+(1|id2),
+         RandDist="gaussian")
> res_corr<-jointfit(RespDist=c("gaussian","binomial"),
+         DataMain=list(data1,data2),
+         MeanModel=list(jm1,jm2),
```

```
+         structure="correlated",Init_Corr=0.1,
+         mord = 0, dord = 1,
+         Maxiter = 200, convergence = 10^-6)
```

The models for the responses `y1` and `y2` are defined by the `DHGLMMODELING()` function by specifying their link functions, linear predictors, and distributions of random effects. The `jointfit` function with `DHGLMMODELING` objects specifies the distribution of `y1` and `y2`, the structure of correlation between random effects and the method of fitting, for example, the order of Laplace approximation.

For the above example, the first response, `y1`, is specified to follow a Gaussian random-effect model with a fixed covariate `x1` and the subject identifier `id1` specifying a random intercept model. The second response, `y2`, is specified to follow a binary-logistic random-effect model with a fixed covariate `x2` and the subject identifier `id2` specifying a random intercept model. The `jointfit` function can fit the multivariate model with two different DHGLMs for response variables. The parameters `mord` and `dord` are the orders of Laplace approximations to fit the mean parameters (`mord` = 0 or 1) and the dispersion parameters (`dord` = 1 or 2), respectively. Higher-order corrections allow accurate approximation but involve expensive computation. The structure specifies the structure of correlation between two random effects, such as independent, shared or correlated (saturated). The `Maxiter` parameter specifies the maximum number of iterations and `convergence` specifies the tolerance of the convergence criterion.

## 7.1 Examples

### *7.1.1 Ethylene glycol data*

Price et al. (1985) presented data from a study on the developmental toxicity of ethylene glycol (EG) in mice. Table 7.1 summarizes the data on malformation (binary response) and fetal weight (continuous response) and shows clear dose-related trends with respect to both. The rates of fetal malformation increase with dose, ranging from 0.3% in the control group to 57% in the highest-dosage (3g/kg/day) group. Fetal weight decreases as dosage increases with the average weight ranging from 0.972g in the control group to 0.704g in the highest-dose group.

To analyze this dataset, Yun and Lee (2004) proposed the following bi-variate HGLM: let $y_{1ij}$ be fetal weights and $y_{2ij}$ an indicator for malformation, obtained from the $i$th dam. Let $y_{ij} = (y_{1ij}, y_{2ij})^t$ be the bivariate responses and $v_i = (w_i, u_i)^t$ be the unobserved random effects

Table 7.1 *Descriptive statistics for the ethylene glycol data*

| Dose (g/kg) | Dams | Live | Malformations No. | % | Weight (g) Mean | (S.D) |
|---|---|---|---|---|---|---|
| 0.00 | 25 | 297 | 1 | (0.34) | 0.972 | (0.0976) |
| 0.75 | 24 | 276 | 26 | (9.42) | 0.877 | (0.1041) |
| 1.50 | 22 | 229 | 89 | (38.86) | 0.764 | (0.1066) |
| 3.00 | 23 | 226 | 129 | (57.08) | 0.704 | (0.1238) |

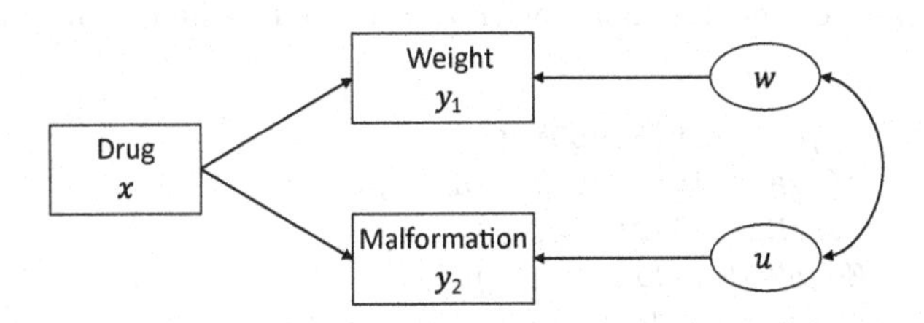


Figure 7.1 *Path diagram for the MDHGLM fitted to the ethylene glycol data.*

for the $i$th cluster. It is assumed that $y_{1ij}$ and $y_{2ij}$ are conditionally independent given $v_i$ (see Figure 7.1). Hence, the following bivariate HGLM is proposed:

$$y_{1ij}|w_i \sim \mathrm{N}(\mu_{ij}, \phi),$$

where $\mu_{ij} = x_{1ij}\beta_1 + w_i$

$$y_{2ij}|u_i \sim Bernoulli(p_{ij}),$$

where $logit(p_{ij}) = x_{2ij}\beta_2 + u_i$, and

$$v_i \sim \mathrm{N}(0, \Sigma), \tag{3}$$

where $\Sigma = \begin{pmatrix} \sigma_1^2 & \rho\sigma_1\sigma_2 \\ \rho\sigma_1\sigma_2 & \sigma_2^2 \end{pmatrix}$ and $-1 < \rho < 1$.

We consider three models; the R-codes for fitting these models are as follows:

```
> data(eg,package="mdhglm")
> jm1<-DHGLMMODELING(Link="identity",
+          LinPred=y1~dose+dose2+(1|litter),
+          RandDist="gaussian")
> jm2<-DHGLMMODELING(Link="logit",
+          LinPred=y2~dose+dose2+(1|litter),
+          RandDist="gaussian")
```

(i) M1: independent random-effects model where $\rho = 0$,

```
> res_ind<-jointfit(RespDist=
+         c("gaussian","binomial"),
+         DataMain=list(eg,eg),
+         MeanModel=list(jm1,jm2),
+         structure="independent")
```

(ii) M2: random-effects model with a saturated variance-covariance matrix, and

```
> res_sat<-jointfit(RespDist=
+         c("gaussian","binomial"),
+         DataMain=list(eg,eg),
+         MeanModel=list(jm1,jm2),
+         structure="correlated")
```

(iii) M3: shared random-effects model where $u_i = \delta w_i$ for some constant $\delta$

```
> res_shared<-jointfit(RespDist=
+         c("gaussian","binomial"),
+         DataMain=list(eg, eg),
+         MeanModel=list(jm1,jm2),
+         structure="shared")
```

The cAIC has values of $-1649.9$, $-1657.0$ and $-1600.1$ for M1, M2, and M3, respectively. Thus, cAIC selects M2 as the best-fitting model. This means that the model with the saturated variance-covariance matrix fits the best, rather than models with the independent or shared random effects. Table 7.2 shows the result for the best-fitting model.

### 7.1.2 *Rheumatoid arthritis data*

The Rheumatoid Arthritis Patients rePort Onset Re-activation sTudy (RAPPORT study) is a longitudinal study that aims to identify an increase in disease activity by self-reported questionnaires. A cohort of 159

Table 7.2 *Fitting results for M2 in the ethylene glycol data*

| | Weight | | | Malformation | | |
|---|---|---|---|---|---|---|
| | Estimate | SE | t-value | Estimate | SE | t-value |
| Intercept | 0.978 | 0.017 | 57.529 | −5.857 | 0.678 | −8.639 |
| Dose | −0.163 | 0.029 | −5.621 | 4.742 | 0.905 | 5.240 |
| Dose$^2$ | 0.025 | 0.009 | 2.778 | −0.885 | 0.245 | −3.612 |
| $\log(\sigma_1^2)$ | −4.957 | 0.219 | | | | |
| $\log(\sigma_2^2)$ | | | | 0.613 | 0.339 | |
| $\rho$ | −0.619 | 0.047 | | | | |
| $\phi$ | −5.190 | 0.046 | | | | |

patients is followed throughout one year. Self-reported questionnaires are provided for patients every three months together with clinical evaluations of patients' disease status. Health assessment questionnaires (HAQ) (20 questions from eight categories) and the Rheumatoid Arthritis Disease Activity Index (RADAI) (five items, e.g., today's disease activity in terms of swollen and tender joints) were used for patients to self-report their functional status. We use binarized versions of HAQ and RADAI, with cut-off points of 0.5 (HAQ) and 2.2 (RADAI). Lower values indicate disease stability, while higher values show increased disease activity.

A clinical examination was recorded using the disease activity score with 28 joint counts (DAS28), which is a composite score that includes for example the swollen joints count. The DAS28 score varies between 0 and 10, and we assume a Gaussian model to approximate its distribution. Information about DAS28, RADAI, and HAQ taken at months 0, 3, 6, 9, and 12 were considered in this example. The analysis includes gender and baseline age of the patients as covariates.

DAS28 is denoted as `y1` in the R code below, HAQ as `y2` and RADAI as `y3`. As covariates we use the intercept, month of measurement, gender and age at the baseline. Therefore each $\boldsymbol{X}_k$, where $k = 1, 2, 3$ has four columns. There are 159 patients in the study; therefore $i = 1, 2, ..., 159$ and $j = 1, 2, ..., 5$ as there are five visits. Note that not all patients gave information for each $k$th response and not all patients were measured at each of the five visits. We consider the following multivariate model with three responses (see Figure 7.2).

$$y_{1ij}|v_{11i}, v_{12i} \sim \mathrm{N}(X_{1ij}\beta_1 + v_{11i} + v_{12i} * time, \phi)$$

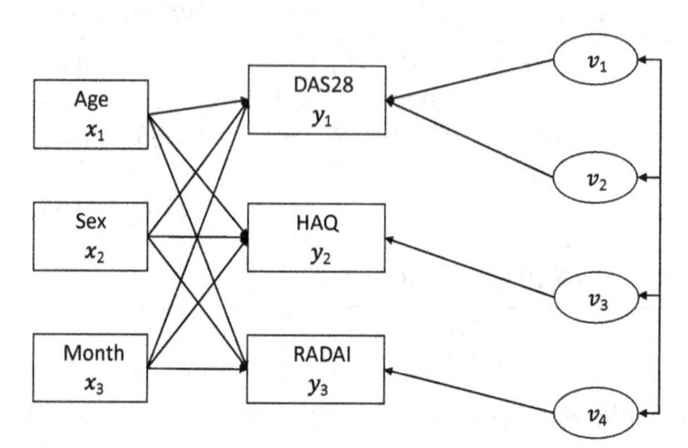


Figure 7.2 *Path diagram for the MDHGLM fitted to the rheumatoid arthritis data.*

$$y_{2ij}|v_{21i} \sim Bernoulli\left(\frac{\exp(X_{2ij}\beta_2 + v_{21i})}{1 + exp(X_{2ij}\beta_2 + v_{21i})}\right)$$

$$y_{3ij}|v_{31i} \sim Bernoulli\left(\frac{\exp(X_{3ij}\beta_3 + v_{31i})}{1 + exp(X_{3ij}\beta_3 + v_{31i})}\right)$$

The model for DAS28 includes a random intercept and slope, while HAQ and RADAI have only random intercepts. Thus, we assume a four-dimensional latent structure:

$$\begin{pmatrix} v_{11i} \\ v_{12i} \\ v_{21i} \\ v_{31i} \end{pmatrix} \sim MVN\left[\begin{pmatrix} 0 \\ 0 \\ 0 \\ 0 \end{pmatrix}, \begin{pmatrix} \lambda_{11} & \rho_1\lambda^*_{11,12} & \rho_2\lambda^*_{11,21} & \rho_3\lambda^*_{11,31} \\ \rho_1\lambda^*_{11,12} & \lambda_{12} & \rho_4\lambda^*_{12,21} & \rho_5\lambda^*_{12,31} \\ \rho_2\lambda^*_{11,21} & \rho_4\lambda^*_{12,21} & \lambda_{21} & \rho_6\lambda^*_{21,31} \\ \rho_3\lambda{*}_{11,31} & \rho_5\lambda^*_{12,31} & \rho_6\lambda^*_{21,31} & \lambda_{31} \end{pmatrix}\right]$$

where $\lambda^*_{ij,lk} = \sqrt{\lambda_{ij}\lambda_{lk}}$.

To fit the above model, the following commands are used to create individual HGLMs:

```
> data(ra,package="mdhglm")
> jm1<-DHGLMMODELING(Link="identity",
+ LinPred=y1~time+age+sex+(1|subject)+(time|subject),
+ RandDist=c("gaussian","gaussian"))
```

```
> jm2<-DHGLMMODELING(Link="logit",
+ LinPred=y2~time+age+sex+(1|subject),
+ RandDist=c("gaussian"))
> jm3<-DHGLMMODELING(Link="logit",
+ LinPred=y3~time+age+sex+(1|subject),
+ RandDist=c("gaussian"))
```

To link them by the common correlation matrix we apply the next procedure:

```
> res<-jointfit(RespDist=
+ c("gaussian","binomial","binomial"),
+ DataMain=list(ra1,ra2,ra3),
+ MeanModel=list(jm1,jm2,jm3),
+ structure="correlated")
```

Random intercepts of HAQ and RADAI show the strongest correlation, 0.85 (Table 7.3). Latent profiles were highly correlated, showing a higher level of HAQ associated with higher level of RADAI score. This indicates that the average probability of disease remission within a subject was shown by the two questionnaire measures over time. Inferences from the independence model (Table 7.4) and the correlated model were somewhat different, which might have been explained by the fact that the average probability of remission is highly correlated in both questionnaires. Age effect for HAQ does not show significance in the correlated model, while it is significant in separate univariate models. Moderate correlations were found between the clinical evaluation response and the self-reported questionnaires which were estimated at 0.61 and 0.63. The random slope of DAS28 was not strongly associated with the random intercepts of any response.

To summarize, the latent intercepts of the three responses play a major role, HAQ and RADAI seem to be equivalent binary measures as to whether disease progresses or remains at its current level, and they might be treated exchangeably. Further, the level of latent DAS28 predicts moderately well the status of a patient obtained from self-assessment.

Table 7.3 *Results of the h-likelihood DHGLM fit: rheumatoid arthritis with correlated models*

| Correlated Joint Model | | | | |
|---|---|---|---|---|
| | | Fixed Effects | | |
| | Estimate | S.E. | z-score | p-value |
| | | DAS28 | | |
| Intercept | 1.773 | 0.413 | 4.293 | <0.0001 |
| age | 0.008 | 0.006 | 1.333 | 0.182 |
| sex=f | 0.760 | 0.198 | 3.838 | <0.0001 |
| time | −0.010 | 0.008 | −1.250 | 0.211 |
| | | HAQ | | |
| Intercept | −7.500 | 2.726 | −2.751 | 0.006 |
| age | 0.073 | 0.042 | 1.730 | 0.084 |
| sex=f | 3.156 | 1.300 | 2.428 | 0.015 |
| time | 0.024 | 0.034 | 0.698 | 0.485 |
| | | RADAI | | |
| Intercept | −2.095 | 0.978 | −2.142 | 0.032 |
| age | 0.018 | 0.015 | 1.200 | 0.230 |
| sex=f | 0.460 | 0.458 | 1.004 | 0.315 |
| time | 0.018 | 0.023 | 0.783 | 0.434 |
| | | Correlation matrix | | |
| | DAS RI | DAS RS | HAQ RI | RADAI RI |
| DAS RI | 1 | −0.159 | 0.614 | 0.626 |
| DAS RS | −0.159 | 1 | 0.117 | 0.095 |
| HAQ RI | 0.614 | 0.117 | 1 | 0.846 |
| RADAI RI | 0.626 | 0.095 | 0.846 | 1 |
| | | Residual Variance | | |
| DAS 28 | | 0.466 | | |

### *7.1.3 National merit scholarship qualifying test data for twins*

Loehlin and Nichols (1976) presented the national merit twins data including extensive questionnaires from 839 adolescent twins. The twins were identified among the roughly 600,000 US high school juniors who took the national merit scholarship qualifying test (NMSQT) in 1962. They were diagnosed as identical (509 pairs) or same-sex fraternal (330 pairs) by a brief mail questionnaire and later completed a 1082-item questionnaire covering a variety of behaviors, attitudes, personality, life experiences, health, vocational preferences, etc., plus the 480-item California psychological inventory. Twins' scores on the NMSQT and their

Table 7.4 *Results of the h-likelihood DHGLM fit: rheumatoid arthritis with independent models*

| Correlated Joint Model | | | | |
|---|---|---|---|---|
| | | Fixed Effecs | | |
| | Estimate | S.E. | z-score | p-value |
| | | DAS28 | | |
| Intercept | 1.697 | 0.411 | 4.129 | <0.0001 |
| age | 0.009 | 0.006 | 1.500 | 0.134 |
| sex=f | 0.799 | 0.197 | 4.056 | <0.0001 |
| time | −0.011 | 0.008 | −1.375 | 0.169 |
| | | HAQ | | |
| Intercept | −9.002 | 2.347 | −3.835 | <0.0001 |
| age | 0.010 | 0.037 | 2.631 | 0.009 |
| sex=f | 3.238 | 1.106 | 2.926 | 0.003 |
| time | 0.023 | 0.033 | 0.692 | 0.489 |
| | | RADAI | | |
| Intercept | −2.283 | 0.964 | −2.367 | 0.018 |
| age | 0.022 | 0.015 | 1.479 | 0.139 |
| sex=f | 0.361 | 0.449 | 0.804 | 0.421 |
| time | 0.015 | 0.023 | 0.663 | 0.508 |
| | | Correlation matrix | | |
| | DAS RI | DAS RS | HAQ RI | RADAI RI |
| DAS RI | 1 | −0.206 | | |
| DAS RS | −0.206 | 1 | | |
| HAQ RI | | | 1 | |
| RADAI RI | | | | 1 |
| | | Residual Variance | | |
| DAS 28 | | 0.453 | | |

five subscales are also included. The 285-item questionnaire filled out by the parent was mainly focused on the life histories and experiences of the twins.

Four NMSQT scores recorded within 0-100 are considered as response variables: English ($y_1$), mathematics ($y_2$), social science ($y_3$) and natural science ($y_4$). The purpose of the study is to examine the effects of five covariates in Table 7.5 on four NMSQT scores.

We consider a multivariate HGLM for four response variables for the $j$th person of the $i$th twin (see Figure 7.3), for $k = 1, \cdots, 4$,

Table 7.5 *Covariate and response variables lists for NMSQT for twins data*

| variables | code | definition |
|---|---|---|
| | | Covariates |
| Gender | $x_1$ | $= 0$ (male), $= 1$ (female) |
| Mother's educational level | $x_2$ | $= 1$ ($\leq$ 8th grade), $= 2$ (part high school), $= 3$ (high school grad), $= 4$ (part college), $= 5$ (college grad), $= 6$ (graduate degree) |
| Father's educational level | $x_3$ | $= 1$ ($\leq$ 8th grade), $= 2$ (part high school), $= 3$ (high school grad), $= 4$ (part college), $= 5$ (college grad), $= 6$ (graduate degree) |
| Family income level | $x_4$ | $= 1$ ($\leq$ \$5000), $= 2$ (\$5000 to \$7499), $= 3$ (\$7500 to \$9999), $= 4$ (\$10000 to \$14999), $= 5$ (\$15000 to \$19999), $= 6$ (\$20000 to \$24999), $= 7$ ($\geq$ \$25000) |
| Zygosity | x5 | $= 0$ (identical), $= 1$ (fraternal) |

$$y_{kij}|v_{ki} \sim \mathrm{N}(X_{ij}\beta_k^{(\mu)} + v_{ki}, \phi_{kij}),$$

where random effects follow multivariate normal distribution

$$\begin{pmatrix} v_{1i} \\ v_{2i} \\ v_{3i} \\ v_{4i} \end{pmatrix} \sim MVN\left[\begin{pmatrix} 0 \\ 0 \\ 0 \\ 0 \end{pmatrix}, \begin{pmatrix} \lambda_{1i} & \rho_1\lambda^*_{1i,2i} & \rho_2\lambda^*_{1i,3i} & \rho_3\lambda^*_{1i,4i} \\ \rho_1\lambda^*_{1i,2i} & \lambda_{2i} & \rho_4\lambda^*_{2i,3i} & \rho_5\lambda^*_{2i,4i} \\ \rho_2\lambda^*_{1i,3i} & \rho_4\lambda^*_{2i,3i} & \lambda_{3i} & \rho_6\lambda^*_{3i,4i} \\ \rho_3\lambda^*_{1i,4i} & \rho_5\lambda^*_{2i,4i} & \rho_6\lambda^*_{3i,4i} & \lambda_{4i} \end{pmatrix}\right],$$

$\lambda^*_{ji,ki} = \sqrt{\lambda_{ji}\lambda_{ki}}$ and $\lambda_{ki} = \exp\beta^{(\lambda)}_{k0}$ is the variance of random effects. To allow heterogeneity between type of zygosity, we consider the model for residual variance

$$\log(\phi_{kij}) = \beta^{(\phi)}_{k0} + \beta^{(\phi)}_{k5} x_{5i}.$$

To fit this multivariate HGLM, the following commands are used. We make individual HGLMs as follows:

```
> data(nmsqt,package="mdhglm")
> jm1<-DHGLMMODELING(Link="identity",
+           LinPred=y1~x1+as.factor(x2)+as.factor(x3)+
+     as.factor(x4)+x5+(1|pairnum),
```

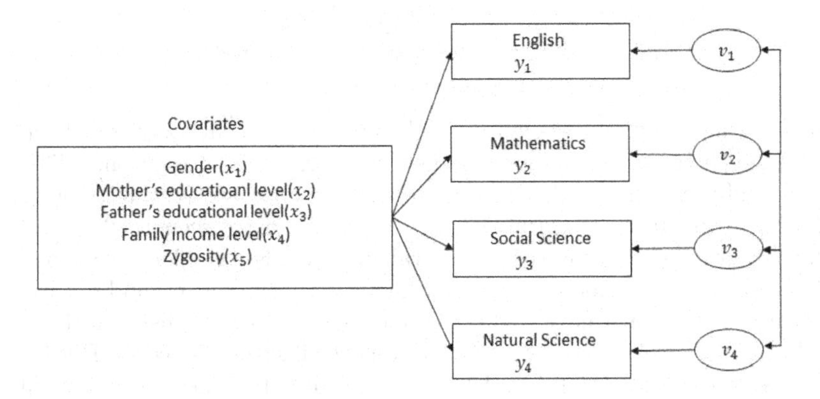


Figure 7.3 *Path diagram for the MDHGLM fitted to the national merit scholarship qualifying test for twins data.*

```
+          RandDist=c("gaussian"))
> jm2<-DHGLMMODELING(Link="identity",
+          LinPred=y1~x1+as.factor(x2)+as.factor(x3)+
+     as.factor(x4)+x5+(1|pairnum),
+          RandDist=c("gaussian"))
> jm3<-DHGLMMODELING(Link="identity",
+          LinPred=y1~x1+as.factor(x2)+as.factor(x3)+
+     as.factor(x4)+x5+(1|pairnum),
+          RandDist=c("gaussian"))
> jm4<-DHGLMMODELING(Link="identity",
+          LinPred=y1~x1+as.factor(x2)+as.factor(x3)+
+     as.factor(x4)+x5+(1|pairnum),
+          RandDist=c("gaussian"))
> jd1<-DHGLMMODELING(Model="dispersion",
+    Link="log", LinPred=phi~x5)
> jd2<-DHGLMMODELING(Model="dispersion",
+    Link="log", LinPred=phi~x5)
> jd3<-DHGLMMODELING(Model="dispersion",
+    Link="log", LinPred=phi~x5)
> jd4<-DHGLMMODELING(Model="dispersion",
+    Link="log", LinPred=phi~x5)
```

To link HGLMs by correlated random effects, we use the following codes:

```
> res1<-jointfit(RespDist=
+          c("gaussian","gaussian","gaussian","gaussian"),
+          DataMain=list(cog,cog,cog,cog),
```

```
+          MeanModel=list(jm1,jm2,jm3,jm4),
+          DispersionModel=list(jm1,jm2,jm3,jm4),
+          structure="correlated")
```

The fitting results are shown in Table 7.6. Random effects of social science and natural science scores show the strongest correlation, 0.738. This indicates that higher scores in social science are associated with higher scores in the natural sciences. Correlation between English and mathematics scores has the lowest value, 0.622. For gender effect, men have higher significant scores on mathematics, social science and natural science, but women have higher significant scores on English. Mother's educational level is not significant at almost all subject's scores. But father's educational level 4 and 6 are significant. If father's educational level is somewhat high, then their offsprings will have a high score too. Family's income level 5 has a significant positive effect and it has the highest estimate. Zygosity is not significant for the mean but, it is positively significant for the residual variance. In dispersion models for residual variances, we see that fraternal twins have greater heterogeneity than identical twins.

Even though this is not shown, the multivariate HGLM model with constant residual variances does not have linear normal probability plots. From Figure 7.4, we can see that the normal probability plots are approximately linear in the absence of outliers. Thus, the fitted model is satisfactory.

### *7.1.4 Vascular cognitive impairment data*

Lee, Nelder, and Pawitan (2017) considered the Vascular Cognitive Impairment (VCI) data. As the elderly population increases, the proportion of strokes, one of the main geriatric diseases, also rises. The VCI measurements are increased among stroke patients, because cognitive function is declined due to stroke. However, through an early intervention based on the VCI, the cognitive function can be improved.

Here, the standardized VCI scores are considered for four domains (response variables): executive ($y1$), memory ($y2$), visuospatial ($y3$), and language ($y4$). The purpose of the study is to examine the effects of ten demographic and ten acute neuroimaging variables on the cognitive function in the ischemic stroke patients in Table 7.7.

First, consider a multivariate HGLM for four response variables for the $t$th visit of the $i$th patient (see Figure 7.5), for $k = 1, \cdots, 4$,

$$y_{kit}|v_{ki} \sim \mathrm{N}(X_{it}\beta_k^{(\mu)} + v_{ki}, \phi_{kit}),$$

Table 7.6 *MDHGLM Results for the national merit scholarship qualifying test for twins data*

| $y_k$ covariate | $y_1$ (english) Estimate | SE | t-value | $y_2$ (mathematics) Estimate | SE | t-value | $y_3$ (social science) Estimate | SE | t-value | $y_4$ (natural science) Estimate | SE | t-value |
|---|---|---|---|---|---|---|---|---|---|---|---|---|
| model for $\mu$ | | | | | | | | | | | | |
| Intercept | 18.273 | 0.715 | 25.538 | 19.947 | 0.894 | 22.298 | 19.760 | 0.742 | 26.621 | 19.779 | 0.840 | 23.541 |
| Gender | 0.9784 | 0.304 | 3.216 | −3.734 | 0.381 | −9.800 | −1.208 | 0.316 | −3.821 | −2.511 | 0.356 | −7.038 |
| Mother's eduation | | | | | | | | | | | | |
| 2 (part high school) | −1.334 | 0.744 | −1.791 | −0.178 | 0.933 | −0.191 | −0.312 | 0.774 | −0.403 | −0.404 | 0.873 | −0.463 |
| 3 (high school grad) | −1.136 | 0.711 | −1.598 | −0.167 | 0.892 | −0.187 | −0.906 | 0.740 | −1.223 | 0.183 | 0.833 | 0.220 |
| 4 (part college) | −0.644 | 0.752 | −0.857 | −0.435 | 0.944 | −0.461 | −0.378 | 0.783 | −0.482 | 0.613 | 0.881 | 0.696 |
| 5 (college grad) | −0.082 | 0.813 | −0.101 | 1.204 | 1.021 | 1.178 | 0.105 | 0.847 | 0.124 | 1.667 | 0.952 | 1.749 |
| 6 (graduate degree) | −0.917 | 1.002 | −0.915 | −0.579 | 1.258 | −0.460 | −0.522 | 1.044 | −0.500 | 0.506 | 1.174 | 0.431 |
| Father's eduation | | | | | | | | | | | | |
| 2 (part high school) | 0.204 | 0.637 | 0.321 | 0.760 | 0.798 | 0.952 | 0.474 | 0.662 | 0.716 | −0.539 | 0.747 | −0.722 |
| 3 (high school grad) | 0.398 | 0.591 | 0.674 | 0.626 | 0.742 | 0.843 | 0.619 | 0.615 | 1.006 | −0.599 | 0.693 | −0.864 |
| 4 (part college) | 1.292 | 0.625 | 2.065 | 1.694 | 0.785 | 2.156 | 1.753 | 0.651 | 2.689 | 0.610 | 0.733 | 0.832 |
| 5 (college grad) | 1.089 | 0.708 | 1.539 | 2.036 | 0.888 | 2.292 | 1.709 | 0.737 | 2.318 | −0.405 | 0.830 | −0.489 |
| 6 (graduate degree) | 2.582 | 0.739 | 3.491 | 3.796 | 0.928 | 4.087 | 2.899 | 0.770 | 3.763 | 1.286 | 0.866 | 1.483 |
| Family income | | | | | | | | | | | | |
| 2 ($5000 to $7499) | 0.286 | 0.541 | 0.529 | 1.371 | 0.677 | 2.025 | 0.188 | 0.561 | 0.335 | 0.590 | 0.636 | 0.928 |
| 3 ($7500 to $9999) | 0.955 | 0.577 | 1.655 | 2.134 | 0.722 | 2.954 | 1.312 | 0.599 | 2.189 | 1.340 | 0.678 | 1.977 |
| 4 ($10000 to $14999) | 0.650 | 0.594 | 1.094 | 2.190 | 0.743 | 2.948 | 1.218 | 0.616 | 1.975 | 1.769 | 0.697 | 2.537 |
| 5 ($15000 to $19999) | 2.016 | 0.717 | 2.811 | 3.902 | 0.897 | 4.349 | 2.071 | 0.744 | 2.783 | 2.454 | 0.841 | 2.914 |
| 6 ($20000 to $24999) | 1.310 | 1.039 | 1.260 | 3.820 | 1.302 | 2.932 | 1.591 | 1.080 | 1.472 | 2.191 | 1.220 | 1.796 |
| 7 ($\geq$ $25000) | 1.071 | 0.856 | 1.251 | 3.253 | 1.072 | 3.034 | 1.325 | 0.889 | 1.490 | 1.368 | 1.004 | 1.362 |
| zygosite | 0.189 | 0.309 | 0.612 | 0.114 | 0.391 | 0.292 | −0.224 | 0.324 | −0.692 | 0.141 | 0.361 | 0.392 |
| model for $\log(\phi)$ | | | | | | | | | | | | |
| Intercept | 1.710 | 0.061 | 28.240 | 2.344 | 0.060 | 39.330 | 1.685 | 0.061 | 27.681 | 2.369 | 0.059 | 40.177 |
| zygosite | 0.442 | 0.095 | 4.641 | 0.603 | 0.093 | 6.493 | 0.734 | 0.095 | 7.758 | 0.233 | 0.093 | 2.493 |
| $\log \lambda_i$ | 2.605 | 0.058 | 45.16 | 2.982 | 0.060 | 49.84 | 2.674 | 0.058 | 46.14 | 2.854 | 0.060 | 47.74 |
| | $\widehat{\rho}_1$ =0.622 | $\widehat{\rho}_2$ =0.683 | $\widehat{\rho}_3$ =0.679 | $\widehat{\rho}_4$ =0.635 | $\widehat{\rho}_5$ =0.688 | $\widehat{\rho}_6$ =0.738 | | | | | | |

Table 7.7 *Covariate lists for vascular cognitive impairment data*

| variables | code | definition |
|---|---|---|
| demographic variables | | |
| Age | $x_1$ | integer of age/10 |
| Gender | $x_2$ | $= 1$ (male), $= 0$ (female) |
| Edu | $x_3$ | $= 0$ (none), $= 1$ (elementary) , $= 2$ (middle), $= 3$ (high), $= 4$ (over college) |
| HTN | $x_4$ | $= 1$ (hypertension), $= 0$ (none) |
| DM | $x_5$ | $= 1$ (diabetes mellitus), $= 0$ (none) |
| Af | $x_6$ | $= 1$ (atrial fibrillation), $= 0$ (none) |
| HxStroke | $x_7$ | $= 1$ (history of stroke), $= 0$ (none) |
| NIHSS | $x_8$ | national institute of health stroke scale score at admission |
| VCINP | $x_9$ | time interval from stroke onset to first K-VCIHS-NP |
| PCI | $x_{10}$ | $= 1$ (IQCODE $\geq 3.6$), $= 0$ (otherwise) |
| neuroimaging variables | | |
| AcuteLeft | $x_{11}$ | Left or bilateral involvement |
| AcuteMulti | $x_{12}$ | lesion multiplicity in acute DWI imaging |
| AcuteCS | $x_{13}$ | cortical involvement of acute lesions |
| ChrCS | $x_{14}$ | cortical involvement of chronic territorial infarction |
| PVWM | $x_{15}$ | Periventricular white matter lesions (PVWM); $= 0$ (PVWN $0, 1$), $= 1$ (PVWN $2, 3$) |
| SCWM | $x_{16}$ | Subcortical white matter lesions (SCWM); $= 0$ (SCWM $0, 1$), $= 1$ (SCWM $2, 3$) |
| LAC | $x_{17}$ | The presence of lacunes |
| CMB | $x_{18}$ | The presence of cerebral microbleeds |
| | | Medial temporal lobe atrophy (MTA); |
| MTA1 | $x_{19}$ | $= 1$ (MTA 2), $= 0$ (not 2) |
| MTA2 | $x_{20}$ | $= 1$ (MTA $3, 4$), $= 0$ (not $3, 4$) |

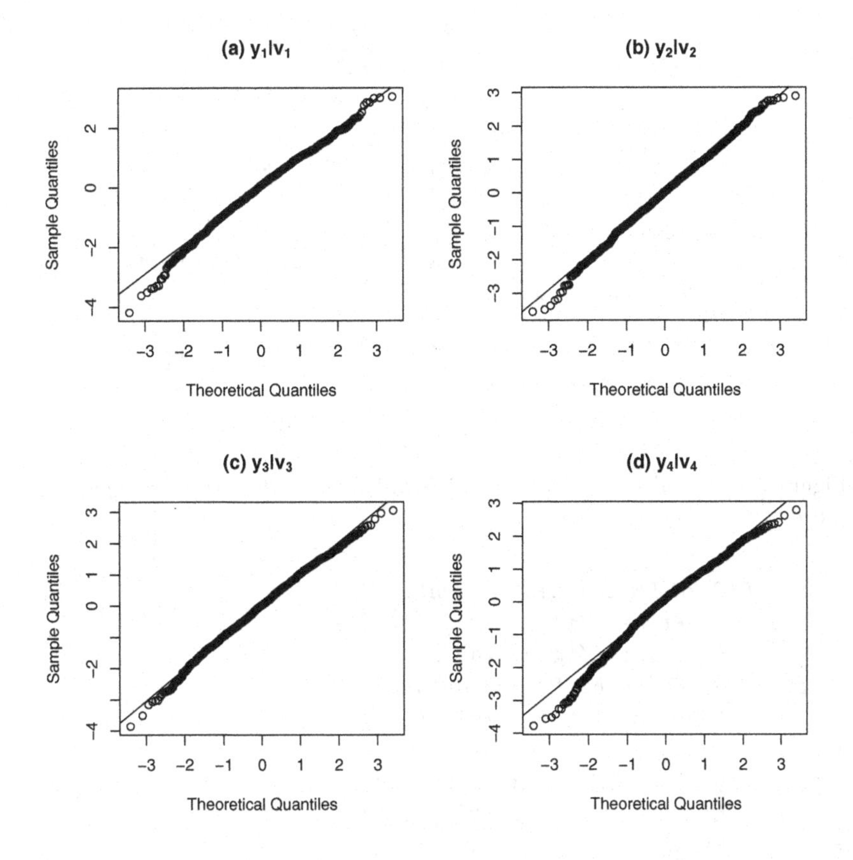


Figure 7.4 *Normal probability plots for (a) $y_1|v_1$, (b) $y_2|v_2$, (c) $y_3|v_3$, and (d) $y_4|v_4$ under the multivariate HGLM on the national merit scholarship qualifying test for twins data.*

where $X_{it}$ are covariates, $\phi_{kit} = \exp(\beta_{k0}^{(\phi)})$ is the residual variance, and the random effects follow a multivariate normal distribution

$$\begin{pmatrix} v_{1i} \\ v_{2i} \\ v_{3i} \\ v_{4i} \end{pmatrix} \sim MVN \left[ \begin{pmatrix} 0 \\ 0 \\ 0 \\ 0 \end{pmatrix}, \begin{pmatrix} \lambda_{1i} & \rho_1\lambda^*_{1i,2i} & \rho_2\lambda^*_{1i,3i} & \rho_3\lambda^*_{1i,4i} \\ \rho_1\lambda^*_{1i,2i} & \lambda_{2i} & \rho_4\lambda^*_{2i,3i} & \rho_5\lambda^*_{2i,4i} \\ \rho_2\lambda^*_{1i,3i} & \rho_4\lambda^*_{2i,3i} & \lambda_{3i} & \rho_6\lambda^*_{3i,4i} \\ \rho_3\lambda^*_{1i,4i} & \rho_5\lambda^*_{2i,4i} & \rho_6\lambda^*_{3i,4i} & \lambda_{4i} \end{pmatrix} \right],$$

$\lambda^*_{ji,ki} = \sqrt{\lambda_{ji}\lambda_{ki}}$ and $\lambda_{ki} = \exp(\beta_{k0}^{(\lambda)})$ is the variance of random effects.

To fit this multivariate HGLM, the following commands are used, where we make individual HGLMs:

```
> data(cog)
```

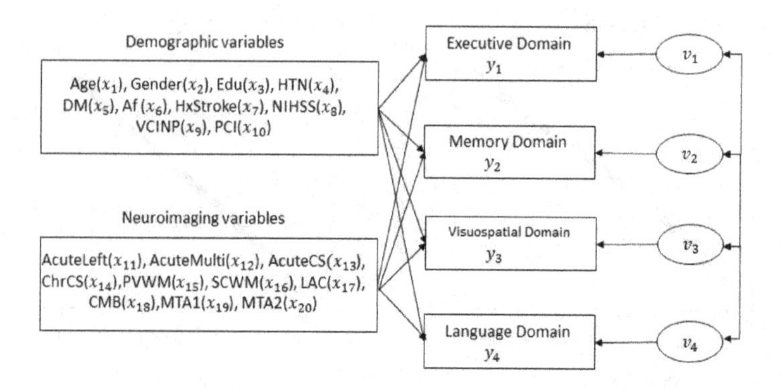


Figure 7.5 *Path diagram for the MDHGLM fitted to the vascular cognitive impairment data.*

```
> jm1<-DHGLMMODELING(Link="identity",
+          LinPred=y1~x1+...+x20+(1|id),
+          RandDist=c("gaussian"))
> jm2<-DHGLMMODELING(Link="identity",
+          LinPred=y2~x1+...+x20+(1|id),
+          RandDist=c("gaussian"))
> jm3<-DHGLMMODELING(Link="identity",
+          LinPred=y3~x1+...+x20+(1|id),
+          RandDist=c("gaussian"))
> jm4<-DHGLMMODELING(Link="identity",
+          LinPred=y4~x1+...+x20+(1|id),
+          RandDist=c("gaussian"))
```

To link HGLMs by correlated random effects, we use the following code:

```
> res1<-jointfit(RespDist=
+          c("gaussian","gaussian","gaussian","gaussian"),
+          DataMain=list(cog,cog,cog,cog),
+          MeanModel=list(jm1,jm2,jm3,jm4),
+          structure="correlated")
```

From Figure 7.6, we see many large outliers. Thus, we consider a multivariate DHGLM (called MDHGLM1) that allows a heavy-tailed distribution for $(y_{1it}|v_{1i}, \cdots, y_{4it}|v_{4i})$ as follows: for $k = 1, \cdots, 4$,

$$\log(\phi_{kit}) = \beta_{k0}^{(\phi)} + v_{ki}^{(\phi)},$$

where $v_{ki}^{(\phi)} \sim \mathrm{N}(0, \alpha_k)$ and $\lambda_{ki} = \exp \beta_{k0}^{(\lambda)}$.

The MDHGLM1 can be fitted by specifying models for dispersions as follows:

```
> jd<-DHGLMMODELING(Model="dispersion",Link="log",
+         LinPred=phi~1+(1|id),
+         RandDist=c("gaussian"))
```

To link DHGLMs by correlated random effects, we use the following code:

```
> res2<-jointfit(RespDist=
+         c("gaussian","gaussian","gaussian","gaussian"),
+         DataMain=list(cog,cog,cog,cog),
+         MeanModel=list(jm1,jm2,jm3,jm4),
+         DispersionModel=list(jd,jd,jd,jd),
+         structure="correlated")
```

We further consider a MDHGLM (called MDHGLM2), also allowing heavy-tailed distribution for $v_{ki}$ as follows: for $k = 1, \cdots, 4$,

$$\log(\phi_{kit}) = \beta_{k0}^{(\phi)} + v_{ki}^{(\phi)} \text{ and } \log(\lambda_{ki}) = \beta_{k0}^{(\lambda)} + v_{ki}^{(\lambda)},$$

where $v_{ki}^{(\phi)} \sim \mathrm{N}(0, \alpha_k)$ and $v_{ki}^{(\lambda)} \sim \mathrm{N}(0, \tau_k)$. The MDHGLM2 can be fitted by specifying models for the means and dispersions as follows:

```
> data(cog)
> jm11<-DHGLMMODELING(Link="identity",
+         LinPred=y1~x1+...+x20+(1|id),
+         RandDist=c("gaussian"),
+   LinkRandVariance="log",
+   LinPredRandVariance=lambda~1+(1|id),
+   RandDistRandVariance="gaussian")
> jm21<-DHGLMMODELING(Link="identity",
+         LinPred=y2~x1+...+x20+(1|id),
+         RandDist=c("gaussian"),
+   LinkRandVariance="log",
+   LinPredRandVariance=lambda~1+(1|id),
+   RandDistRandVariance="gaussian")
> jm31<-DHGLMMODELING(Link="identity",
+         LinPred=y3~x1+...+x20+(1|id),
+         RandDist=c("gaussian"),
+   LinkRandVariance="log",
+   LinPredRandVariance=lambda~1+(1|id),
```

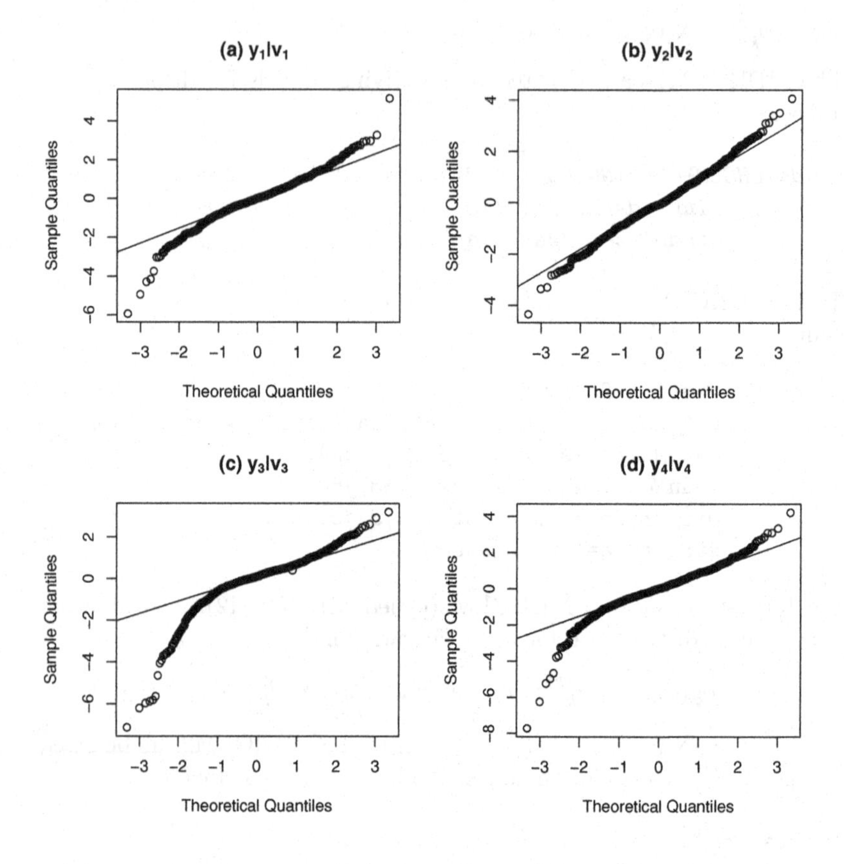


Figure 7.6 *Normal probability plots for (a) $y_1|v_1$, (b) $y_2|v_2$, (c) $y_3|v_3$, and (d) $y_4|v_4$ under the multivariate HGLM on the vascular cognitive impairment data.*

```
+   RandDistRandVariance="gaussian")
> jm41<-DHGLMMODELING(Link="identity",
+          LinPred=y4~x1+...+x20+(1|id),
+          RandDist=c("gaussian"),
+   LinkRandVariance="log",
+   LinPredRandVariance=lambda~1+(1|id),
+   RandDistRandVariance="gaussian")
```

To link DHGLMs by correlated random effects, we use the following code:

```
> res3<-jointfit(RespDist=
+         c("gaussian","gaussian","gaussian","gaussian"),
```

```
+         DataMain=list(cog,cog,cog,cog),
+         MeanModel=list(jm11,jm21,jm31,jm41),
+         DispersionModel=list(jd,jd,jd,jd),
+         structure="correlated")
```

The fitted results from the three multivariate models are in Tables 7.8-7.10. cAIC selects MDHGLM2 (cAIC=10437.4) as the best-fitting model among models we considered, because cAIC for the multivariate HGLM (cAIC=13260.0) and MDHGLM1 (cAIC=10548.1) are larger. From Figures 7.7 and 7.8, we see that most outliers in the multivariate HGLMs disappear by using MDHGLM1; MDHGLM2 has similar plots to MDHGLM1. From the normal probability plots for $v_{ki}^{(\lambda)}$ in Figures 7.9 and 7.10, the DHGLM2 is prefered to the DHGLM1 because $\hat{v}_{ki}^{(\lambda)}$ leans more toward the line. Thus, we select the MDHGLM2 as the final model, which gives robust estimators against outliers as well as robustness against misspecification of distributional assumptions on random effects. From Table 7.10 we see that MDHGLM2 tends to have the smallest p-values, i.e., having the most efficient analysis. Thus, it is important to find the better fitting model.

From the results under MDHGLM2, we can see that the random effects of executive and memory domains show the strongest correlation, 0.584. This indicates that a higher score in the executive domain is associated with higher score in the memory domain. Correlations between memory and visuospatial domains have the lowest value, 0.210. For all domains, NIHSS score and group of PCI=1 of demographic variables and higher MTA group (3-4) of neuroimaging variables give significant negative effect. These results indicate that higher NIHSS score, PCI=1 and MTA 3-4 groups lower the scores of cognitive function.

### *7.1.5 Mother's stress and children's morbidity data*

Consider the longitudinal data set from mother's stress and children's morbidity study (MSCM) in the R package mmm (Asar and Ilk, 2014). In this MSCM study, 167 mothers and their preschool children were enrolled for 28 days. Investigation of the serial dependence structures of the two longitudinal responses suggested a weak correlation structure for the period of days 1~16. Therefore, only the period of days 17~28 is considered in this dataset.

The bivariate binary responses were mother's stress status $y_{1it}$ (stress; 0 = absence, 1 = presence) and children's illness status $y_{2it}$ (illness; 0 = absence, 1 = presence) for the $t$th visit of the $i$th family. Asar

Table 7.8 *Results under multivariate HGLM for the vascular cognitive impairment data*

| $y_k$ covariate | $y_1$ (executive) Estimate | SE | p-value | $y_2$ (memory) Estimate | SE | p-value | $y_3$ (visuospatial) Estimate | SE | p-value | $y_4$ (language) Estimate | SE | p-value |
|---|---|---|---|---|---|---|---|---|---|---|---|---|
| Intercept | −0.908 | 0.395 | 0.021 | −0.705 | 0.297 | 0.018 | −1.649 | 0.712 | 0.021 | −0.25 | 0.399 | 0.531 |
| Age | 0.029 | 0.053 | 0.578 | 0.115 | 0.041 | 0.005 | 0.216 | 0.096 | 0.024 | 0.049 | 0.052 | 0.343 |
| Male | −0.222 | 0.178 | 0.214 | −0.852 | 0.138 | 0.000 | −0.658 | 0.285 | 0.021 | −0.394 | 0.182 | 0.030 |
| Edu | 0.038 | 0.015 | 0.012 | −0.011 | 0.012 | 0.349 | −0.029 | 0.024 | 0.224 | 0.010 | 0.015 | 0.507 |
| HTN | −0.158 | 0.162 | 0.329 | −0.111 | 0.134 | 0.407 | −0.351 | 0.258 | 0.173 | −0.38 | 0.167 | 0.022 |
| DM | −0.040 | 0.145 | 0.781 | −0.080 | 0.117 | 0.493 | −0.069 | 0.230 | 0.764 | −0.069 | 0.144 | 0.633 |
| Af | −0.014 | 0.201 | 0.945 | −0.242 | 0.156 | 0.120 | −0.202 | 0.328 | 0.539 | 0.259 | 0.208 | 0.214 |
| HxStroke | −0.233 | 0.172 | 0.175 | −0.294 | 0.137 | 0.031 | −0.292 | 0.283 | 0.302 | 0.135 | 0.178 | 0.448 |
| NIHSS | −0.071 | 0.014 | 0.000 | −0.032 | 0.010 | 0.002 | −0.088 | 0.023 | 0.000 | −0.050 | 0.015 | 0.001 |
| VCINP | −0.062 | 0.018 | 0.000 | 0.019 | 0.016 | 0.230 | −0.062 | 0.034 | 0.069 | 0.011 | 0.017 | 0.526 |
| PCI | −1.151 | 0.160 | 0.000 | −0.654 | 0.122 | 0.000 | −1.190 | 0.253 | 0.000 | −1.006 | 0.167 | 0.000 |
| AcuteLeft | −0.105 | 0.136 | 0.441 | −0.289 | 0.112 | 0.010 | 0.689 | 0.217 | 0.002 | −0.166 | 0.144 | 0.248 |
| AcuteMulti | −0.154 | 0.159 | 0.333 | −0.054 | 0.132 | 0.685 | 0.047 | 0.252 | 0.853 | −0.058 | 0.166 | 0.724 |
| AcuteCS | −0.073 | 0.152 | 0.633 | 0.046 | 0.130 | 0.724 | −0.168 | 0.244 | 0.491 | 0.181 | 0.175 | 0.299 |
| ChrCS | 0.011 | 0.207 | 0.957 | 0.252 | 0.164 | 0.125 | 0.625 | 0.321 | 0.051 | −0.206 | 0.215 | 0.339 |
| PVWM | 0.002 | 0.162 | 0.990 | −0.088 | 0.128 | 0.492 | 0.146 | 0.255 | 0.567 | −0.242 | 0.169 | 0.152 |
| SCWM | −0.342 | 0.205 | 0.096 | 0.081 | 0.152 | 0.595 | −0.373 | 0.314 | 0.235 | 0.083 | 0.203 | 0.685 |
| LAC | −0.233 | 0.158 | 0.142 | −0.067 | 0.125 | 0.592 | −0.408 | 0.245 | 0.096 | −0.255 | 0.165 | 0.122 |
| CMB | −0.122 | 0.154 | 0.429 | −0.204 | 0.128 | 0.109 | 0.114 | 0.249 | 0.647 | 0.141 | 0.158 | 0.371 |
| MTA1 | −0.141 | 0.183 | 0.441 | −0.103 | 0.148 | 0.488 | −0.142 | 0.297 | 0.634 | −0.266 | 0.187 | 0.153 |
| MTA2 | −0.779 | 0.219 | 0.000 | −0.458 | 0.175 | 0.009 | −1.141 | 0.358 | 0.001 | −0.808 | 0.229 | 0.000 |
| $\log \beta_{k0}^{(\lambda)}$ | 0.329 | 0.075 | | −0.167 | 0.077 | | 1.117 | 0.081 | | 0.422 | 0.075 | |
| $\log \beta_{k0}^{(\phi)}$ | −0.448 | 0.043 | | −0.673 | 0.044 | | 0.983 | 0.045 | | −0.662 | 0.045 | |
| | $\widehat{\rho}_1$ =0.496 | $\widehat{\rho}_2$ =0.697 | $\widehat{\rho}_3$ =0.557 | $\widehat{\rho}_4$ =0.238 | $\widehat{\rho}_5$ =0.463 | $\widehat{\rho}_6$ =0.355 | | | | | | |

Table 7.9 *Results under MDHGLM1 for the vascular cognitive impairment data*

| $y_k$ covariate | $y_1$ (executive) Estimate | SE | p-value | $y_2$ (memory) Estimate | SE | p-value | $y_3$ (visuospatial) Estimate | SE | p-value | $y_4$ (language) Estimate | SE | p-value |
|---|---|---|---|---|---|---|---|---|---|---|---|---|
| Intercept | −1.009 | 0.378 | 0.008 | −0.700 | 0.326 | 0.032 | −1.197 | 0.564 | 0.034 | −0.228 | 0.36 | 0.526 |
| Age | 0.027 | 0.049 | 0.576 | 0.113 | 0.044 | 0.009 | 0.096 | 0.076 | 0.208 | 0.032 | 0.045 | 0.475 |
| Male | −0.212 | 0.168 | 0.208 | −0.848 | 0.136 | 0.000 | −0.652 | 0.228 | 0.004 | −0.385 | 0.171 | 0.024 |
| Edu | 0.041 | 0.015 | 0.005 | −0.011 | 0.012 | 0.353 | −0.007 | 0.020 | 0.713 | 0.013 | 0.015 | 0.387 |
| HTN | −0.129 | 0.158 | 0.414 | −0.107 | 0.128 | 0.402 | −0.299 | 0.215 | 0.164 | −0.379 | 0.161 | 0.018 |
| DM | −0.046 | 0.145 | 0.752 | −0.066 | 0.117 | 0.572 | −0.087 | 0.196 | 0.659 | −0.062 | 0.147 | 0.674 |
| Af | 0.022 | 0.199 | 0.912 | −0.259 | 0.161 | 0.108 | −0.052 | 0.273 | 0.849 | 0.264 | 0.202 | 0.191 |
| HxStroke | −0.247 | 0.181 | 0.172 | −0.291 | 0.145 | 0.045 | −0.300 | 0.245 | 0.221 | 0.144 | 0.183 | 0.433 |
| NIHSS | −0.071 | 0.014 | 0.000 | −0.032 | 0.011 | 0.004 | −0.084 | 0.019 | 0.000 | −0.047 | 0.014 | 0.001 |
| VCINP | −0.027 | 0.013 | 0.038 | 0.014 | 0.013 | 0.274 | −0.001 | 0.022 | 0.964 | 0.014 | 0.012 | 0.244 |
| PCI | −1.148 | 0.16 | 0.000 | −0.653 | 0.128 | 0.000 | −1.062 | 0.217 | 0.000 | −0.98 | 0.162 | 0.000 |
| AcuteLeft | −0.098 | 0.138 | 0.475 | −0.298 | 0.112 | 0.008 | 0.673 | 0.187 | 0.000 | −0.119 | 0.14 | 0.394 |
| AcuteMulti | −0.170 | 0.160 | 0.289 | −0.049 | 0.129 | 0.704 | 0.041 | 0.217 | 0.851 | −0.082 | 0.163 | 0.612 |
| AcuteCS | −0.073 | 0.162 | 0.651 | 0.046 | 0.131 | 0.725 | −0.082 | 0.219 | 0.707 | 0.196 | 0.164 | 0.233 |
| ChrCS | −0.023 | 0.209 | 0.910 | 0.254 | 0.168 | 0.130 | 0.407 | 0.282 | 0.149 | −0.225 | 0.212 | 0.289 |
| PVWM | 0.002 | 0.159 | 0.989 | −0.079 | 0.129 | 0.540 | 0.234 | 0.216 | 0.279 | −0.234 | 0.162 | 0.148 |
| SCWM | −0.346 | 0.186 | 0.063 | 0.074 | 0.150 | 0.621 | −0.218 | 0.252 | 0.389 | 0.100 | 0.189 | 0.596 |
| LAC | −0.214 | 0.157 | 0.172 | −0.065 | 0.126 | 0.603 | −0.462 | 0.211 | 0.028 | −0.255 | 0.158 | 0.107 |
| CMB | −0.128 | 0.157 | 0.412 | −0.203 | 0.126 | 0.108 | 0.056 | 0.212 | 0.792 | 0.138 | 0.159 | 0.387 |
| MTA1 | −0.14 | 0.183 | 0.446 | −0.099 | 0.147 | 0.501 | −0.219 | 0.247 | 0.375 | −0.272 | 0.186 | 0.143 |
| MTA2 | −0.755 | 0.226 | 0.001 | −0.457 | 0.178 | 0.010 | −0.913 | 0.309 | 0.003 | −0.742 | 0.229 | 0.001 |
| $\log \beta_{k0}^{(\lambda)}$ | 0.368 | 0.079 | | −0.114 | 0.081 | | 0.886 | 0.082 | | 0.426 | 0.078 | |
| $\log \beta_{k0}^{(\phi)}$ | −1.131 | 0.046 | | −1.110 | 0.045 | | 0.092 | 0.045 | | −1.336 | 0.045 | |
| $\log \alpha_k$ | −0.980 | 0.226 | | −1.070 | 0.191 | | −0.933 | 0.515 | | −1.044 | 0.254 | |
| | $\widehat{\rho}_1$ =0.420 | $\widehat{\rho}_2$ =0.591 | $\widehat{\rho}_3$ =0.489 | $\widehat{\rho}_4$ =0.227 | $\widehat{\rho}_5$ =0.400 | $\widehat{\rho}_6$ =0.310 | | | | | | |

Table 7.10 *Results under MDHGLM2 for the vascular cognitive impairment data*

| $y_k$ covariate | $y_1$ (executive) Estimate | SE | p-value | $y_2$ (memory) Estimate | SE | p-value | $y_3$ (visuospatial) Estimate | SE | p-value | $y_4$ (language) Estimate | SE | p-value |
|---|---|---|---|---|---|---|---|---|---|---|---|---|
| Intercept | −0.995 | 0.351 | 0.005 | −0.679 | 0.301 | 0.024 | −1.091 | 0.530 | 0.040 | −0.193 | 0.331 | 0.560 |
| Age | 0.018 | 0.046 | 0.702 | 0.113 | 0.041 | 0.006 | 0.080 | 0.072 | 0.268 | 0.027 | 0.043 | 0.532 |
| Male | −0.214 | 0.150 | 0.154 | −0.879 | 0.122 | 0.000 | −0.662 | 0.209 | 0.002 | −0.441 | 0.149 | 0.003 |
| Edu | 0.048 | 0.013 | 0.000 | −0.01 | 0.011 | 0.358 | 0.000 | 0.018 | 0.987 | 0.020 | 0.013 | 0.116 |
| HTN | −0.116 | 0.141 | 0.408 | −0.095 | 0.115 | 0.412 | −0.277 | 0.195 | 0.157 | −0.378 | 0.141 | 0.007 |
| DM | −0.107 | 0.129 | 0.407 | −0.093 | 0.104 | 0.371 | −0.168 | 0.180 | 0.351 | −0.079 | 0.129 | 0.538 |
| Af | 0.002 | 0.180 | 0.990 | −0.333 | 0.145 | 0.021 | −0.032 | 0.252 | 0.898 | 0.233 | 0.178 | 0.190 |
| HxStroke | −0.271 | 0.161 | 0.093 | −0.265 | 0.129 | 0.039 | −0.335 | 0.225 | 0.136 | 0.135 | 0.161 | 0.404 |
| NIHSS | −0.069 | 0.012 | 0.000 | −0.030 | 0.010 | 0.002 | −0.081 | 0.018 | 0.000 | −0.042 | 0.012 | 0.001 |
| VCINP | −0.026 | 0.013 | 0.040 | 0.013 | 0.012 | 0.286 | 0.000 | 0.022 | 1.000 | 0.015 | 0.012 | 0.196 |
| PCI | −1.122 | 0.143 | 0.000 | −0.674 | 0.114 | 0.000 | −1.049 | 0.201 | 0.000 | −1.006 | 0.143 | 0.000 |
| AcuteLeft | −0.110 | 0.123 | 0.371 | −0.288 | 0.100 | 0.004 | 0.605 | 0.171 | 0.000 | −0.080 | 0.123 | 0.514 |
| AcuteMulti | −0.168 | 0.144 | 0.243 | −0.007 | 0.116 | 0.951 | 0.002 | 0.198 | 0.993 | −0.113 | 0.143 | 0.431 |
| AcuteCS | −0.107 | 0.145 | 0.462 | 0.054 | 0.117 | 0.646 | −0.051 | 0.201 | 0.801 | 0.209 | 0.145 | 0.147 |
| ChrCS | −0.077 | 0.187 | 0.681 | 0.239 | 0.150 | 0.110 | 0.348 | 0.259 | 0.179 | −0.152 | 0.188 | 0.420 |
| PVWM | 0.028 | 0.143 | 0.845 | −0.118 | 0.115 | 0.304 | 0.273 | 0.197 | 0.166 | −0.254 | 0.142 | 0.073 |
| SCWM | −0.306 | 0.167 | 0.067 | 0.070 | 0.133 | 0.599 | −0.219 | 0.231 | 0.344 | 0.027 | 0.165 | 0.871 |
| LAC | −0.222 | 0.139 | 0.111 | −0.094 | 0.112 | 0.401 | −0.446 | 0.193 | 0.021 | −0.246 | 0.139 | 0.078 |
| CMB | −0.102 | 0.139 | 0.464 | −0.171 | 0.112 | 0.126 | 0.063 | 0.194 | 0.743 | 0.148 | 0.139 | 0.287 |
| MTA1 | −0.134 | 0.164 | 0.417 | −0.170 | 0.131 | 0.196 | −0.158 | 0.226 | 0.486 | −0.300 | 0.163 | 0.065 |
| MTA2 | −0.723 | 0.204 | 0.000 | −0.479 | 0.157 | 0.002 | −0.877 | 0.288 | 0.002 | −0.681 | 0.203 | 0.001 |
| $\log \beta_{k0}^{(\lambda)}$ | 0.136 | 0.078 | | −0.360 | 0.075 | | 0.812 | 0.077 | | 0.191 | 0.077 | |
| $\log \tau_k$ | −1.401 | 0.809 | | −1.350 | 0.791 | | −1.285 | 0.317 | | −1.290 | 0.563 | |
| $\log \beta_{k0}^{(\phi)}$ | −1.139 | 0.045 | | −1.129 | 0.045 | | 0.058 | 0.045 | | −1.347 | 0.045 | |
| $\log \alpha_k$ | −0.982 | 0.226 | | −1.073 | 0.191 | | −0.929 | 0.434 | | −1.040 | 0.256 | |
| | $\widehat{\rho}_1$ =0.420 | $\widehat{\rho}_2$ =0.584 | $\widehat{\rho}_3$ =0.487 | $\widehat{\rho}_4$ =0.210 | $\widehat{\rho}_5$ =0.400 | $\widehat{\rho}_6$ =0.301 | | | | | | |

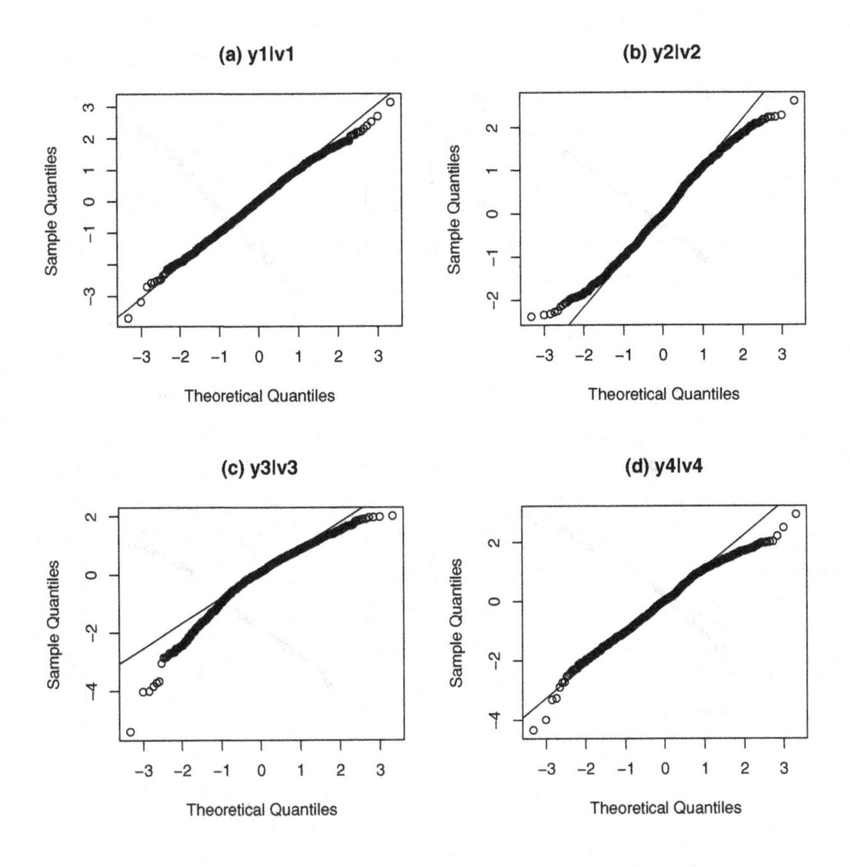


Figure 7.7 *Normal probability plots for ((a) $y_1|v_1$, (b) $y_2|v_2$, (c) $y_3|v_3$, and (d) $y_4|v_4$ under the MDHGLM1 on the vascular cognitive impairment data.*

and Ilk (2013) considered ten between-subject covariates: (i) marriage status (married; 0 = other, 1 = married), (ii) highest education level (education; 0 = < high school, 1 = $\geq$ high school), (iii) employment status (employed; 0 = unemployed, 1 = employed) of mother, (iv) race (race; 0 = white, 1 = non-white), (v) gender (csex; 0 = male, 1 = female) of the children, (vi) health statuses of children, (vii) mothers at baseline (chlth and mhlth, respectively; 0 = very poor/poor, 1 = fair, 2 = good, 3 = very good), (viii) household size (housize; 0 = 2-3 people, 1 = more than 3 people), (ix) the average response values $y_{1ij}$ of the $1 \sim 16$ days (bstress), and (x) that of $y_{2ij}$ (billness). A within-subject covariate is the study time (week= $(\text{day}_{ij} - 22)/7$).

To analyze this dataset, we use the bivariate Bernoulli HGLMs as shown

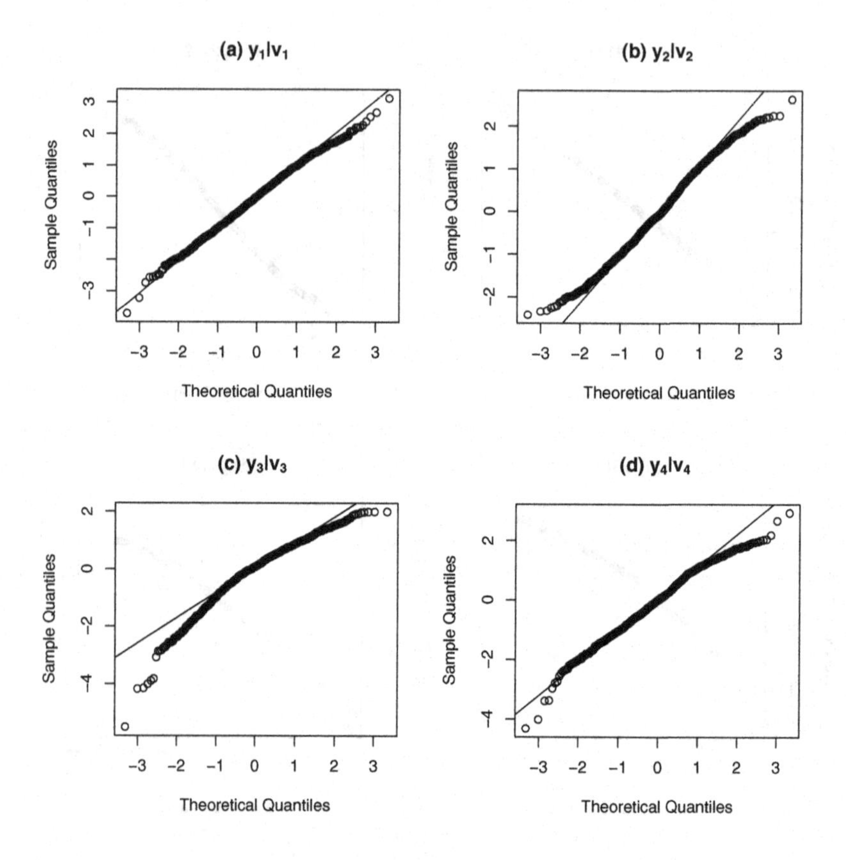


Figure 7.8 *Normal probability plots for (a)* $y_1|v_1$*, (b)* $y_2|v_2$*, (c)* $y_3|v_3$*, and (d)* $y_4|v_4$ *under the MDHGLM2 on the vascular cognitive impairment data.*

in Figure 7.11: Let $y_{ij} = (y_{1ij}, y_{2ij})^t$ be the bivariate binary responses and $v_i^{(\mu)} = (w_i^{(\mu)}, u_i^{(\mu)})^t$ be the unobserved random effects for the $i$th cluster.

$$\begin{aligned} y_{1ij}|v_i^{(\mu)} &\sim Bernoulli(\frac{\exp(X_{it}\beta_1^{(\mu)} + w_i^{(\mu)})}{1+\exp(X_{it}\beta_1^{(\mu)} + w_i^{(\mu)})}), \\ y_{2ij}|v_i^{(\mu)} &\sim Bernoulli(\frac{\exp(X_{it}\beta_2^{(\mu)} + u_i^{(\mu)})}{1+\exp(X_{it}\beta_2^{(\mu)} + u_i^{(\mu)})}) \end{aligned} \tag{7.1}$$

where $v_i^{(\mu)} \sim \mathrm{N}(0, \Sigma_i)$ with $\Sigma_i = \begin{pmatrix} \lambda_{1i} & \rho\sqrt{\lambda_{1i}\lambda_{2i}} \\ \rho\sqrt{\lambda_{1i}\lambda_{2i}} & \lambda_{2i} \end{pmatrix}$ and $-1 < \rho < 1$. Thus, given $v_i^{(\mu)}$, $y_{1ij}$ and $y_{2ij}$ are independent.

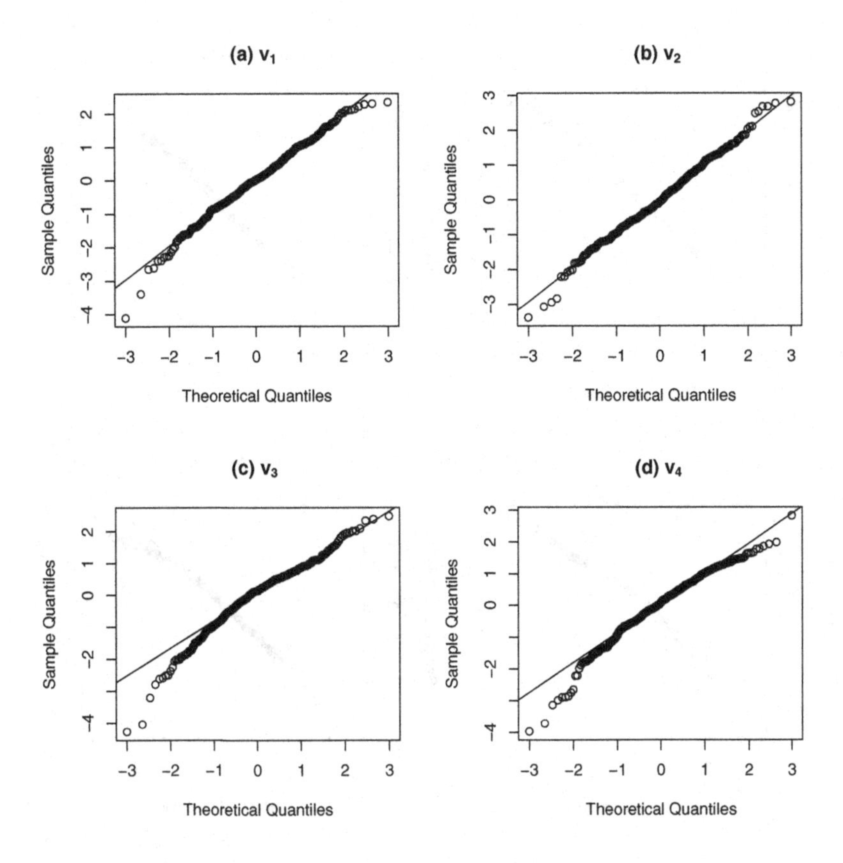


Figure 7.9 *Normal probability plots for (a) $v_1$, (b) $v_2$, (c) $v_3$, and (d) $v_4$ under the MDHGLM1 on the vascular cognitive impairment data.*

We first consider three models with $\log(\lambda_{1i}) = \beta_{10}^{(\lambda)}$ and $\log(\lambda_{2i}) = \beta_{20}^{(\lambda)}$ :

(i) M1: independent model, having $\rho = 0$,

(ii) M2: random-effects model with a saturated variance-covariance matrix,

(iii) M3: shared random-effects model, having $u_i^{(\mu)} = \delta w_i^{(\mu)}$ for some constant $\delta$.

The R-codes for fitting these models are as follows:

```
> data(motherStress,package="mmm")
> jm1<-DHGLMMODELING(Link="logit",
+          LinPred=stress~married+education
```

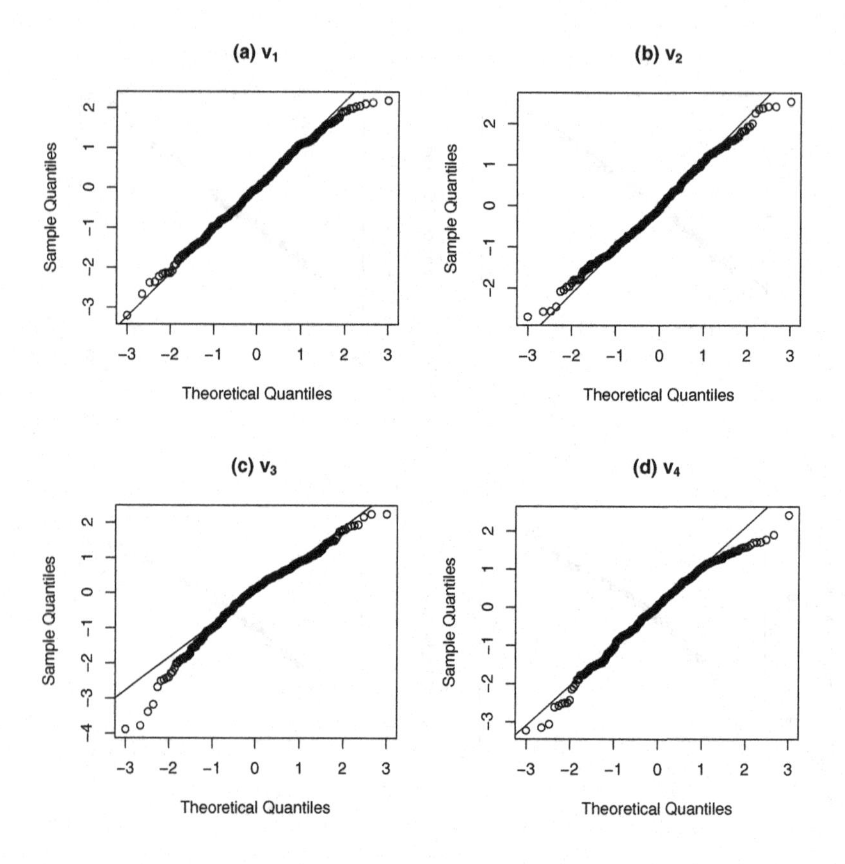


Figure 7.10 *Normal probability plots for (a) $v_1$, (b) $v_2$, (c) $v_3$, and (d) $v_4$ under the MDHGLM2 on the vascular cognitive impairment data.*

```
+    +employed+race+csex+chlth
+    +mhlth+housize+bstress+billness
+    +week+(1|id),
+          RandDist="gaussian")
> jm2<-DHGLMMODELING(Link="logit",
+          LinPred=stress~married+education
+    +employed+race+csex+chlth
+    +mhlth+housize+bstress+billness
+    +week+(1|id),
+          RandDist="gaussian")
```

(i) M1: independent model,

```
> res_ind<-jointfit(RespDist=
```

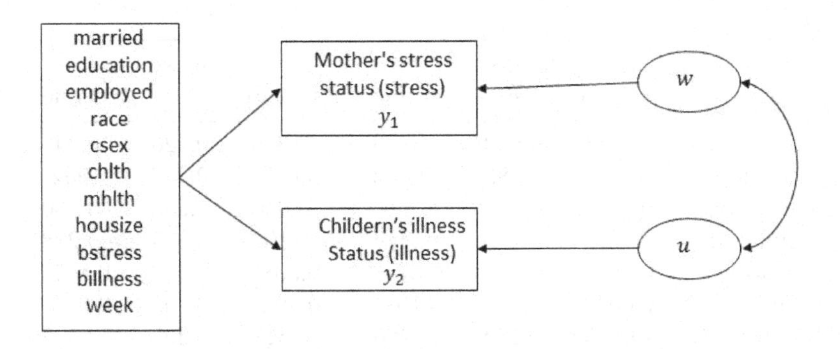


Figure 7.11 *Path diagram for the MDHGLM fitted to the mother's stress and children's morbidity data.*

```
+         c("binomial","binomial"),
+         DataMain=list(motherStress,motherStress),
+         MeanModel=list(jm1,jm2),
+         structure="independent")
```

(ii) M2: random-effect model with a saturated covariance matrix,

```
> res_sat<-jointfit(RespDist=
+         c("binomial","binomial"),
+         DataMain=list(motherStress,motherStress),
+         MeanModel=list(jm1,jm2),
+         structure="correlated")
```

(iii) M3: shared random-effect model,

```
> res_shared<-jointfit(RespDist=
+         c("binomial","binomial"),
+         DataMain=list(motherStress,motherStress),
+         MeanModel=list(jm1,jm2),
+         structure="shared")
```

These three models have nested structure; M1$\subset$M3$\subset$M2. The cAIC has values of 2653.7, 2428.9 and 2517.3 for M1, M2, and M3, respectively. Thus, cAIC selects the full model M2 among three models. Table 7.11 shows the fitting result from M2.

In binary data, GLMMs are sensitive to a distributional assumption of

Table 7.11 *Fitting results for M2 in the mother's stress and children's morbidity data*

| | Stress | | | Illness | | |
|---|---|---|---|---|---|---|
| | Estimate | SE | t-value | Estimate | SE | t-value |
| Intercept | −2.744 | 0.609 | −4.506 | −2.336 | 0.526 | −4.441 |
| married | −0.025 | 0.283 | −0.088 | 0.695 | 0.356 | 1.952 |
| education | 0.490 | 0.315 | 1.556 | −0.046 | 0.355 | −0.130 |
| employed | −0.796 | 0.325 | −2.449 | −0.108 | 0.375 | −0.288 |
| race | 0.114 | 0.302 | 0.377 | 0.308 | 0.351 | 0.877 |
| csex | 0.054 | 0.274 | 0.197 | 0.191 | 0.309 | 0.618 |
| chlth | −0.245 | 0.189 | −1.296 | −0.513 | 0.205 | −2.502 |
| mhlth | −0.248 | 0.168 | −1.476 | 0.096 | 0.196 | 0.490 |
| housize | −0.050 | 0.309 | −0.162 | −0.909 | 0.352 | −2.582 |
| bstress | 5.101 | 0.982 | 5.195 | 0.105 | 1.186 | 0.089 |
| billness | 0.598 | 0.949 | 0.630 | 2.424 | 1.135 | 2.136 |
| week | −0.480 | 0.164 | −2.927 | −0.243 | 0.169 | −1.438 |
| $\beta_{10}^{(\lambda)}$ | 0.307 | 0.320 | | | | |
| $\beta_{20}^{(\lambda)}$ | | | | 0.775 | 0.369 | |
| $\rho$ | 0.364 | 0.154 | | | | |

random effects, which is difficult to identify. Thus, we consider the robust bivariate DHGLM by allowing random effects in the variance for random effects in model (7.1):

(iv) M4: the same as M2, but having $\log(\lambda_{1i}) = \beta_{10}^{(\lambda)} + w_i^{(\lambda)}$ and $\log(\lambda_{2i}) = \beta_{20}^{(\lambda)} + u_i^{(\lambda)}$ where $w_i^{(\lambda)} \sim \mathrm{N}(0, \tau_1)$ and $u_i^{(\lambda)} \sim \mathrm{N}(0, \tau_2)$

R-code for M4 is as follows:

```
> jm11<-DHGLMMODELING(Link="logit",
+         LinPred=stress~married+education
+   +employed+race+csex+chlth
+   +mhlth+housize+bstress+billness
+   +week+(1|id),
+         RandDist="gaussian",
+   LinkRandVariance="log",
+   LinPredRandVariance=lambda~1+(1|id),
+   RandDistRandVariance="gaussian")

> jm21<-DHGLMMODELING(Link="logit",
+         LinPred=stress~married+education
```

```
+     +employed+race+csex+chlth
+     +mhlth+housize+bstress+billness
+     +week+(1|id),
+          RandDist="gaussian",
+     LinkRandVariance="log",
+     LinPredRandVariance=lambda~1+(1|id),
+     RandDistRandVariance="gaussian")

> res_dhglm<-jointfit(RespDist=
+          c("binomial","binomial"),
+          DataMain=list(motherStress,motherStress),
+          MeanModel=list(jm11,jm21),
+          structure="correlated")
```

The cAIC has value of 2103.9 for M4, Thus, cAIC selects M4 as the best-fitting model. Table 7.12 shows the fitting results from model M4. We also use the GEE method (Asar and Ilk, 2013) to fit the marginal model,

$$
\begin{aligned}
y_{1ij} &\sim Bernoulli\left(\frac{\exp(X_{it}\beta_1^{(\mu)})}{1+\exp(X_{it}\beta_1^{(\mu)})}\right), \\
y_{2ij} &\sim Bernoulli\left(\frac{\exp(X_{it}\beta_2^{(\mu)})}{1+\exp(X_{it}\beta_2^{(\mu)})}\right)
\end{aligned}
$$

with the working correlation matrix whose elements are

$$
\begin{aligned}
\mathrm{Corr}(y_{1ij}, y_{1i'j'}) &= \mathrm{Corr}(y_{1ij}, y_{2i'j'}) = \mathrm{Corr}(y_{2ij}, y_{2i'j'}) \\
&\phantom{=} 1 \text{ for } i=i' \text{ and } j=j' \\
&= \rho_2 \text{ for } i=i' \text{ and } j\neq j' \\
&\phantom{=} 0 \text{ for } i\neq i'
\end{aligned}
$$

This marginal model can be fitted by using the following package mmm.

```
> fit<-mmm(formula=cbind(stress, illness)~
+      married+education
+     +employed+race+csex+chlth
+     +mhlth+housize+bstress+billness
+     +week,
+     id=motherStress$id,
+     data=motherStress,
+     family=binomial(link=logit),
+     corStruct="exchangeable")
```

Because differences between parameter estimates from HGLM (M2) and

DHGLM (M4) are quite different, we rely on DHGLM results, which are less sensitive to the distributional assumption on random effects. Compared with HGLM, the DHGLM gives closer estimates to those from the marginal GEE model, even though DHGLM models conditional means, while GEE models marginal models. Both GEE and DHGLM are known to be less sensitive to the distributional assumption on random effects.

### *7.1.6 Primary biliary cirrhosis data*

We used a longitudinal data set in the R package JM (Komarek, 2015) from a Mayo Clinic trial on 312 patients with primary biliary cirrhosis (PBC) conducted in 1974-1984. There are 1 to 5 visits per subject (coded as "id") performed at time of months ($x$). At each visit, measurements of three response variables are observed: continuous logarithmic bilirubin ($y1$), discrete platelet count ($y2$) and dichotomous indication of blood vessel malformations ($y3$). Komarek (2015) used 260 subjects known to be alive at 910 days of follow-up, and only the longitudinal measurements by this point will be considered. The corresponding data are available as a PBC910 dataset in the mixAK package.

With this dataset, we consider a multivariate model for three response variables with a covariate $x$ for the $t$th visit of the $i$th patient (see Figure 7.12),

$$\begin{aligned}
&y_{1it}|v_{1i} \sim \mathrm{N}(\mu_{1it}, \phi_{1i}) \\
&\quad \text{with } \mu_{1it} = \beta_{10}^{(\mu)} + \beta_{11}^{(\mu)} x_{it} + v_{1i}^{(\mu)} \text{ and } \log(\phi_{1i}) = \beta_{10}^{(\phi)} + \beta_{11}^{(\phi)} x_{it} \\
&y_{2it}|v_{1i} \sim \mathrm{N}(\mu_{2it}, \phi_{2i}) \\
&\quad \text{with } \mu_{2it} = \beta_{20}^{(\mu)} + \beta_{21}^{(\mu)} x_{it} + v_{2i}^{(\mu)} \text{ and } \log(\phi_{2i}) = \beta_{20}^{(\phi)} + \beta_{21}^{(\phi)} x_{it} \\
&y_{3it}|v_{1i} \sim Bernoulli(p_{3it}) \\
&\quad \text{with } \log(p_{3it}/(1-p_{3it})) = \beta_{30}^{(\mu)} + \beta_{31}^{(\mu)} x_{it} + v_{3i}^{(\mu)}
\end{aligned}$$

where the random effects follow multivariate normal distribution:

$$\begin{pmatrix} v_{1i}^{(\mu)} \\ v_{2i}^{(\mu)} \\ v_{3i}^{(\mu)} \end{pmatrix} \sim MVN \left[ \begin{pmatrix} 0 \\ 0 \\ 0 \end{pmatrix}, \begin{pmatrix} \lambda_1 & \rho_1 \lambda_{1,2}^* & \rho_2 \lambda_{1,3}^* \\ \rho_1 \lambda_{1,2}^* & \lambda_2 & \rho_3 \lambda_{2,3}^* \\ \rho_2 \lambda_{1,3}^* & \rho_3 \lambda_{2,3}^* & \lambda_3 \end{pmatrix} \right]$$

where $\lambda_{j,k}^* = \sqrt{\lambda_j \lambda_k}$.

To fit this multivariate model, the following commands are used. First we create individual HGLMs as follows:

```
> data(PBC910,package="mixAK")
```

Table 7.12 *Fitting results for M4 and GEE in the mother's stress and children's morbidity data*

| | Stress | | | Illness | | |
|---|---|---|---|---|---|---|
| | Estimate | SE | t-value | Estimate | SE | t-value |
| | bi-variate DHGLM | | | | | |
| Intercept | −2.237 | 0.476 | −4.700 | −1.796 | 0.506 | −3.549 |
| married | −0.008 | 0.243 | −0.033 | 0.513 | 0.287 | 1.787 |
| education | 0.370 | 0.247 | 1.498 | −0.051 | 0.301 | −0.169 |
| employed | −0.677 | 0.273 | −2.480 | −0.209 | 0.331 | −0.631 |
| race | 0.013 | 0.270 | 0.048 | 0.103 | 0.256 | 0.402 |
| csex | −0.027 | 0.234 | −0.115 | 0.053 | 0.277 | 0.191 |
| chlth | −0.253 | 0.159 | −1.591 | −0.436 | 0.177 | −2.463 |
| mhlth | −0.186 | 0.159 | −1.170 | 0.036 | 0.170 | 0.212 |
| housize | 0.053 | 0.276 | 0.192 | −0.603 | 0.276 | −2.185 |
| bstress | 4.010 | 0.803 | 4.994 | 0.077 | 1.031 | 0.075 |
| billness | 0.791 | 0.758 | 1.044 | 2.270 | 0.894 | 2.539 |
| week | −0.456 | 0.163 | −2.798 | −0.207 | 0.205 | −1.010 |
| $\beta_{10}^{(\lambda)}$ | 0.137 | 0.336 | | | | |
| $\log(\tau_1)$ | −0.131 | 0.077 | | | | |
| $\beta_{20}^{(\lambda)}$ | | | | 0.591 | 0.273 | |
| $\log(\tau_2)$ | | | | −0.236 | 0.086 | |
| $\rho$ | 0.391 | 0.181 | | | | |
| | Marginal model using GEE | | | | | |
| Intercept | −2.138 | 0.415 | −5.149 | −1.576 | 0.478 | −3.297 |
| married | −0.005 | 0.236 | −0.021 | 0.499 | 0.266 | 1.876 |
| education | 0.364 | 0.225 | 1.615 | −0.055 | 0.287 | −0.193 |
| employed | −0.647 | 0.250 | −2.589 | −0.218 | 0.329 | −0.663 |
| race | −0.015 | 0.242 | −0.062 | 0.020 | 0.244 | 0.080 |
| csex | −0.043 | 0.220 | −0.196 | 0.019 | 0.246 | 0.076 |
| chlth | −0.261 | 0.133 | −1.962 | −0.400 | 0.156 | −2.561 |
| mhlth | −0.172 | 0.123 | −1.393 | 0.025 | 0.166 | 0.152 |
| housize | 0.061 | 0.239 | 0.257 | −0.564 | 0.259 | −2.183 |
| bstress | 3.894 | 0.711 | 5.477 | 0.061 | 0.981 | 0.062 |
| billness | 0.858 | 0.708 | 1.212 | 2.176 | 0.752 | 2.893 |
| week | −0.429 | 0.162 | −2.648 | −0.194 | 0.216 | −0.901 |
| $\rho_2$ | 0.070 | | | | | |

```
> jm1<-DHGLMMODELING(Model="mean",Link="identity",
+         LinPred=lbili~month+(1|id),
+         RandDist=c("gaussian"))
> jm2<-DHGLMMODELING(Model="mean",Link="identity",
+         LinPred=platelet~month+(1|id),
```

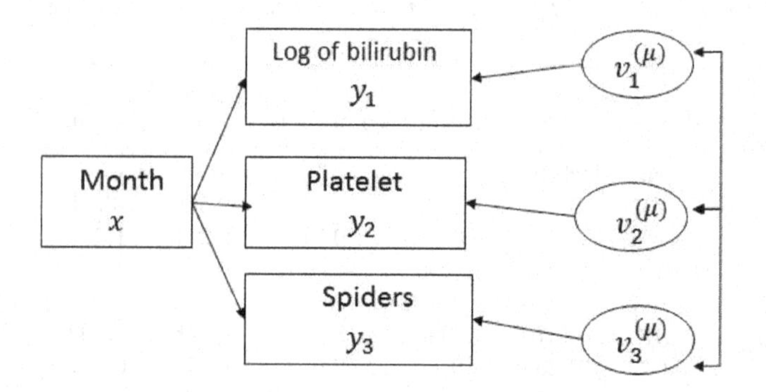


Figure 7.12 *Path diagram for the MDHGLM fitted to the primary biliary cirrhosis data.*

```
+          RandDist=c("gaussian"))
> jm3<-DHGLMMODELING(Model="mean",Link="logit",
+          LinPred=spiders~month+(1|subject),
+          RandDist=c("gaussian"))
```

To link HGLMs by correlated random effects, we use the following codes:

```
> res<-jointfit(RespDist=
+          c("gaussian","gaussian","binomial"),
+          DataMain=list(PBC910,PBC910,PBC910),
+          MeanModel=list(jm1,jm2,jm3),
+          structure="correlated")
```

From Figures 7.13 for $y_1$ and 7.14 for $y_2$ under the multivariate HGLM, we see that many large outliers exist. Thus, we can consider DHGLMs allowing heavy-tailed distributions for $y_1$ and $y_2$ as follows:

$$\begin{aligned}\log(\phi_{1i}) &= \beta_{10}^{(\phi)} + \beta_{11}^{(\phi)} x_{it} + v_{1i}^{(\phi)} \text{ with } v_{1i}^{(\phi)} \sim \text{N}(0, \alpha_1)\\ \log(\phi_{2i}) &= \beta_{20}^{(\phi)} + \beta_{21}^{(\phi)} x_{it} + v_{1i}^{(\phi)} \text{ with } v_{2i}^{(\phi)} \sim \text{N}(0, \alpha_2).\end{aligned}$$

This DHGLM can be fitted by specifying model for dispersion as follows:

```
> jd1<-DHGLMMODELING(Model="dispersion",Link="log",
+          LinPred=phi~month+(1|id),
+          RandDist=c("gaussian"))
> jd2<-DHGLMMODELING(Model="dispersion",Link="log",
```

```
+         LinPred=phi~month+(1|id),
+         RandDist=c("gaussian"))
> jd3<-DHGLMMODELING(Model="dispersion")
```

To link HGLMs by correlated random effects, we use the following codes:

```
> res<-jointfit(RespDist=
+         c("gaussian","gaussian","binomial"),
+         DataMain=list(PBC910,PBC910,PBC910),
+         MeanModel=list(jm1,jm2,jm3),
+    DispersionModel=list(jd1,jd2,jd3),
+         structure="correlated")
```

The fitting results are shown in Table 7.13. cAIC shows that DHGLM (cAIC=13068.1) is better fit than HGLM (cAIC=19776.5). From Figures 7.15 for $y_1$ and 7.16 for $y_2$ under the multivariate DHGLM, we see that most outliers in multivariate HGLM disappear allowing heavy-tailed distribution for $y_1$ and $y_2$. The estimates under the multivariate DHGLM are somewhat different than those from the multivariate HGLM. Thus, we select DHGLM which gives robust estimators against outliers.

From the DHGLM results, we see that random effects of platelet and spiders show the strongest correlation, 0.50. Correlations between logarithm of bilirubin and platelet (spiders) have lower values, -0.16 (-0.10). For the mean, Month has a positively (negatively) significant effect for the mean of logarithm of bilirubin and spiders (platelet). And it also gives a positively significant effect for the variance of logarithm of bilirubin and platelet.

### *7.1.7 Modeling missing data mechanism for the schizophrenic behavior data*

With this example, we show how to analyze the dataset presented in Chapter 6 with missingness. Rubin and Wu (1997) analyzed schizophrenic behavior data from an eye-tracking experiment with a visual target moving back and forth along a horizontal line on a screen.

Here, we first show how to analyze the data, assuming missingness at random which means that the probability of being missing does not depend on the values of missing data (unobserved random variables). Because abrupt changes among repeated responses were peculiar to schizophrenics, we proposed using a DHGLM with

$$\begin{aligned} y_{ij} &= \beta_0 + x_{1ij}\beta_1 + x_{2ij}\beta_2 + t_j\beta_3 + sch_i\beta_4 \\ &\quad + sch_i \cdot x_{1ij}\beta_5 + sch_i \cdot x_{2ij}\beta_6 + v_i + e_{ij} \end{aligned}$$

Table 7.13 *MDHGLM Results for the primary biliary cirrhosis data*

| $y_k$ | $y_1$ (logarithm of bilirubin) | | | $y_2$ (platelet) | | | $y_3$ (spiders) | | |
|---|---|---|---|---|---|---|---|---|---|
| covariate | Estimate | SE | p-value | Estimate | SE | p-value | Estimate | SE | p-value |
| | multivariate HGLM | | | | | | | | |
| | model for $\mu$ | | | | | | | | |
| Intercept | 0.3140 | 0.0598 | 0.0000 | 264.4598 | 5.5651 | 0.0000 | −3.2913 | 0.7976 | 0.0000 |
| Month | 0.0083 | 0.0014 | 0.0000 | −1.0255 | 0.1978 | 0.0000 | 0.0284 | 0.0134 | 0.0353 |
| | model for $\log(\phi)$ | | | | | | | | |
| Intercept | −2.2624 | 0.0856 | 0.0000 | 7.6654 | 0.0847 | 0.0000 | | | |
| Month | 0.0288 | 0.0060 | 0.0000 | 0.0205 | 0.0060 | 0.0001 | | | |
| | model for $\log(\lambda)$ | | | | | | | | |
| $\log\lambda_k$ | −0.2405 | 0.0901 | 0.0076 | 8.8040 | 0.0930 | 0.0000 | 4.3020 | 1.0513 | 0.0000 |
| | $\widehat{\rho_1} = -0.1705$ | $\widehat{\rho_2} = 0.4873$ | $\widehat{\rho_3} = -0.0887$ | | | | | | |
| | multivariate DHGLM | | | | | | | | |
| | model for $\mu$ | | | | | | | | |
| Intercept | 0.3341 | 0.0566 | 0.0000 | 264.3980 | 5.3515 | 0.0000 | −3.1305 | 0.7813 | 0.0000 |
| Month | 0.0047 | 0.0011 | 0.0000 | −1.2520 | 0.1441 | 0.0000 | 0.0271 | 0.0126 | 0.0315 |
| | model for $\log(\phi)$ | | | | | | | | |
| Intercept | −2.870 | 0.0664 | 0.0000 | 7.1248 | 0.0675 | 0.0000 | | | |
| Month | 0.0350 | 0.0043 | 0.0000 | 0.0137 | 0.0044 | 0.0018 | | | |
| | model for $\log(\lambda)$ | | | | | | | | |
| $\log\lambda_k$ | −0.2368 | 0.0894 | 0.0081 | 8.7830 | 0.09147 | 0.0000 | 4.1073 | 1.0417 | 0.0000 |
| | $\widehat{\rho_1} = -0.1611$ | $\widehat{\rho_2} = 0.5013$ | $\widehat{\rho_3} = -0.0975$ | | | | | | |
| | model for $\log(\alpha)$ | | | | | | | | |
| $\log\alpha_k$ | −1.0460 | 0.2147 | 0.0000 | −0.9948 | 0.2377 | 0.0000 | | | |

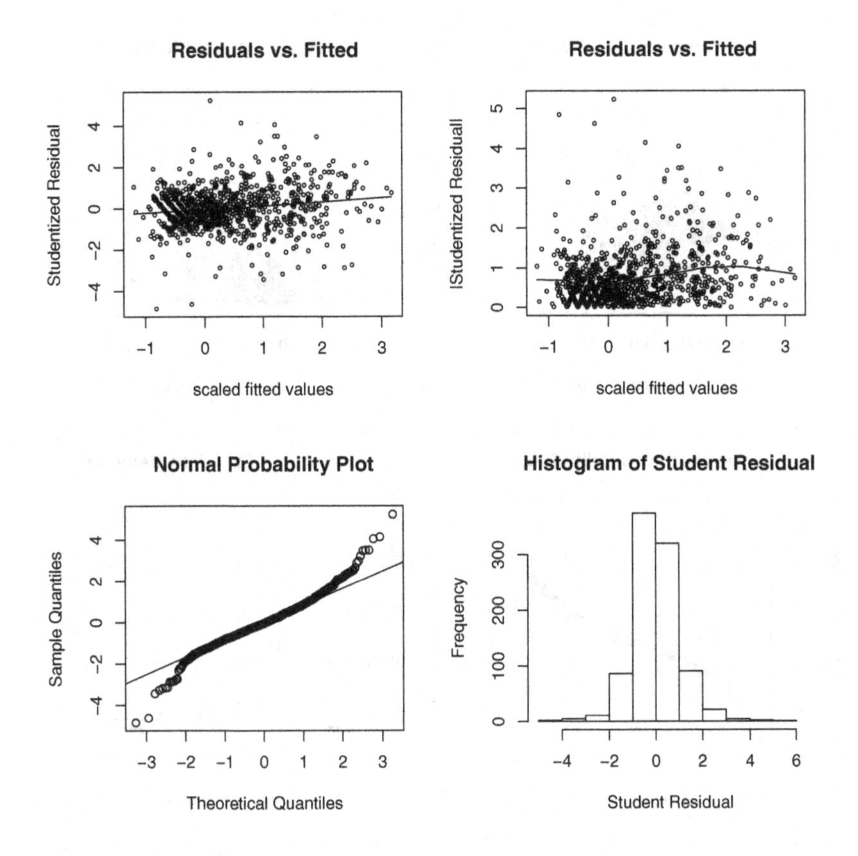


Figure 7.13 *Model checking plots for multivariate HGLM of* $y_1$ *on the primary biliary cirrhosis data.*

where $v_i \sim \mathrm{N}(0, \alpha)$ is the random effect, $e_{ij} \sim \mathrm{N}(0, \phi_i)$, $sch_i$ is equal to 1 if a subject is schizophrenic and 0 otherwise, $t_j$ is the measurement time, $x_{1ij}$ is the effect of PS vs. CS, $x_{2ij}$ is the effect of TR vs. the average of CS and PS, and

$$\log(\phi_i) = \gamma_0 + sch_i\beta_1^{(\phi)} + sch_i v_i^{(\phi)},$$

where $b_i \sim \mathrm{N}(0,\ \tau)$ are the dispersion random effects. We call this model DI (DHGLM with ignorable missingness).

In this dataset, we have another response, $\delta_{ij}$ (indicator variables: 1 if the data is missing and 0 otherwise). At first we ignored missingness using the DI model after discussion with the psychologists. However, according to the physicians missingness could be caused by eye blinks,

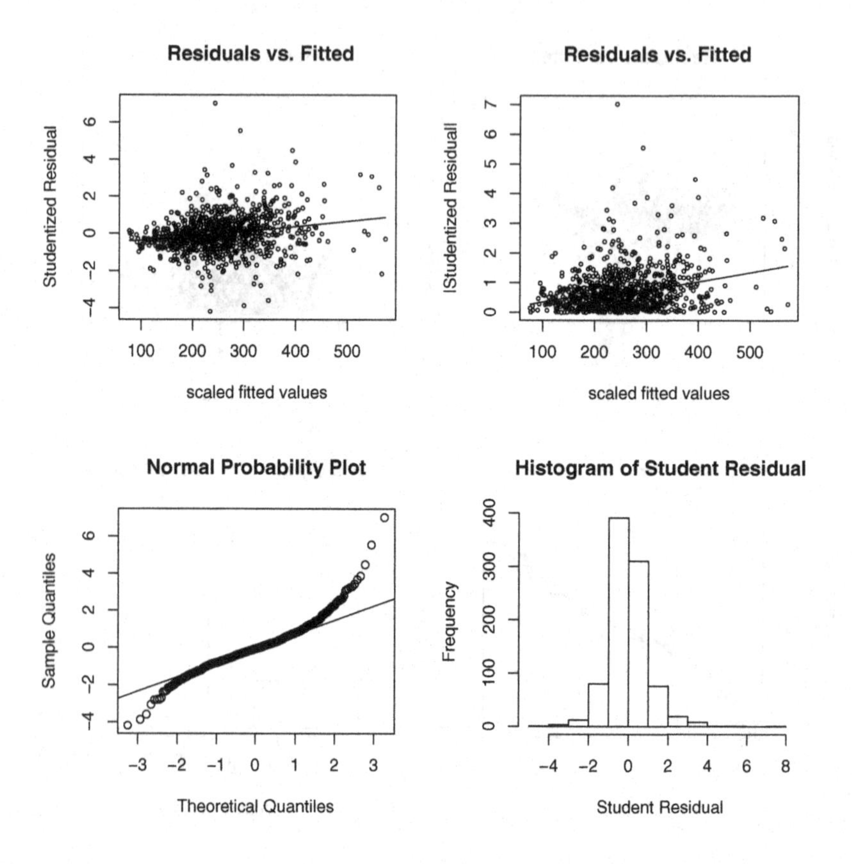


Figure 7.14 *Model checking plots for multivariate DHGLM of $y_2$ on the primary biliary cirrhosis data.*

which are related to eye movements (responses) (Goossens and Opstal, 2000); this leads to the following model for missing data:

$$\begin{aligned}\eta &= \Phi^{-1}(p_{ij}) = \delta_0 + x_{1ij}\delta_1 + x_{2ij}\delta_2 + sex_i\delta_3 + sch_i\delta_4 \\ &\quad + sex_i \cdot x_{1ij}\delta_5 + sex_i \cdot x_{2ij}\delta_6 + sex_i \cdot sch_i\delta_7 + \rho y_{ij}^*,\end{aligned}$$

where $p_{ij} = P(\delta_{ij} = 1)$.

We can consider the model DI as well as DN (DN: DHGLM with non-ignorable missingness) with the probit model having two responses: $y1$ for a continuous response and $y2$ for a missing indicator. Here, R-codes using mdhglm are as follows.

```
> # read variables for continuous response
```

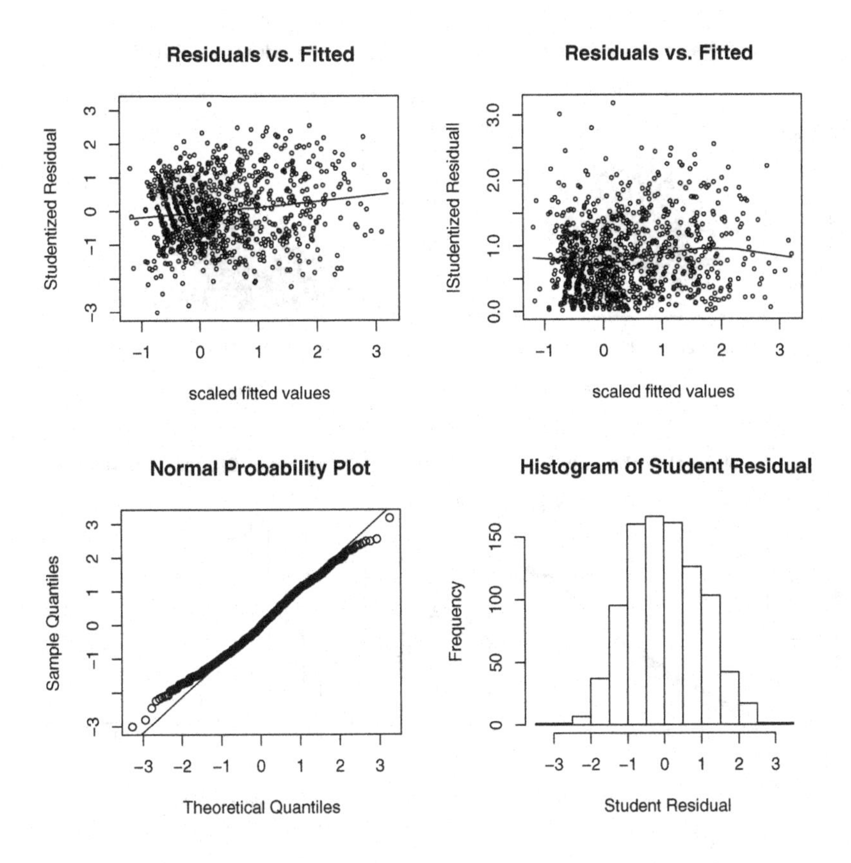


Figure 7.15 *Model checking plots for multivariate DHGLM of $y_1$ on the primary biliary cirrhosis data.*

```
> data(ech_data1)
> # read variables for missingness
> data(ech_data2)
> jm1m<-DHGLMMODELING(Model="mean",Link="identity",
+     LinPred=y1~x1+x2+time+sch+sch*x1+sch*x2+(1|subject),
+     RandDist="gaussian")
> jm1d<-DHGLMMODELING(Model="dispersion",Link="log",
+     LinPred=phi~sch+(sch|subject),
+     RandDist="gaussian")
> jm2m<-DHGLMMODELING(Model="mean",Link="probit",
+     LinPred=y2~x1+x2+time+sch+sch*x1+sch*x2+(1|subject),
+     RandDist="gaussian")
> jm2d<-DHGLMMODELING(Model="dispersion")
```

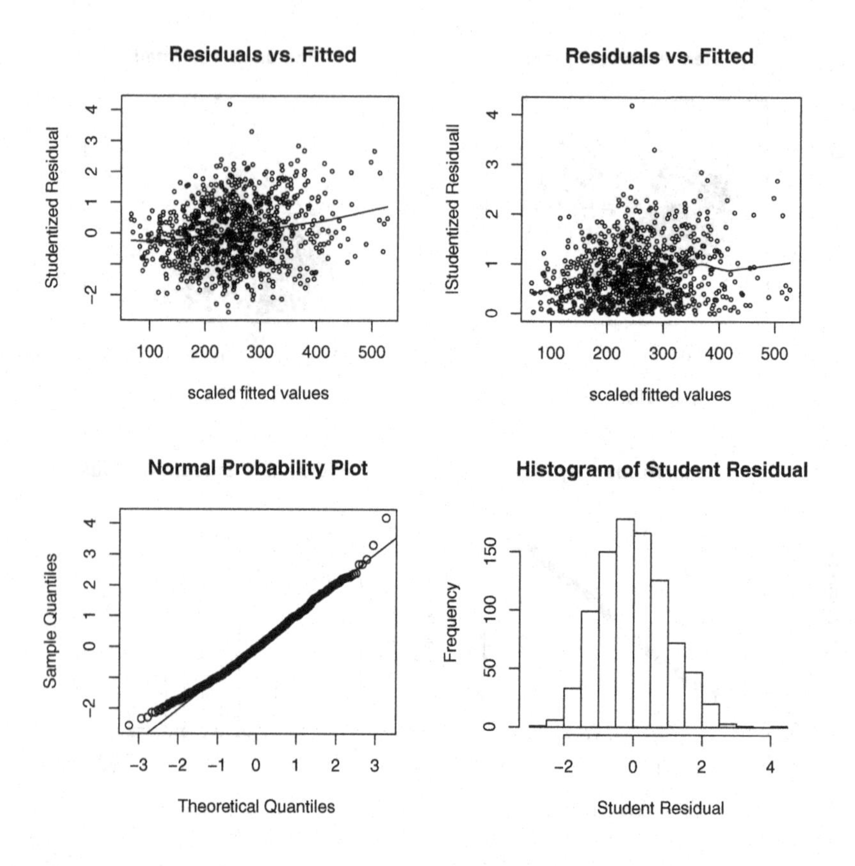


Figure 7.16 *Model checking plots for multivariate DHGLM of* $y_2$ *on the primary biliary cirrhosis data.*

(i) DI: DHGLM with ignorable missingness where $\rho = 0$,

```
> res_ind<-jointfit(RespDist=
+          c("gaussian","binomial"),
+  DataMain=list(sch_data1,sch_data2),
+          MeanModel=list(jm1m,jm2m),
+          DispersionModel=list(jm1d,jm2d),
+  structure="independent")
```

(ii) DN: DHGLM with non-ignorable missingness where $\rho \neq 0$ and

```
> res_selection<-jointfit(RespDist=
+          c("gaussian","binomial"),
+  DataMain=list(sch_data1,sch_data2),
```

```
+         MeanModel=list(jm1m,jm2m),
+         DispersionModel=list(jm1d,jm2d),
+  structure="selection")
```

From Table 7.14, we see that the time effect is not significant in the probit model. The analyses from DHGLMs with and without a model for missingness are in Table 7.14. The negative value of $\hat{\rho}$ supports the physicians' opinions that lower values of the response are more likely to be missing at each cycle. However, the conclusions concerning non-ignorable missingness depend crucially on untestable distributional assumptions; thus, sensitivity analysis has been recommended. Fortunately, the analysis of the responses in these data indicates that they are not sensitive to the assumptions about the heavy tails or the missing mechanism (Yun and Lee, 2006).

### *7.1.8 Binary factor model for law school admission data*

Factor analysis and structural equation model (SEM) are widely used in many fields of science. Lee, Nelder, and Pawitan (2017) considered law school admission data of Bock and Lieberman (1970), consisting of six items $y_{ij}$ taking 1 (Correct) or 0 (Not correct) for law school admission test with $n = 350$ subjects. Let $\pi_{ij} = P(y_{ij} = 1|\boldsymbol{v}_i)$ and consider a binary two-factor model with the path diagram in Figure 7.17.

$$\text{logit}(\boldsymbol{\pi}_i) = \boldsymbol{\beta}_0 + \boldsymbol{\Lambda}\boldsymbol{v}_i,$$

where $\boldsymbol{\pi}_i = (\pi_{i1}, \cdots, \pi_{i6})^T$ and $\boldsymbol{\beta}_0 = (\beta_{01}, \cdots, \beta_{06})^T$ be the vector of $\pi_{ij}$'s and $\beta_{0j}$'s, respectively,

$$\Lambda^T = \begin{pmatrix} 1 & \lambda_2 & \lambda_3 & 0 & 0 & 0 \\ 0 & 0 & 0 & 1 & \lambda_5 & \lambda_6 \end{pmatrix}$$

and $\boldsymbol{v}_i = (v_{1i}, v_{2i})^T$ follow the bivariate normal distribution with the covariance matrix

$$\text{cov}(\boldsymbol{v}_i) = \begin{pmatrix} \gamma_{11} & \gamma_{12} \\ \gamma_{12} & \gamma_{22} \end{pmatrix}.$$

Here, R-codes using mdhglm for fitting factor model are as follows.

```
> data(factor,packge="mdhglm")
> jm1<-DHGLMMODELING(Model="mean",Link="logit",
+     LinPred=y1~1+(1|subject),
+     RandDist="gaussian")
> jm2<-DHGLMMODELING(Model="mean",Link="logit",
+     LinPred=y2~1+(1|subject),
```

Table 7.14 *Results from the MDHGLM fit in the schizophrenic behavior data.*

| Part | Parameter | DI | | | DN | | |
|---|---|---|---|---|---|---|---|
| | | est. | s.e. | t-value | est. | s.e. | t-value |
| $y_1$ | Int | 0.811 | 0.014 | 59.29 | 0.802 | 0.014 | 55.89 |
| | $x_1$ | 0.006 | 0.005 | 1.42 | 0.004 | 0.005 | 0.96 |
| | $x_2$ | −0.121 | 0.005 | −24.63 | −0.121 | 0.005 | −24.40 |
| | $time$ | −0.002 | 0.000 | −5.23 | −0.002 | 0.000 | −5.97 |
| | $sch$ | −0.036 | 0.019 | −1.85 | −0.051 | 0.020 | −2.51 |
| | $sch \cdot x_1$ | −0.023 | 0.006 | −3.54 | −0.022 | 0.007 | −3.33 |
| | $sch \cdot x_2$ | −0.004 | 0.007 | −0.59 | −0.005 | 0.007 | −0.63 |
| $y_2$ | Int | | | | 2.148 | 0.231 | 9.30 |
| | $x_1$ | | | | 0.065 | 0.062 | 1.05 |
| | $x_2$ | | | | −0.276 | 0.071 | −3.86 |
| | $sex$ | | | | −0.085 | 0.072 | −1.18 |
| | $sch$ | | | | −0.072 | 0.054 | −1.30 |
| | $sex \cdot x_1$ | | | | −0.171 | 0.123 | −1.39 |
| | $sex \cdot x_2$ | | | | −0.379 | 0.128 | −2.97 |
| | $sex \cdot sch$ | | | | −0.284 | 0.103 | −2.76 |
| | $y^*$ | | | | −3.704 | 0.296 | −12.53 |
| $phi$ | $\lambda_1$ | 0.089 | | | 0.093 | | |
| | $\gamma_0$ | −5.320 | | | −5.287 | | |
| | $\gamma_1$ | 0.149 | | | 0.241 | | |
| | $\lambda_2$ | 0.583 | | | 0.738 | | |
| cAIC | | −2143.4 | | | −2155.4 | | |
| elapsed time (sec.) | | 327.6 | | | 453.1 | | |

```
+    RandDist="gaussian")
> jm3<-DHGLMMODELING(Model="mean",Link="logit",
+    LinPred=y3~1+(1|subject),
+    RandDist="gaussian")
> jm4<-DHGLMMODELING(Model="mean",Link="logit",
+    LinPred=y4~1+(1|subject),
+    RandDist="gaussian")
> jm5<-DHGLMMODELING(Model="mean",Link="logit",
+    LinPred=y5~1+(1|subject),
+    RandDist="gaussian")
> jm6<-DHGLMMODELING(Model="mean",Link="logit",
```

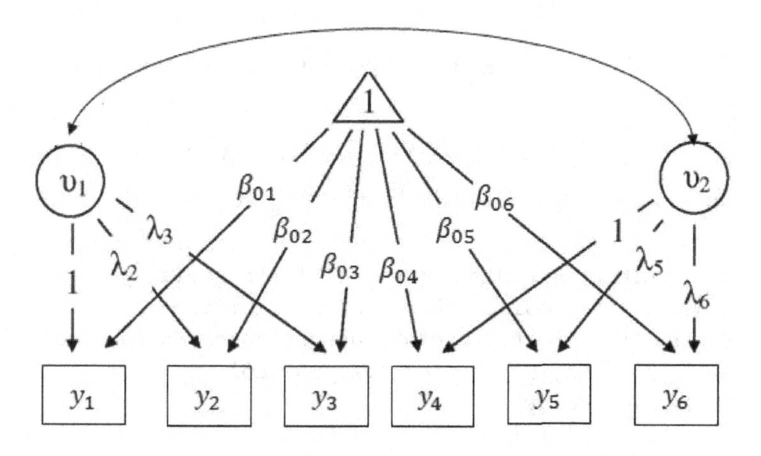


Figure 7.17 *Path diagram for the binary two-factor model*

```
+     LinPred=y6~1+(1|subject),
+     RandDist="gaussian")
```

To link them by the factor model we apply the next procedure:

```
> res<-jointfit(RespDist=c("binomial","binomial",
+ "binomial","binomial","binomial","binomial",),
+ DataMain=list(factor,factor,factor,factor,factor,factor),
+ MeanModel=list(jm1,jm2,jm3,jm4,jm5,jm6),
+ structure="factor",
+ factor=c(1,1,1,2,2,2),
+ REML=TRUE,order=2)
```

When the `REML` option is specified TRUE (or FALSE), REML (or ML) procedures are implemented. The `order` option (1 or 2) specifies order of Laplace approximation. Fitting results are in Table 7.15. Here REML(1) is based on the first-order Laplace approximation equations (3.3) and (3.4) in Chapter 3. and REML(2) is based on the second-order Laplace approximation. Via a small simulation study, Noh et al. (2017) showed that REML(1) has the least mean squares errors while REML(2) has the least biases. The goodness-of-fit test based on the deviance and the degrees of freedom in Section 2.2.2 shows no lack of fit. In this two-factor model, using REML(2) the correlation between $v_{1i}$ and $v_{2i}$ is estimated as $0.997 = 0.47/\sqrt{0.39 \times 0.57}$.

We also consider one factor model, which is equivalent to assume the correlation between $v_{1i}$ and $v_{2i}$ being $\pm 1$,

$$\text{logit}(\boldsymbol{\pi}_i) = \boldsymbol{\beta}_0 + \boldsymbol{\Lambda} w_{1i},$$

where

$$\Lambda^T = \begin{pmatrix} 1 & \lambda_2 & \lambda_3 & \lambda_4 & \lambda_5 & \lambda_6 \end{pmatrix}$$

and $w_{1i} \sim \text{N}(0, \gamma_{11})$. This can be fitted by

```
> res<-jointfit(RespDist=c("binomial","binomial",
+ "binomial","binomial","binomial","binomial",),
+ DataMain=list(factor,factor,factor,factor,factor,factor),
+ MeanModel=list(jm1,jm2,jm3,jm4,jm5,jm6),
+ structure="factor",
+ factor=c(1,1,1,1,1,1),
+ REML=TRUE,order=2)
```

In Table 7.15, this one-factor model has $cAIC = 2,371.7$ which is less than $cAIC = 2,548.6$ of the previous two-factor model. Thus, cAICs clearly prefers the one-factor model.

Table 7.15 *Parameter estimates (SE) with one-factor and two-factor models for law school admission data*

| | two-factor model | | one-factor model | |
|---|---|---|---|---|
| | REML(1) | REML(2) | REML(1) | REML(2) |
| $\lambda_2$ | 0.48(0.12) | 0.50(0.13) | 0.45(0.13) | 0.48(0.14) |
| $\lambda_3$ | 0.44(0.10) | 0.46(0.11) | 0.40(0.10) | 0.42(0.10) |
| $\lambda_4$ | - | - | 0.79(0.13) | 0.85(0.14) |
| $\lambda_5$ | 0.51(0.11) | 0.53(0.11) | 0.41(0.10) | 0.42(0.10) |
| $\lambda_6$ | 0.42(0.10) | 0.42(0.10) | 0.32(0.09) | 0.33(0.10) |
| $\gamma_{11}$ | 0.39(0.11) | 0.39(0.13) | 0.49(0.10) | 0.51(0.11) |
| $\gamma_{12}$ | 0.46(0.08) | 0.47(0.08) | - | - |
| $\gamma_{22}$ | 0.56(0.08) | 0.57(0.10) | - | - |
| $\beta_{01}$ | −1.39(0.08) | −1.42(0.09) | −1.38(0.08) | −1.39(0.08) |
| $\beta_{02}$ | −0.52(0.05) | −0.50(0.05) | −0.50(0.05) | −0.49(0.05) |
| $\beta_{03}$ | −0.12(0.07) | −0.11(0.07) | −0.12(0.07) | −0.11(0.07) |
| $\beta_{04}$ | −0.65(0.05) | −0.67(0.05) | −0.63(0.05) | −0.66(0.05) |
| $\beta_{05}$ | −1.07(0.06) | −1.06(0.06) | −1.10(0.07) | −1.09(0.07) |
| $\beta_{06}$ | −0.91(0.05) | −0.93(0.06) | −0.93(0.05) | −0.94(0.05) |
| cAIC | 2548.6 | | 2371.7 | |

## 7.2 Implementation in the mdhglm package

The mdhglm package is used to fit MDHGLMs in which multivariate responses may vary and follow the DHGLM. A variety of distributions and link functions for both the response variables and the random effects are allowed. Fixed and random effects can also be fitted in both the mean and the dispersion components. To call the fitting function jointfit, models for the mean and dispersion must be specified by DHGLMMODELING object preferably created by calling the DHGLMMODELING function which is defined in the Chapter 6. The structure of that we use is as follows:

```
jointfit(RespDist=c("gaussian", "gaussian"),
BinomialDen=NULL,DataMain, MeanModel=,DispersionModel=,
PhiFix=NULL,LamFix=NULL,
mord=1,dord=1,Maxiter=200,convergence=1e-06,
structure="correlated",Init_Corr=,EstimateCorrelation=TRUE)
```

- RespDist: The distribution of the response is set by the option RespDist. The user can set it to: "gaussian" (default), "binomial", "poisson", or "gamma".
- BinomialDen: When RespDist="binomial", one should use the option BinomialDen to specify the denominator for the binomial distribution. This should be NULL (default) or a numeric vector of length equal to the length of DataMain. When specified as BinomialDen=NULL and RespDist="binomial", the denominator is 1.
- DataMain: The option DataMain determines the data frame to be used (non-optional).
- MeanModel: For the mean model, this option requires DGHLMMODELING object which should be specified by the option Model="mean".
- DispersionModel: For the over-dispersion model, this option requires DGHLMMODELING object which should be specified by the option Model="dispersion".
- PhiFix, LamFix: Two options that determine whether the over-dispersion parameters (phi) and random-effect variance (lambda) are to be estimated or maintained constant. Specifying defaults such as PhiFix =NULL (or LamFix =NULL) implies that phi (or lambda) is to be estimated. If not, phi (or lambda) is fixed at a value specified by PhiFix (or LamFix).
- mord: The option mord specifies the order of Laplace approximation to the marginal likelihood for fitting the mean parameters. The choice is either 0 or 1 (default).

- dord: The option dord specifies the order of Laplace approximation to the adjusted profile likelihood for fitting the dispersion parameters. The choice is either 1 (default) or 2.
- Maxiter: Maxiter determines the maximum number of iterations for estimating all parameters. The default number is 200 iterations.
- convergence: Setting this option determines the criterion for convergence, which is computed as the absolute difference between the values of all the estimated parameters in the previous and current iterations. The default criterion is 1e-06.
- structure: Setting structure of correlations. The choice is "correlated" (default), "shared", "indep", "selection" or "factor."
- factor: Vector indicating factor model. Structure should be specified to be "factor".

## 7.3 Exercises

1. Fit bivariate GLMMs, using the code below where two response variables y1 and y2 are simulated as Gaussian and Poisson, respectively.

a) What is the simulated covariance between the random effects for the two parts of the model? What is the estimated covariance between the random effects?

b) Perform a simulation study with 100 replicates and investigate if there is any bias in the parameter estimates.

```
rm( list = ls() )
library( mvtnorm )
library( mdhglm )
################
#Simulations
set.seed( 1234 )
k <- 50 #No. of clusters
m <- 20 #No. of obs per cluster
N <- m*k #Length of each response variable
Z <- diag(k)%x%rep(1,m)
id <- factor( rep(1:k, each=m) )
sigma <- matrix(c(1,0.5,0.5,1), ncol=2)
u <- rmvnorm(n=k, mean = c(0,0), sigma = sigma)
x <- rbinom(N, 1, 0.5)
y1 <- 2 - 3*x + Z%*%u[,1] + rnorm(N)
eta2 <- Z%*%u[,2]
```

```
y2 <- rpois( length(eta2), exp(eta2) )
sim_data <- data.frame(y1=y1, y2=y2, id=id)
```

2. The LSAT dataset in the R package ltm (Rizopoulos, 2013) has five binary items $\mathbf{y}_i = (y_{i1}, \cdots, y_{i5})^T$ taking values 1 (Correct) or 0 (Not correct) with $n = 1,000$ subjects for law school admission data. Fit a two-factor logistic model as follows,

$$\text{logit}(\boldsymbol{\pi}_i) = \boldsymbol{\beta}_0 + \boldsymbol{\Lambda} \boldsymbol{v}_i,$$

where $\boldsymbol{\pi}_i = (\pi_{i1}, \cdots, \pi_{i5})^T$ and $\boldsymbol{\beta}_0 = (\beta_{01}, \cdots, \beta_{05})^T$ are the vector of $\pi_{ij}$'s and $\beta_{0j}$'s, respectively, and

$$\Lambda^T = \begin{pmatrix} 1 & \lambda_2 & \lambda_3 & 0 & 0 \\ 0 & 0 & 0 & 1 & \lambda_5 \end{pmatrix}$$

and $\boldsymbol{v}_i = (v_{1i}, v_{2i})^T$ follow the bivariate normal distribution with the covariance matrix

$$\text{cov}(\boldsymbol{v}_i) = \begin{pmatrix} \gamma_{11} & \gamma_{12} \\ \gamma_{12} & \gamma_{22} \end{pmatrix}.$$

CHAPTER 8

# Survival analysis

In this chapter we study the analysis of incomplete data, caused by censoring in event-time survival data. Cox's proportional hazards model is widely used for the analysis of survival data. This model is interesting due to its semi-parametric nature, where the baseline hazards assume non-parametric but treatment effects are modeled parametrically. To handle, such a semi-parametric structure, the partial likelihood has been proposed. It can be fitted using Poisson GLM algorithms, but it is slow due to a large number of nuisance parameters caused by non-parametric baseline hazards. Thus, a fast algorithm, overcoming this drawback, is needed for proportional hazards models.

Frailty models with a non-parametric baseline hazard extend proportional hazards model by allowing random effects in hazards and have been widely adopted for the analysis of survival data (Hougaard, 2000; Duchateau and Janssen, 2008). Using h-likelihood theory we can show that Poisson HGLM algorithms can be used to fit these models. However, this algorithm is again slow because the number of nuisance parameters in non-parametric baseline hazards increases quickly with the number of observations. Ha, Lee, and Song (2001) showed that with the h-likelihood it is easy to eliminate nuisance parameters by using a plug-in method and a fast estimation algorithm can thereby be used.

Either a log-normal or gamma distribution can be used as the frailty distribution; therefore, normal and log-gamma distribution can be adopted for the log frailties. The h-likelihood provides a computationally efficient procedure (Ha, Jeong, and Lee, 2017).

Previously, frailty models have been implemented in several R functions such as the `coxph()` function in the survival package (Therneau and Lumley, 2015) and the `coxme()` function in the coxme package (Therneau, 2015), based on the penalized partial likelihood (PPL), the `phmm()` function in the phmm package (Donohue and Xu, 2013), based on a Monte Carlo EM (MCEM) method, and the `frailtyPenal()` function in the frailtypack package,(Gonzalez et al., 2012) based on a penalized marginal likelihood. The phmm package fits only one-component frailty models, although it does allow multivariate frailties. The `coxme` function can fit the multi-component model as shown in the CGD example below. The frailtyHL package is used to fit frailty models with a non-parametric baseline hazard, providing estimates of fixed effects, random effects, and variance components as well as their standard errors. In addition, it provides a statistical test for the variance components of frailties and also three AIC criteria for the model selection.

## 8.1 Examples

We first consider survival analyses using frailty models (Box 7). The frailtyHL package makes it possible to

1. fit models when the log-frailty distribution is not necessarily normal, and

2. estimate variance components when the frailty structure is shared or nested.

### *8.1.1 Kidney infection data*

To demonstrate differences of various estimation methods on small cluster size $n_i \equiv 2$, we use the kidney infection data (McGilchrist and Aisbett, 1991) whose state diagram is shown in Figure 8.1. The data consist of times until the first and second recurrences ($n_i \equiv 2$ ) of kidney infection in 38 ($q = 38$) patients using a portable dialysis machine. Each survival time (`time`) is the time until infection since the insertion of the catheter. The survival times for the same patient are likely to be correlated because of a shared frailty describing the common patient's effect. The catheter is later removed if infection occurs and can be removed for other reasons, which we regard as censoring; about 24 per cent of the data was censored.

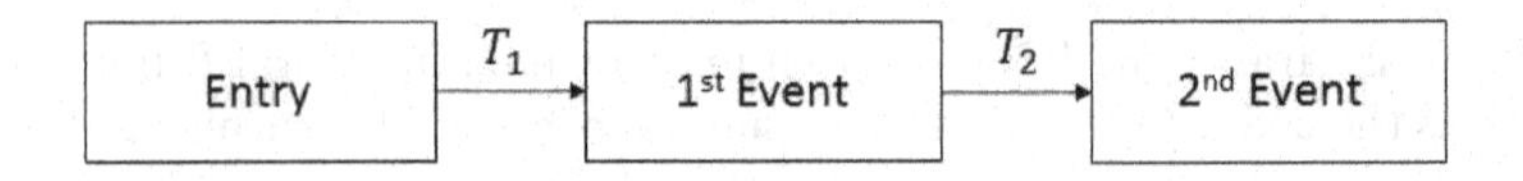


Figure 8.1 *State diagram for the kidney infection data*

We fit frailty models with two covariates, the sex (1 = male; 2 = female)

Table 8.1 *Comparison of different estimation methods for the kidney infection data*

| Method | Sex<br>$\hat{\beta}_1$ (SE) | Age<br>$\hat{\beta}_2$ (SE) | Patient<br>$\hat{\alpha}$ (SE) |
|---|---|---|---|
| log-normal model | | | |
| HL(0,1) | −1.380 | 0.005 | 0.535 |
| | (0.431) | (0.012) | (0.338) |
| HL(1,1) | −1.414 | 0.005 | 0.545 |
| | (0.432) | (0.012) | (0.340) |
| `coxph` | −1.388 | 0.005 | 0.551 |
| (Breslow) | (0.441) | (0.012) | ( − ) |
| `coxph` | −1.411 | 0.005 | 0.569 |
| (Efron) | (0.445) | (0.013) | ( − ) |
| `coxme` | −1.332 | 0.005 | 0.440 |
| (Breslow) | (0.414) | (0.012) | ( − ) |
| `coxme` | −1.355 | 0.004 | 0.456 |
| (Efron) | (0.417) | (0.012) | ( − ) |
| `phmm` | −1.329 | 0.004 | 0.378 |
| | (0.452) | (0.012) | ( − ) |
| gamma model | | | |
| HL(0,2) | −1.691 | 0.007 | 0.561 |
| | (0.483) | (0.013) | (0.280) |
| HL(1,2) | −1.730 | 0.007 | 0.570 |
| | (0.485) | (0.013) | (0.281) |
| `coxph` | −1.557 | 0.005 | 0.398 |
| (Breslow) | (0.456) | (0.012) | ( − ) |
| `coxph` | −1.587 | 0.005 | 0.412 |
| (Efron) | (0.461) | (0.012) | ( − ) |

and age of each patient, using the functions (`frailtyHL`, `coxph`, `coxme` and `phmm`) in the four packages. The results are summarized in Table 8.1. In PPL procedures (`coxph` and `coxme`), the Breslow method provides a slightly smaller estimate for $\alpha$ than the Efron method.

**Box 7: Frailty models**

Suppose that data consist of right censored time-to-event observations from $q$ subjects (or clusters), with $n_i$ observations each; $i = 1, \ldots, q$. Let $T_{ij}$ be the survival time for the $j$th observation of the $i$th subject; $j = 1, \ldots, n_i$. Here, $n = \sum_i n_i$ is the total sample size and $n_i$ is the cluster size. Let $C_{ij}$ be the corresponding censoring time and let $y_{ij} = \min\{T_{ij}, C_{ij}\}$ and $\delta_{ij} = I(T_{ij} \leq C_{ij})$ be the observable random variables, where $I(\cdot)$ is the indicator function. Given the common unobserved frailty for the $i$th subject $u_i$, the conditional hazard function of $T_{ij}$ is of the form

$$\lambda_{ij}(t|u_i) = \lambda_0(t) \exp(x_{ij}^T \boldsymbol{\beta}) u_i, \tag{8.1}$$

where $\lambda_0(\cdot)$ is an unspecified baseline hazard function and $\boldsymbol{\beta} = (\beta_1, \ldots, \beta_p)^T$ is a vector of regression parameters for the fixed covariates $x_{ij}$. Here, the term $x_{ij}^T \boldsymbol{\beta}$ does not include an intercept term because of identifiability. Assume that the frailties $u_i$ are i.i.d. random variables with a frailty parameter $\alpha$; the gamma and log-normal frailty models assume gamma and log-normal distributions for $u_i$, respectively. That is, our package allows
(i) gamma frailty with $E(u_i) = 1$ and $\text{var}(u_i) = \alpha$, and
(ii) log-normal frailty having $v_i = \log u_i \sim \text{N}(0, \alpha)$.
In multi-component frailty models,

$$\boldsymbol{X}\beta + \boldsymbol{Z}^{(1)}\boldsymbol{v}^{(1)} + \boldsymbol{Z}^{(2)}\boldsymbol{v}^{(2)} + \cdots + \boldsymbol{Z}^{(k)}\boldsymbol{v}^{(k)}, \tag{8.2}$$

where $\boldsymbol{X}$ is the $n \times p$ model matrix, $\boldsymbol{Z}^{(r)}$ $(r = 1, 2, \ldots, k)$ are the $n \times q_r$ model matrices corresponding to the $q_r \times 1$ frailties $\boldsymbol{v}^{(r)}$, and $\boldsymbol{v}^{(r)}$ and $\boldsymbol{v}^{(l)}$ are independent for $r \neq l$. Notice here that $\boldsymbol{Z}^{(r)}$ is a matrix having indicator values such that $Z_{st}^{(r)} = 1$ if observation $s$ is a member of subject (or cluster) $t$ in the $r$th frailty component and 0 otherwise (Therneau and Grambsch, 2000). For example, the CGD data (Fleming and Harrington, 1991) have a multilevel structure in which patients nested within centers have recurrent infection times. Later, we analyze these data using model (8.2) with $k = 2$. Here, the frailty structures are:

$$\begin{aligned} \boldsymbol{v}^{(1)}&: \text{center frailty} \sim \text{N}(0, \alpha_1 \boldsymbol{A}_1), \\ \boldsymbol{v}^{(2)}&: \text{patient frailty} \sim \text{N}(0, \alpha_2 \boldsymbol{A}_2), \end{aligned}$$

where $\boldsymbol{A}_r = I_r$ $(r = 1, 2)$ are the $q_r \times q_r$ identity matrices, and $q_1$ and $q_2$ are the number of centers and patients, respectively.

In the log-normal frailty, REML procedures (fraityHL() and coxph()) give larger estimate for $\alpha$ than ML procedures (`coxme` and `phmm`). However, both ML estimates from `coxme` and `phmm` are somewhat different because cluster size is small, $n_i \equiv 2$ for all $i$. For gamma frailty, `coxph()` uses the ML procedure so that they give smaller estimate for $\alpha$ than the h-likelihood procedures. Compared with the h-likelihood methods, PPL methods are computationally more efficient, but could have larger biases (Ha et al., 2010). For further discussions in survival analysis, see Ha, Jeong, and Lee (2017).

The standard shared frailty model assumes that censoring times are independent of event times within clusters. When competing risks are present this assumption is no longer reasonable, since subjects who experience a competing event are censored informatively.

Traditionally two legitimate and identifiable quantities to analyze competing risks data have been the cause-specific hazard function and the cumulative incidence function (Kalbfleisch and Prentice, 1980). Here we consider the cause-specific hazard frailty model which is a generalization of the shared frailty model that allows for competing risks as well as independent censoring (Box 8). The cause-specific hazard regression model can be directly fitted via the frailtyHL package allowing competing events.

### *8.1.2 Log-normal and gamma frailty models on rat data*

The dataset presented by Mantel et al. (1977) is based on a tumorigenesis study of 50 ($q = 50$) litters of female rats. For each litter, one rat was selected to receive the drug and the other two rats were placebo-treated controls ($n_i \equiv 3$). Here each litter is treated as a cluster. The survival time (`time`) is the time to development of tumor, measured in weeks. Death before occurrence of tumor yields a right-censored observation; forty rats developed a tumor, leading to censoring of about 73 per cent. The survival times for rats in a given litter may be correlated due to a random effect representing shared genetic or environmental effects.

We fit frailty models with one covariate, the rx (1 = drug; 0 = placebo), using `frailtyHL()`. Below, we present the code and results for the log-normal frailty model using HL(1,1). The output from the R code shows that the effect of rx is significant (t-value = 2.823 with p-value = 0.005). That is, the rx group has a significantly higher risk than the control group. Here, the variance estimate of the frailty is $\hat{\alpha} = 0.4272$ (with SE = 0.4232).

Note that although we report the SE of the $\alpha$, one should not use it for testing the absence of frailty $\alpha = 0$ (Vaida and Xu, 2000). Such a null hypothesis is on the boundary of the parameter space, so that the critical value of an asymptotic $(\chi^2(0)+\chi^2(1))/2$ distribution is 2.71 at 5% significant level (Lee, Nelder, and Pawitan, 2017; Ha, Sylvester, Legrand, and MacKenzie, 2011). The difference in deviance $-2p_{\beta,v}(h_p)$ between Cox's PHM without frailty and log-normal frailty model is $364.15 - 362.56 = 1.59 (< 2.71)$, indicating that the frailty effect is non-significant, i.e., $\alpha = 0$. Here, the results from fitting Cox's PHM without frailty are available by adding the two arguments `varfixed=TRUE` and `varinit=c(0)` in the `frailtyHL()`: see below.

```
> library(survival)
> data(rats)
> rat_cox <- frailtyHL(Surv(time,status)
+ ~rx+(1|litter), rats,
+ varfixed=TRUE, varinit=c(0))
iteration :
         4
convergence :
  4.801639e-09
[1] "converged"
[1] "Results from the log-normal frailty model"
[1] "Number of data : "
[1] 150
[1] "Number of event : "
[1] 40
[1] "Model for conditional hazard : "
Surv(time, status) ~ rx + (1 | litter)
[1] "Method : HL(0,1)"
[1] "Estimates from the mean model"
   Estimate Std. Error t-value  p-value
rx   0.8982     0.3174    2.83 0.004655
[1] "Estimates from the dispersion model"
       Estimate Std. Error
litter "0"      "NULL"
       -2h0  -2*hp -2*p_b,v(hp)
[1,] 363.69 363.69    364.15
       cAIC   mAIC   rAIC
[1,] 365.69 365.69 364.15
> rat_ln11 <- frailtyHL(Surv(time,status)~rx+(1|litter),
+ rats, RandDist="Normal", mord=1, dord=1)
iteration :
        87
convergence :
   9.97616e-07
```

```
[1] "converged"
[1] "Results from the log-normal frailty model"
[1] "Number of data : "
[1] 150
[1] "Number of event : "
[1] 40
[1] "Model for conditional hazard : "
Surv(time, status) ~ rx + (1 | litter)
[1] "Method : HL(1,1)"
[1] "Estimates from the mean model"
   Estimate Std. Error t-value  p-value
rx   0.9107     0.3226   2.823 0.004754
[1] "Estimates from the dispersion model"
       Estimate Std. Error
litter   0.4272     0.4232
       -2h0  -2*hp -2*p_v(hp) -2*p_b,v(hp)
[1,] 335.97 397.36     362.14     362.56
       cAIC   mAIC   rAIC
[1,] 362.22 366.14 364.56
```

The code and results for the gamma frailty model using HL(1,2) are provided below. The output from the R code shows that these results are similar to those of the log-normal frailty, particularly for estimation of $\beta$.

```
> rat_g12 <- frailtyHL(Surv(time,status)~rx+(1|litter),
+ rats, RandDist="Gamma", mord=1, dord=2)
iteration :
        170
convergence :
  9.567765e-07
[1] "converged"
[1] "Results from the gamma frailty model"
[1] "Number of data : "
[1] 150
[1] "Number of event : "
[1] 40
[1] "Model for conditional hazard : "
Surv(time, status) ~ rx + (1 | litter)
[1] "Method : HL(1,2)"
[1] "Estimates from the mean model"
   Estimate Std. Error t-value  p-value
rx   0.9126     0.3236    2.82 0.004806
[1] "Estimates from the dispersion model"
```

```
        Estimate Std. Error
litter   0.5757     0.5977
      -2h0  -2*hp  -2*p_v(hp) -2*p_b,v(hp) -2*s_b,v(hp)
[1,] 331.60 413.85   365.35    365.77       362.12
       cAIC   mAIC   rAIC
[1,] 365.3 369.35 367.77
```

For the selection of a model between non-nested models such as log-normal and gamma frailty models, we may use three Akaike information criterion (AIC) criteria (Lee, Nelder, and Pawitan, 2017; Ha, Lee, and MacKenzie, 2007; Donohue, Overholser, Xu, and Vaida, 2011) based on conditional likelihood, marginal likelihood and restricted likelihood, respectively, defined by

$$\begin{aligned} \text{cAIC} &= -2h_0 + 2\text{df}_\text{c}, \\ \text{mAIC} &= -2p_v(h_p) + 2\text{df}_\text{m}, \\ \text{rAIC} &= -2p_{\beta,v}(h_p) + 2\text{df}_\text{r}, \end{aligned}$$

where $h_0 = \ell_0^*$ (8.5) in Section 8.4, and
$\text{df}_\text{c} = \text{trace}\{D^{-1}(h_p, (\beta, v))D(h_0, (\beta, v))\}$ is an effective degrees of freedom adjustment for estimating the fixed and random effects, computed using the Hessian matrices $D(h_p, (\beta, v)) = -\partial^2 h_p/\partial(\beta, v)^2$ and

$D(h_0, (\beta, v)) = -\partial^2 h_0/\partial(\beta, v)^2$. Note here that $\text{df}_\text{m}$ is the number of fixed parameters and $\text{df}_\text{r}$ is the number of dispersion parameters (Ha et al., 2007). Thus we select a model to minimize the AIC values among models. In the dataset, for log-normal frailty we have cAIC=362.22, mAIC=366.14, and rAIC=364.56, and for gamma frailty cAIC=365.30, mAIC=369.35, and rAIC=367.77. All three AICs above indicate that the log-normal frailty model is better than the gamma frailty model.

### *8.1.3 Multilevel frailty models on CGD infection data*

The CGD dataset presented by Fleming and Harrington (1991) consists of a placebo-controlled randomized trial of gamma interferon (rIFN-g) in the treatment of chronic granulomatous disease (CGD). 128 patients (`id`) from 13 centers ($q_1 = 13, q_2 = 128$) were tracked for around 1 year. The number of patients (i.e., cluster size) per center ranged from 4 to 26. The survival times (`tstop-tstart`) are the recurrent infection times of each patient from the different centers. Censoring occurred at the last observation for all patients, except one, who experienced a serious infection on the date he left the study; in the CGD study about 63 per cent of the data were censored. The recurrent infection times for a given patient are likely to be correlated. However, each patient belongs to one of the 13 centers; hence, the correlation may also be attributed to a center effect.

We fit a multilevel log-normal frailty model with two frailties and a single covariate, treatment (rIFN-g, placebo), using `frailtyHL()`. Here, the two frailties are random center and patient effects. The code and results using HL(1,1) are provided below. The output shows that the effect of treatment is significant (t-value = -3.476 with p-value $< 0.001$), indicating that rIFN-g significantly reduces the rate of serious infection in CGD patients. The estimate of variance of patient frailty ($\hat{\alpha}_2 = 1.00235$) is considerably larger than variance of center frailty ($\hat{\alpha}_1 = 0.02987$), indicating that the random-patient effect is more heterogeneous.

```
> library(survival)
> data(cgd)
> cgd11 <- frailtyHL(Surv(tstop-tstart,status)
+ ~treat+(1|center)+(1|id), cgd,
+ RandDist="Normal",mord=1, dord=1)
iteration :
        157
convergence :
  9.336249e-07
[1] "converged"
[1] "Results from the log-normal frailty model"
[1] "Number of data : "
[1] 203
[1] "Number of event : "
[1] 76
[1] "Model for conditional hazard : "
Surv(tstop-tstart,status)~treat+(1|center)+(1|id)
[1] "Method : HL(1,1)"
[1] "Estimates from the mean model"
            Estimate Std. Error t-value   p-value
treatrIFN-g   -1.184     0.3407  -3.476 0.0005085
[1] "Estimates from the dispersion model"
       Estimate Std. Error
center  0.02986     0.1572
id      1.00235     0.5089
       -2h0  -2*hp -2*p_v(hp) -2*p_b,v(hp)
[1,] 603.3 853.66     692.63       692.95
       cAIC   mAIC   rAIC
[1,] 684.92 698.63 696.95
```

Ignoring important random components may lead to invalid conclusions

(Ha et al., 2007). For testing the need for a random component ($\alpha_1 = 0$ or $\alpha_2 = 0$), we again use the deviance, $-2p_{\beta,v}(h_p)$, and fit the following four models including Cox's model and three log-normal frailty models,

M1: Cox's model without frailty ($\alpha_1 = 0, \alpha_2 = 0$) has $-2p_{\beta,v}(h_p) =$ 707.48,

M2: model without patient effect ($\alpha_2 = 0, \alpha_1 > 0$) has $-2p_{\beta,v}(h_p) =$ 703.66,

M3: model without center effect ($\alpha_1 = 0, \alpha_2 > 0$) has $-2p_{\beta,v}(h_p) =$ 692.99, and

M4: multilevel model above requiring both patient and center effects ($\alpha_1 > 0, \alpha_2 > 0$) has $-2p_{\beta,v}(h_p) = 692.95$.

For fitting of M2 and M3, we use the HL(1,1) method. The deviance difference between M3 and M4 is 692.99 – 692.95 = 0.04 , which is not significant at a 5% level ($\chi^2_{0.10}(1) = 2.71$), indicating the absence of the random-center effects, i.e., $\alpha_1 = 0$. The deviance difference between M2 and M4 is 703.66-692.95=10.71, indicating that the random-patient effects are necessary, i.e., $\alpha_2 > 0$. In addition, the deviance difference between M1 and M3 is 707.48-692.99=14.49, indicating that the random-patient effects are indeed necessary with or without random-center effects. All of the three criteria (cAIC, mAIC and rAIC) also choose M3 among M1-M4 in the CGD dataset (not shown).

### *8.1.4 Log-normal and gamma frailty models on bladder cancer data*

Therneau and Lumley (2015) reported data on recurrences of bladder cancer, which were used to demonstrate methodology for recurrent event modeling (Wei et al., 1989). `Bladder2` in the survival package is the dataset from 85 subjects who were assigned to either thiotepa or placebo, and reports up to four recurrences for any patient. The start variable is start of interval (0 or previous recurrence time) in month and stop variable is the tumor recurrence or censoring time in month. The event variable is 1 for recurrence and 0 for everything else (including death for any reason). We fit frailty models with three covariates, the rx (1 = placebo; 2 = thiotepa), the number (initial number of tumors) and the size (cm of largest initial tumor) using `frailtyHL()`. The code and results for the log-normal frailty model using HL(1,1) are shown below. The output from the R code shows the effects of rx (t-value = −1.655 with p-value = 0.098) and number (t-value = 2.559 with p-value = 0.005). That is the thiotepa treatment has a marginally significant lower recurrent risk than in the placebo group controlling initial number of tumors. For testing the absence of frailty $\alpha = 0$, the difference in deviance $-2p_{\beta,v}(h)$ between Cox's PHM and log-normal frailty model is 1029.4−1024.1=5.3 (>2.71), indicating that the frailty effect is significant (p=0.011).

```
> library(survival)
> data(bladder)

> bladder_cox <- frailtyHL(Surv(stop-start,event)
+ ~rx+number+size+(1|id), bladder2,
+ varfixed=TRUE, varinit=c(0))

iteration :
          5
convergence :
  4.48187e-12
[1] "converged"
[1] "Results from the Cox model"
[1] "Number of data : "
[1] 178
[1] "Number of event : "
[1] 112
[1] "Model for conditional hazard : "
Surv(stop - start, event) ~ rx + number + size + (1 | id)
[1] "Method : HL(0,1)"
[1] "Estimates from the mean model"
       Estimate Std. Error t-value  p-value
rx     -0.36743    0.20263  -1.813 0.069784
number  0.15523    0.04904   3.166 0.001548
size   -0.02006    0.06799  -0.295 0.768017
[1] "Estimates from the dispersion model"
   Estimate Std. Error
id "0"       "NULL"
       -2h0  -2*hp -2*p_b,v(hp)
[1,] 1020.3 1020.3       1029.4
       cAIC   mAIC   rAIC
[1,] 1026.3 1026.3 1029.4

> bladder11 <- frailtyHL(Surv(stop-start,event)
+ ~rx+number+size+(1|id), bladder2,
+ RandDist="Normal",mord=1, dord=1)

iteration :
         84
```

```
convergence :
  8.858652e-07
[1] "converged"
[1] "Results from the log-normal frailty model"
[1] "Number of data : "
[1] 178
[1] "Number of event : "
[1] 112
[1] "Model for conditional hazard : "
Surv(stop - start, event) ~ rx + number + size + (1 | id)
[1] "Method : HL(1,1)"
[1] "Estimates from the mean model"
        Estimate Std. Error  t-value  p-value
rx     -0.454174    0.27440 -1.65514 0.097896
number  0.204903    0.07140  2.86984 0.004107
size   -0.004441    0.09143 -0.04857 0.961261
[1] "Estimates from the dispersion model"
   Estimate Std. Error
id   0.5058     0.2933
       -2h0  -2*hp -2*p_v(hp) -2*p_b,v(hp)
[1,] 950.02 1075.7     1016.8       1024.1
     cAIC   mAIC   rAIC
[1,] 1013 1024.8 1026.1
```

For the gamma frailty model using HL(1,2), the code and results are presented below. The output shows that these results are slightly different to those of the log-normal frailty, particularly for estimation of $\beta$. AICs indicate that log-normal or gamma frailty models are better than Cox's PHM, which gives evidence of existence of the frailty. Between log-normal and gamma frailty models, AICs indicate that the log-normal frailty model is better than the gamma frailty model.

```
> bladder_g12 <- frailtyHL(Surv(stop-start,event)
+ ~rx+number+size+(1|id),bladder2,
+ RandDist="Gamma", mord=1, dord=2)

iteration :
         143
convergence :
  9.518445e-07
[1] "converged"
[1] "Results from the gamma frailty model"
[1] "Number of data : "
[1] 178
```

```
[1] "Number of event : "
[1] 112
[1] "Model for conditional hazard : "
Surv(stop - start, event) ~ rx + number + size + (1 | id)
[1] "Method : HL(1,2)"
[1] "Estimates from the mean model"
        Estimate Std. Error  t-value  p-value
rx     -0.486833    0.29345 -1.65902 0.097112
number  0.221928    0.08105  2.73815 0.006179
size   -0.001715    0.09831 -0.01745 0.986079
[1] "Estimates from the dispersion model"
   Estimate Std. Error
id   0.6906     0.3876
      -2h0  -2*hp -2*p_v(hp) -2*s_v(hp)
[1,]  942.09 1105.3 1022.5     1015.9
      -2*p_b,v(hp) -2*s_b,v(hp)
         1029.2        1022.6
        cAIC   mAIC   rAIC
[1,] 1018.9 1023.9 1024.6
```

### 8.1.5 *Grouped duration model on smoke onset data*

Suppose that $T_i$ is the duration time until occurrence of event for the $i$th individual. However, the duration, $T_i$, is not observed exactly, but we have information that the event happened in a specific interval. As shown in Figure 8.2, the durations are observed at the $t$th time point $a_t$ $(t = 1, \cdots, r)$ with the initial time point being taken as zero, i.e., $a_0 = 0$. Conditioned on the $i$th individual having survived until the $(t-1)$th time point at $a_{t-1}$, the binary random variable $d_{it}$ was one if the $i$th individual experienced event during the $t$th time interval, and was zero otherwise. We considered the binary variable $d_{it}$ as the response variable with the corresponding $x_{it}$, which were observed at the $(t-1)$st time point $a_{t-1}$.

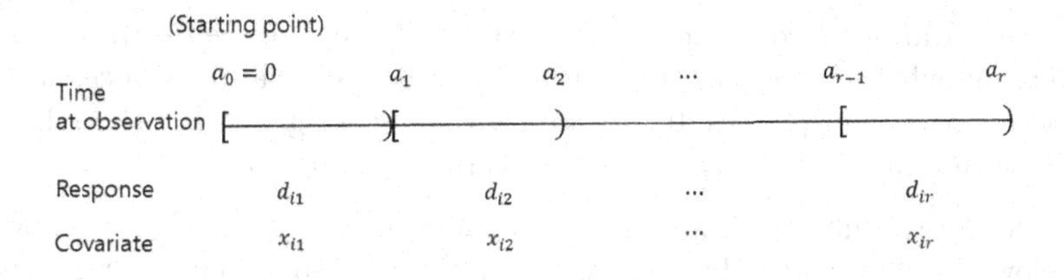


Figure 8.2 *Structure of grouped duration data.*

Suppose that $T_i$ follows the frailty model as below. Given the random effect $v_i$, the conditional hazard rate at time $T_i = u$ for $a_{t-1} \leq u < a_t$ with $t = 1, \cdots, r$ of the form

$$\lambda(u|v_i) = \lambda_0(u)\exp(x_{it}^T\beta + v_i)$$

where $\lambda_0(\cdot)$ is the baseline hazard function, $\beta$ are regression coefficients of covariates of interests, $x_{it}$ are risk factors observed over multiple time points $t = 1, \cdots, r$, and $v_i$ are frailties of individuals. Ha, Jeong, and Lee (2017) showed that the responses $d_{it}$ follow the Bernoulli HGLM with the complementary-log-log link

$$\log(-\log(1 - p_{it})) = \gamma_t + x_{it}^T\beta + v_i,$$

where $p_{it} = Pr(d_{it} = 1|v_i)$ and $\gamma_t = \log\int_{a_{t-1}}^{a_t}\lambda_0(u)du$.

As an example, we consider the SmokeOnset dataset in the `mixor` package. For 1,556 students in the Los Angels area, onset of smoking is observed at each of three timepoints $a_1$, $a_2$, and $a_3$ where $a_1$ is starting time for investigation, $a_2$ is the 1-year follow-up and $a_3$ is the 2-year follow-up. These event times are grouped at the three intervals $[0, a_1)$, $[a_1, a_2)$, and $[a_2, a_3)$.

In the dataset, we have the variable "smkonset", coded by i when the event occurs in the ith time interval and the variable "event", coded 1 if he/she smoked, 0 otherwise. For each student, we generate the following four responses :

i) $d_{i1} = 1$ if "smkonset"=1 and "event"=1, i.e., if he/she started smoking at intervals $[0, a_1)$,

ii) $(d_{i1}\ d_{i2}) = (0, 1)$ if "smkonset"=2 and "event"=1, i.e., if he/she started smoking at intervals $[a_1, a_2)$,

iii) $(d_{i1}, d_{i2}, d_{i3}) = (0\ 0\ 1)$ if "smkonset"=3 and "event"=1, i.e., if he/she started smoking at intervals $[a_2, a_3)$,

iv) $(d_{i1}\ d_{i2}\ d_{i3}) = (0\ 0\ 0)$ if "smkonset"=3 and "event"=0, i.e., if he/she had not smoked until $a_3$, this data are right-censored.

Three covariates are considered "SexMale" (1: Male, 0: Female), "cc" indicating whether the school was randomized to a social-resistance classroom curriculum (1: Yes, 0: No) and "tv" indicating whether the school was randomized to a media intervention (1: Yes, 0: No).

Codes and results for the Cox and log-normal frailty models are shown below. First, we test the necessity of frailty. Because the difference in deviance $-2p_{\beta,v}(h)$ between Cox's PHM and log-normal frailty model is $40189.8 - 40123.6 = 66.2 (> 2.71)$, the frailty is necessary. From the output below, male has higher risk for smoking than female and schools with cc or tv give lower risk for smoking to their students.

```
> library(mixor)
> data(SmokeOnSet)
# Fit Cox model
> smoke_cox <- frailtyHL(Surv(smokonset,event)
+ ~SexMale+cc+tv+(1|school), SmokeOnset,
+ grouped=TRUE,
+ varfixed=TRUE, varinit=c(0))

iteration :
          3
convergence :
   2.77905e-09
[1] "converged"
[1] "Results from the Cox model"
[1] "Number of data : "
[1] 1556
[1] "Number of event : "
[1] 634
[1] "Model for conditional hazard : "
frailtyHL(Surv(smokonset,event)~SexMale+cc+tv+(1|school)
[1] "Method : HL(1,1)"
[1] grouped : TRUE
[1] "Estimates from the mean model"
                      Estimate  Std. Error   z value P(>|z|)
(Intercept)        -206.013467    0.653472 -315.2613  <1e-16
gamma2             -315.45364     0.862413 -365.7807  <1e-16
gamme3                0.87654     0.007643  114.6853  <1e-16
SexMale              21.963457    0.315437   69.6287  <1e-16
cc                  -18.661435    0.050312 -370.9140  <1e-16
tv                   -6.347921    0.043137 -147.1573  <1e-16
[1] "Estimates from the dispersion model"
     Estimate Std. Error
school "0"      "NULL"

 "== Likelihood Function Values and Condition AIC =="
-2 log(likelihood)             : 39523.6
-2 log(restricted likelihhod) : 40189.8
cAIC                           : 37736.0

# Fit grouped duration model
> smoke_11 <- frailtyHL(Surv(smokonset,event)
```

```
+ ~SexMale+cc+tv+(1|school), SmokeOnset,
+ grouped=TRUE)

iteration :
          25
convergence :
   9.72315e-07
[1] "converged"
[1] "Results from the log-normal frailty model"
[1] "Number of data : "
[1] 1556
[1] "Number of event : "
[1] 634
[1] "Model for conditional hazard : "
frailtyHL(Surv(smokonset,event)~SexMale+cc+tv+(1|school)
[1] "Method : HL(1,1)"
[1] grouped : TRUE
[1] "Estimates from the mean model"
                      Estimate  Std. Error   z value P(>|z|)
(Intercept)        -276.105404    0.683146 -404.1674  <1e-16
gamma2             -327.818555    1.000000 -327.8186  <1e-16
gamme3                1.523108    0.000000       Inf  <1e-16
SexMale              25.632504    0.337942   75.8489  <1e-16
cc                  -20.002003    0.060507 -330.5743  <1e-16
tv                   -7.116128    0.060361 -117.8932  <1e-16

[1] "Estimates from the dispersion model"
   Estimate Std. Error
school   0.1645    0.8112

 "== Likelihood Function Values and Condition AIC =="
-2 log(likelihood)             : 39457.1
-2 log(restricted likelihhod) : 40123.6
cAIC                           : 37563.0
```

## 8.2 Competing risk models

In various fields including clinical trials and reliability tests, competing risks data commonly arise when an occurrence of an event precludes other types of events from being observed.

**Box 8: Competing risk models**

Suppose there are $i = 1, \cdots, q$ clusters where each cluster has $j = 1, \cdots, n_i$ observations. For the $j$th observation in the $i$th cluster, let $T_{ijk}, k = 1, \ldots, K$, denote time to type $k$ event and let $C_{ij}$ denote independent censoring time. Then, the observed event time would be $y_{ij} = \min(T_{ij1}, T_{ij2}, \ldots, T_{ijK}, C_{ij})$ and define the event indicator $\delta_{ijk} = I(y_{ij} = T_{ijk})$. The cause-specific hazard function conditional on the log-frailty $v_i = (v_{i1}, v_{i2}, \ldots, v_{iK})$ for the $j$th observation in cluster $i$ who failed from cause $k$ ($k = 1 \cdots K$) is

$$\lambda_{ijk}(t|v_i) = \lambda_{0k}(t) \exp(x_{ij}^T \beta_k + v_{ik}), \tag{8.3}$$

where $\lambda_{0k}(t)$ is the unspecified baseline hazard function for event type $k$, $\beta_k = (\beta_{k1}, \beta_{k2}, \ldots, \beta_{kp})^T$ is a $p \times 1$ vector of fixed parameters for event type $k$ and fixed covariates $x_{ij}$ and $v_{ik}$ ($k = 1, \cdots, K$) are the random effect for type $k$ event in cluster $i$.

Consider $K = 2$ in the model (8.3). Specifically, event times from cause 1 would follow a cause-specific proportional hazards model

$$\lambda_{ij1}(t|v_i) = \lambda_{01}(t) \exp(x_{ij}^T \beta_1 + v_{i1}),$$

and event times from cause 2 would follow similarly a model

$$\lambda_{ij2}(t|v_i) = \lambda_{02}(t) \exp(x_{ij}^T \beta_2 + v_{i2}),$$

where $v_{i1}$ and $v_{i2}$ might be correlated. In the traditional cause-specific analysis, patients who failed from cause 2 are treated as censored for the analysis of type 1 events, which ignores a potential correlation between $v_{i1}$ and $v_{i2}$.

The competing risk models can be fitted by using the `jmfit` function in the jointdhglm package. For the model (8.3) with $K = 2$, Christian et al. (2016) and Xue and Brookmeyer (1996) considered a variation of the bivariate frailty model to allow for different, but correlated random effects $v_{i1}$ and $v_{i2}$. Huang and Wolfe (2002) considered a joint model with $v_{i1} = v_i$ and $v_{i2} = \gamma v_i$ in the models above, where $\gamma$ is a real-valued correlation parameter. If $\gamma > 0$ [$\gamma < 0$], both event rates are positively [negatively] associated; a cluster with higher frailty in type 1 event will experience an earlier [delayed] type 2 event, respectively. When $\gamma = 1$, the effect of the frailty is identical for both events. When $\gamma = 0$, type 2 event rate $\lambda_{ij2}(t|v_i)$ does not depend on $v_i$ and is noninformative for the type 1 event rate $\lambda_{ij1}(t|v_i)$, that is the two rates are not associated.

### *8.2.1 Competing risks model on simulated data*

Competing risks data usually arise when an occurrence of a competing event prevents the occurrence of the event of interest. Treating the competing event as a censoring can lead to biased results (Pepe and Mori, 1993). For fitting competing-risks frailty model, we used a simulated dataset, `cdata`, which was generated in the R package crrSC (Zhou et al., 2012, 2015). The dataset consists of a data frame with 200 observations and 4 variables. The ID is the identification number of 100 clusters with each cluster size 2. The $x$ is a binary covariate generated with probability of 0.5. The `time` is event time with status, `status`. Here, the `status` is coded as "1" for the event of interest (112 observations), "2" for the 2 competing events (47 observations), and "0" for censoring (41 observations).

Consider the cause-specific hazard frailty model as shown in Figure 8.3 (Ha, Jeong, and Lee, 2017). Given a shared log-frailty $v_i$, the conditional hazard function $\lambda_{ijk}$ for the $j$th observation in the $i$th cluster that failed from cause $k$ ($k = 1, 2$) can be expressed as

$$\begin{aligned} \lambda_{ij1}(t_i|v_i) &= \lambda_{01}(t)\exp(\beta_1 x_{ij} + v_i), \\ \lambda_{ij2}(t_i|v_i) &= \lambda_{02}(t)\exp(\beta_2 x_{ij} + \gamma v_i), \end{aligned}$$

where $\lambda_{0k}(t)$ is an unspecified baseline hazard function for cause $k$, $\beta_k$ is regression parameters for cause $k$ and $v_i \sim \mathrm{N}(0, \sigma^2)$. If $\gamma > 0$ [$\gamma < 0$], a cluster with higher frailty in a Type 1 event will experience an earlier [delayed] Type 2 event, respectively (Huang and Wolfe, 2002).

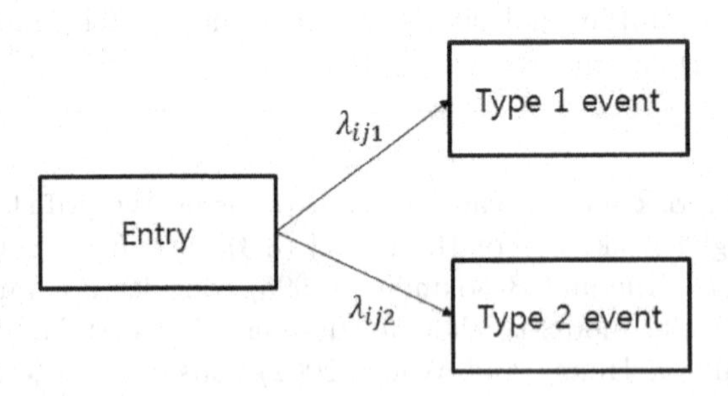


Figure 8.3 *Path diagram for the competing risk frailty model.*

We may view the competing risk model as k-variate frailty models. We developed the jointdhglm package to fit such models. The output from the R code shows that the effect of $x$ is not significant for type 1 events (t-value = 0.678 with p-value = 0.498), but it is positively significant for Type 2 events (t-value = 2.371 with p-value = 0.018). The estimate of shared parameter $\widehat{\gamma} = -1.218$ shows a negative association between two events.

```
> library(crrSC)
> data(cdata,package="crrSC")
> cdata$x<-cdata$z
> cdata$time<-cdata$ftime
> cdata$status<-cdata$fstatus
> jm1<-jointmodeling(Model="mean",
+ RespDist="FM",Link="log",
+ LinPred=Surv(time,status==1)~x+(1|ID),
+ RandDist="gaussian")
> jm2<-jointmodeling(Model="mean",
+ RespDist="FM",Link="log",
+ LinPred=Surv(time,status==2)~x+(1|ID),
+ RandDist="gaussian")
> data_surv<-cdata
> res<-jmfit2(jm1,jm2,data_surv)

    "Estimates from the mean model"
[1] LinPred=Surv(ftime,fstatus==1)~x+(1|ID)
       Estimate Std. Error t-value p-value
z       0.13244  0.1952     0.6783  0.49756
[1] LinPred=Surv(ftime,fstatus==2)~x+(1|ID)
       Estimate Std. Error t-value p-value
z       0.74953  0.3161     2.3711  0.0177

[1] "Estimates for logarithm of variance of random effect"
      Estimate
ID     0.2193
[1] "Estimates for the shared parameter"
        Estimate
gamma   -1.2177
```

## 8.3 Comparison with alternative R procedures

Here, we show results from three functions (coxph, coxme and phmm) for the log-normal frailty model, and the coxph function for the gamma frailty model. Below we present the codes and compare the results for some of the above examples.

### *8.3.1 Rat data*

The codes of `coxph`, `coxme` and `phmm` for fitting log-normal frailty model are, respectively, as follows:

```
>coxph(Surv(time, status) ~ rx +
+ frailty(litter,dist= gauss),
+ method = "breslow", rats)

> coxme(Surv(time, status) ~ rx + (1|litter),
+ ties="breslow", rats)

> fit.phmm<-phmm(Surv(time, status) ~ rx+(1| litter),
+ rats, Gbs = 2000, Gbsvar = 3000, VARSTART = 1,
+ NINIT = 10, MAXSTEP = 200, CONVERG=90)
```

Table 8.2 summarizes the estimation results. As expected from a somewhat larger cluster size $n_i \equiv 3$, the results are similar. Here we find that both ML procedures, `coxme` with Efron method and `phmm`, give similar estimate for $\alpha$.

Next, the code of `coxph` for fitting gamma frailty model is:

```
> coxph(Surv(time, status) ~ rx +
+ frailty(litter,dist=gamma),
+ method = "breslow", rats)
```

The results of `frailtyHL`(HL(0,2), HL(1,2)) and `coxph` for gamma frailty are also presented in Table 8.2. For the estimation of $\beta$ both results are similar, but for $\alpha$ they are somewhat different. That is, our REML estimate ($\hat{\alpha} = 0.576$) using HL(1,2) is somewhat larger than the PPL estimates ($\hat{\alpha} = 0.474$ with Breslow method and $\hat{\alpha} = 0.499$ with Efron method) using the `coxph` function which gives ML estimate.

### *8.3.2 CGD infection data*

The code of the `coxme` function for the fitting multilevel log-normal frailty model is:

```
> coxme(Surv(tstop-tstart,status)~
+ treat+(1|center)+(1|id), ties="breslow" cgd)
```

The results of `frailtyHL`(HL(0,1), HL(1,1)) and `coxme` are summarized in Table 8.3. Again, the results of HL and PPL methods are similar because the cluster sizes (the number of patients from different centers) are somewhat large, ranging from 4 to 26.

Table 8.2 *Comparison of different estimation methods for the rat data*

| Method | Rx<br>$\hat{\beta}$ (SE) | Litter<br>$\hat{\alpha}$ (SE) |
|---|---|---|
| log-normal model | | |
| HL(0,1) | 0.906 (0.323) | 0.427 (0.423) |
| HL(1,1) | 0.911 (0.323) | 0.427 (0.423) |
| coxph (Breslow) | 0.905 (0.322) | 0.395 ( – ) |
| coxph (Efron) | 0.913 (0.323) | 0.412 ( – ) |
| coxme (Breslow) | 0.905 (0.322) | 0.406 ( – ) |
| coxme (Efron) | 0.913 (0.323) | 0.426 ( – ) |
| phmm | 0.920 (0.326) | 0.449 ( – ) |
| gamma model | | |
| HL(0,2) | 0.908 (0.324) | 0.575 (0.598) |
| HL(1,2) | 0.913 (0.324) | 0.576 (0.598) |
| coxph (Breslow) | 0.906 (0.323) | 0.474 ( – ) |
| coxph (Efron) | 0.914 (0.323) | 0.499 ( – ) |

### *8.3.3 Bladder cancer data*

The codes of coxph, coxme and phmm for the fitting log-normal frailty model are, respectively, as follows:

```
> coxph(Surv(stop-start, event) ~ rx+number+size
+ +frailty(id,dist=gauss),
+ method = "breslow", bladder2)

> coxme(Surv(stop-start, event) ~rx+number+size
+ +(1|id),ties="breslow", bladder2)

> fit.phmm<-phmm(Surv(stop-start, event)
+ ~ rx+number+size+(1|id),
```

Table 8.3 *Comparison of different estimation methods for the CGD data*

| Method | Treat $\hat{\beta}$ (SE) | Center $\hat{\alpha}_1$ (SE) | Patient $\hat{\alpha}_2$ (SE) |
|---|---|---|---|
| log-normal model | | | |
| HL(0,1) | −1.074 | 0.026 | 0.982 |
| | (0.335) | (0.153) | (0.501) |
| HL(1,1) | −1.184 | 0.030 | 1.002 |
| | (0.341) | (0.157) | (0.509) |
| `coxme` | −1.074 | 0.033 | 0.939 |
| (Breslow) | (0.333) | ( − ) | ( − ) |
| `coxme` | −1.074 | 0.032 | 0.947 |
| (Efron) | (0.333) | ( − ) | ( − ) |

```
+ bladder2, Gbs = 2000, Gbsvar = 3000, VARSTART = 1,
+ NINIT = 10, MAXSTEP = 200, CONVERG=90)
```

Table 8.4 summarizes the estimation results. The HL(1,1) estimates are close to those of the MCEM method using `phmm`. However, other approximate methods using `coxph` and `coxme` give smaller estimates for $\beta$ and $\alpha$. HL(1,1) is the best approximate method for marginal likelihood.

Next, the code of `coxph` for fitting the gamma frailty model is:

```
> coxph(Surv(stop-start, event) ~ rx+number+size
+ +frailty(id,dist=gamma),
+ method = "breslow", bladder2)
```

The results of `frailtyHL`(HL(0,2), HL(1,2)) and `coxph` for gamma frailty are also presented in Table 8.4. For $\alpha$, our REML estimate ($\hat{\alpha} = 0.69$) using HL(1,2) is somewhat larger than the PPL estimates ($\hat{\alpha} = 0.45$ with Breslow method and $\hat{\alpha} = 0.64$ with Efron method) using the `coxph` function which gives ML estimate. These differences affect the estimates of $\beta$.

## 8.4 H-likelihood theory for the frailty model

Here, we explain the fitting of the frailty models and interpretation of the results of the analyses. Furthermore, our results are compared with those using different R packages such as survival, coxme and phmm. We

Table 8.4 *Comparison of different estimation methods for the bladder cancer data*

| Method | rx<br>$\hat{\beta}$ (SE) | number<br>$\hat{\beta}$ (SE) | size<br>$\hat{\beta}$ (SE) | id<br>$\hat{\alpha}$ (SE) |
|---|---|---|---|---|
| log-normal model | | | | |
| HL(0,1) | −0.43(0.27) | 0.20(0.07) | −0.003(0.09) | 0.49(0.29) |
| HL(1,1) | −0.45(0.27) | 0.20(0.07) | −0.004(0.09) | 0.51(0.29) |
| `coxph` (Breslow) | −0.42(0.26) | 0.19(0.07) | −0.005(0.09) | 0.40( − ) |
| `coxph` (Efron) | −0.44(0.28) | 0.21(0.07) | 0.0007(0.09) | 0.57( − ) |
| `coxme` (Breslow) | −0.42(0.26) | 0.18(0.07) | −0.006(0.09) | 0.38( − ) |
| `coxme` (Efron) | −0.44(0.28) | 0.21(0.07) | 0.001(0.09) | 0.58( − ) |
| `phmm` | −0.48(0.31) | 0.22(0.08) | −0.0004(0.14) | 0.68( − ) |
| gamma model | | | | |
| HL(0,2) | −0.45(0.29) | 0.21(0.08) | −0.0003(0.10) | 0.67(0.39) |
| HL(1,2) | −0.49(0.29) | 0.22(0.08) | −0.002(0.10) | 0.69(0.39) |
| `coxph` (Breslow) | −0.43(0.27) | 0.20(0.07) | −0.005(0.09) | 0.45( − ) |
| `coxph` (Efron) | −0.45(0.29) | 0.22(0.08) | 0.002(0.10) | 0.64( − ) |

use the conventional notation in survival analysis, where $\lambda()$ denotes the hazard function.

The model represented by (8.1) is known as a shared or one-component frailty model (Hougaard, 2000). The implementation for these models in `frailtyHL()` is given in the examples below. The model (8.2) is called the multilevel frailty model (Yau, 2001; Ha et al., 2007). In the CGD infection data example below, we show application of `frailtyHL()` fitting multi-level models.

The h-likelihood gives a straightforward way of handling non-parametric baseline hazards. For frailty model (8.1), the h-likelihood is defined by

$$h = h(\boldsymbol{\beta}, \lambda_0, \alpha) = \ell_0 + \ell_1, \tag{8.4}$$

where $\ell_0 = \sum_{ij} \log f(y_{ij}, \delta_{ij}|u_i; \boldsymbol{\beta}, \lambda_0) = \delta_{ij}\{\log \lambda_0(y_{ij}) + \eta_{ij}\}$

$-\sum_{ij} \Lambda_0(y_{ij}) \exp(\eta_{ij})$ is the sum of conditional log densities for $y_{ij}$ and $\delta_{ij}$ given $u_i$, and $\ell_1 = \sum_i \log f(v_i; \alpha)$ is the sum of log densities for

$v_i = \log u_i$ with parameter $\alpha$. Here, $\eta_{ij} = x_{ij}^T\boldsymbol{\beta} + v_i$ is the linear predictor for the hazards, and $\Lambda_0(t) = \int_0^t \lambda_0(k)dk$ is the baseline cumulative hazard function.

The functional form of $\lambda_0(t)$ in (8.1) is unknown; hence, following Breslow (1972), we consider $\Lambda_0(t)$ to be a step function with jumps at the observed event times:

$$\Lambda_0(t) = \sum_{k:y_{(k)}\leq t} \lambda_{0k}$$

where $y_{(k)}$ is the $k$th ($k = 1, \ldots, l$) smallest distinct event time among the $y_{ij}$'s, and $\lambda_{0k} = \lambda_0(y_{(k)})$. However, the dimension of $\lambda_0 = (\lambda_{01}, \ldots, \lambda_{0l})^T$ increases with the sample size $n$. For inference, Ha, Lee, and Song (2001) proposed the use of the profile h-likelihood with $\lambda_0$ eliminated, $r^* \equiv h|_{\lambda_0=\widehat{\lambda}_0}$, given by

$$r^* = r^*(\boldsymbol{\beta}, \alpha) = \ell_0^* + \ell_1, \tag{8.5}$$

where $\ell_0^* = \sum_{ij} \log f^*(y_{ij}, \delta_{ij}|u_i; \boldsymbol{\beta}) = \sum_{ij} f(y_{ij}, \delta_{ij}|u_i; \boldsymbol{\beta}, \widehat{\lambda}_0)$ does not depend on $\lambda_0$, and

$$\widehat{\lambda}_{0k}(\boldsymbol{\beta}, \boldsymbol{v}) = \frac{d_{(k)}}{\sum_{(i,j)\in R_{(k)}} \exp(\eta_{ij})},$$

are solutions of the estimating equations, $\partial h/\partial\lambda_{0k} = 0$, for $k = 1, \ldots, l$. Here, $d_{(k)}$ is the number of events at $y_{(k)}$ and $R_{(k)} = R(y_{(k)}) = \{(i, j) : y_{ij} \geq y_{(k)}\}$ is the risk set at $y_{(k)}$. Therneau and Grambsch (2000) and Ripatti and Palmgren (2000) proposed an h-likelihood (8.4), while using the partial likelihood (Breslow, 1974) for $\ell_0$. They call it the penalized partial likelihood (PPL) $h_p$, defined by

$$h_p(\boldsymbol{\beta}, \boldsymbol{v}, \alpha) = \sum_{ij} \delta_{ij}\eta_{ij} - \sum_k d_{(k)} \log\Big\{ \sum_{ij\in R_{(k)}} \exp(\eta_{ij})\Big\} + \ell_1.$$

Furthermore, Ha, Lee, and Song (2001) and Ha et al. (2010) have shown that $r^*$ is proportional to the PPL $h_p$,

$$\begin{aligned} r^* &= \sum_k d_{(k)} \log \widehat{\lambda}_{0k} + \sum_{ij} \delta_{ij}\eta_{ij} - \sum_k d_{(k)} + \ell_1 \\ &= h_p + \sum_k d_{(k)}\{\log d_{(k)} - 1\}, \end{aligned}$$

where $\sum_k d_{(k)}\{\log d_{(k)} - 1\}$ is a constant which does not depend upon unknown parameters. Notice here that PPL $h_p$ does not depend on nuisance parameters $\lambda_0$. Thus, the h-likelihood procedure for HGLMs of Lee and Nelder (1996, 2001a) can be extended to frailty models based on $h_p$ (Ha et al., 2010).

Table 8.5 *Estimation criteria for h-likelihood (HL(`mord`, `dord`)), PPL (`coxph`, `coxme`) and marginal likelihood (`phmm`) for log-normal (LN) and gamma frailty models (FMs)*

| | Criterion | | Literature |
|---|---|---|---|
| Method | $\boldsymbol{\beta}$ | $\alpha$ | |
| HL(0,1) | $h_p$ | $p_{\boldsymbol{\beta},\boldsymbol{v}}(h_p)$ | Ha and Lee (2003) |
| HL(0,2) | $h_p$ | $s_{\boldsymbol{\beta},\boldsymbol{v}}(h_p)$ | Ha and Lee (2003) |
| HL(1,1) | $p_{\boldsymbol{v}}(h_p)$ | $p_{\boldsymbol{\beta},\boldsymbol{v}}(h_p)$ | Ha et al. (2011) |
| HL(1,2) | $p_{\boldsymbol{v}}(h_p)$ | $s_{\boldsymbol{\beta},\boldsymbol{v}}(h_p)$ | Ha et al. (2011) |
| `coxph` | $h_p$ | $p_{\boldsymbol{\beta},\boldsymbol{v}}(h_p)$ | Therneau and Lumley (2015) for LN FM |
| `coxph` | $h_p$ | $m$ | Therneau and Lumley (2015) for gamma FM |
| `coxme` | $h_p$ | $p_{\boldsymbol{v}}(h_p)$ | Therneau (2015) for LN FM |
| `phmm` | $m$ | $m$ | Donohue and Xu (2013) for LN FM |

### *8.4.1 Review of estimation procedures*

Lee and Nelder (1996, 2001a) have proposed the use of the Laplace approximation based on the h-likelihood when the marginal likelihood, $m = \log\{\int \exp(h) d\boldsymbol{v}\}$, is hard to obtain. Historical developments for frailty models are listed in Table 8.5. To reduce the bias in small cluster sizes, higher-order approximations for the mean ($\boldsymbol{\beta}$) and the frailty's variance ($\alpha$) have been developed. The lower-order approximation is computationally efficient, but could have large biases when cluster sizes are small (Ha and Lee, 2003, 2005; Ha et al., 2010).

The h-likelihood procedures use the Breslow method for handling tied event times, while the PPL procedures allow the Efron method. For estimating $\boldsymbol{\beta}$, the h-likelihood methods allow the Laplace approximation

$p_v(h_p)$, but the PPL procedures do not. For estimating $\alpha$, the PPL methods use adjusted profile h-likelihoods $p_{\boldsymbol{v}}(h_p)$ and $p_{\boldsymbol{\beta},\boldsymbol{v}}(h_p)$ which give approximate maximum likelihood (ML) and restricted maximum likelihood (REML) estimators, respectively, whereas the h-likelihood method uses the restricted likelihood (based upon the first-order Laplace approximation $p_{\boldsymbol{\beta},\boldsymbol{v}}(h_p)$ or the second-order Laplace approximation $s_{\boldsymbol{\beta},\boldsymbol{v}}(h_p)$) for REML estimators.

In particular, for the log-normal frailty the PPL method (coxph) uses the existing codes in linear mixed models so that it ignores the $\partial\hat{v}/\partial\alpha$ term in solving the score equation $\partial p_{\boldsymbol{\beta},\boldsymbol{v}}(h_p)/\partial\alpha = 0$; this can lead to an underestimation of the parameters, especially when the number of subjects $q$ is large or censoring is high (Lee et al., 2017; Ha et al., 2010). Moreover, for gamma frailty models $p_{\boldsymbol{v}}(h_p)$ often gives non-negligible bias, so that Therneau and Grambsch (2000) recommended the use of the marginal likelihood $m$ for $\alpha$ (Ha et al., 2010).

The PPL method is implemented using the `coxph` or `coxme` function for the log-normal frailty model, and the `coxph` function for the gamma frailty model. Here, for the log-normal frailty the `coxph` and `coxme` functions provide the REML and ML estimators, respectively, and for the gamma frailty the `coxph` function gives an ML estimator. For comparison, we present the Breslow and Efron methods for handling ties in survival times for the `coxph` and `coxme` functions; Therneau and Lumley (2015) recommended the Efron method. For the log-normal frailty the ML estimator using $m$ is obtained via the `phmm` function, but care must be taken to ensure the MCEM algorithm has converged (Donohue and Xu, 2013). However, the ML estimator can be biased when the number of nuisance parameters increases with the sample size (Ha et al., 2010).

The `frailtyHL()` function provides estimators from the four HL estimation criteria summarized in Table 8.5. If the orders in `mord` and `dord` are increased, the bias of estimator is reduced, but computationally intensive due to the computation of extra terms. We recommend the use of HL(1,1) for log-normal frailty and that of HL(1,2) for gamma frailty. However, for log-normal frailty HL(0,1) and for gamma frailty HL(0,2) often perform well if $\alpha$ is not large. Note that the variance matrices of $\hat{\boldsymbol{\tau}} = (\hat{\boldsymbol{\beta}}, \hat{\boldsymbol{v}})$ and $\hat{\alpha}$ are directly obtained from $\{-\partial^2 h_p/\partial\boldsymbol{\tau}^2\}^{-1}$ and $\{-\partial^2 p_{\boldsymbol{\beta},\boldsymbol{v}}(h_p)/\partial\alpha^2\}^{-1}$, respectively; the `frailtyHL()` package provides the standard errors (SEs) of $\hat{\boldsymbol{\beta}}$ as well as $\hat{\alpha}$. For the use of standard errors of $\hat{\boldsymbol{v}} - \boldsymbol{v}$, see Lee and Ha (2010), Lee et al. (2011c) and Ha et al. (2011).

*8.4.2 Fitting algorithm*

Suppose that HL(0,1) is used. The fitting algorithm is as follows:

Step 1: Obtain initial estimates ($\hat{\boldsymbol{\beta}} = 0, \hat{\boldsymbol{v}} = 0$) and $\hat{\alpha} = 0.1$ of $(\boldsymbol{\beta}, \boldsymbol{v})$ and $\alpha$.

Step 2: Given $\hat{\alpha}$, new estimates $(\hat{\boldsymbol{\beta}}, \hat{\boldsymbol{v}})$ are obtained by solving the joint estimating equations $\partial h_p/\partial(\boldsymbol{\beta}, \boldsymbol{v}) = \{\partial h/\partial(\boldsymbol{\beta}, \boldsymbol{v})\}|_{\lambda_0=\hat{\lambda}_0} = 0$; then, given $(\hat{\boldsymbol{\beta}}, \hat{\boldsymbol{v}})$, new estimates $\hat{\alpha}$ are obtained by solving $\partial p_{\boldsymbol{\beta},\boldsymbol{v}}(h_p)/\partial\alpha = 0$.

Step 3: Repeat Step 2 until the maximum absolute difference between the previous and current estimates for $(\boldsymbol{\beta}, \boldsymbol{v})$ and $\alpha$ is less than $10^{-6}$.

After convergence, we compute the estimates of the standard errors of $\hat{\boldsymbol{\beta}}$ and $\hat{\alpha}$.

## 8.5 Running the frailtyHL package

The main function is `frailtyHL()`. For instance,

```
> result<-frailtyHL(formula=Surv(time,status)~x+(1|id),
+ RandDist="Normal", mord=0, dord=1,
+ Maxiter=200, convergence=10^-6,
+ varfixed=FALSE, varinit=c(0.1))
```

fits a log-normal frailty model. Here `formula` is an R formula object, with the response on the left of a $\sim$ operator, and the terms for the fixed and random effects on the right. The response is a survival object as returned by the `Surv` function (Therneau and Lumley, 2015). Here, `time` and `status` denote survival time and censoring indicator having 1 (or 0) for uncensored (or censored) observation, respectively, where `x` denotes a fixed covariate and `id` denotes the subjects for a normally distributed log frailty. The expression `(1|id)` (or `(x|id)`) specifies random intercept (or random slope) model. The `mord` and `dord` are the orders of Laplace approximations to fit the mean parameters (`mord`=0 or 1) and the dispersion parameters (`dord`=1 or 2), respectively. The `Maxiter` specifies the maximum number of iterations and the `convergence` specifies the convergence criterion. If `varfixed` is specified as TRUE (or FALSE), the value of one or more of the variance terms for the frailties is fixed (or estimated) with starting value (e.g., 0.1) given in the `varinit`.

## 8.6 Exercises

1. The colon data are one of the first successful trials of adjuvant chemotherapy for colon cancer (Therneau and Lumley, 2015). Levamisole is a low-toxicity compound previously used to treat worm infestations in animals; 5-FU is a moderately toxic (as these things go) chemotherapy agent. There are two records per person, one for recurrence and one for death. Fit a frailty model by using `frailtyHL`, `coxph`, `coxme` and `phmm` and compare their results.

```
data(colon, package="survival")
# id : id
# study : 1 for all patients
# rx : 1=Treatment
# sex : 1=male
# age : in years
# obstruct : obstruction of colon by tumor
# perfor : perforation of colon
# adhere : adherence to nearby organs
# nodes : number of lymph nodes with detectable cancer
# time : days until event or censoring
# status : censoring status
# differ : differentiation of tumor
# (1=well, 2=moderate, 3=poor)
# extent : Extent of local spread (1=submucosa, 2=muscle,
# 3=serosa, 4=contiguous structures)
# surg : time from surgery to registration (0=short, 1=long)
# node4 : more than 4 positive lymph nodes
# etype : event type (1=recurrence,2=death)
```

In the `coxme` package (Therneau, 2015), we can obtain dataset which is a simulated survival dataset for investigating random center effects. To make it realistic, the number of centers and their sizes are based on an EORTC cancer trial. This dataset has 2323 observations and the following 4 variables. Fit frailty model by using `frailtyHL`, `coxph`, `coxme` and `phmm` and compare their results.

```
data(eortc, package="coxme")
# y : survival time
# uncens : 0=alive, 1=dead
# center : enrolling center, a number from 1 to 37
# trt : treatment arm, 0 or 1
```

2. The colorectal dataset in the (Gonzalez et al., 2012) packages considered randomly chosen 150 patients from the original 410 patients with metastatic colorectal cancer randomized into two therapeutic strategies: combination and sequential. This dataset contains times of observed appearances of new lesions censored by a terminal event (death or right-censoring) with baseline characteristics (treatment arm, age, WHO performance status and previous resection). Fit frailty model by using `frailtyHL`, `coxph`, `coxme` and `phmm` and compare their results.

```
data(colorectal,package="frailtypack")
# id : identification of each subject
# time0 : start of interval
# (0 or previous recurrence time)
# time1 : recurrence or censoring time
# new.lesions : Appearance of new lesions status
#  (0=censored or no event, 1=new lesions)
# treatment : 1=sequential (S), 2=combination (C)
# age : Age at baseline
# (1: <50 years, 2: 50-69 years, 3: >69 years)
# who.PS : WHO performance status at baseline
# (1=status 0, 2=status 1, 3=status 2)
# prev.resection : Previous resection of the primate tumor?
#  (0=No, 1=Yes)
# state : death indicator (0=alive, 1=dead)
# gap.time : interocurrence time or censoring time
```

3. The okiss dataset in the R packages compeir has a random sub-sample of 1,000 patients from ONKO-KISS which is the part of the surveillance program of the German National Reference Centre for Surveillance of Hospital-Acquired Infections (Grambauer and Neudecker, 2011). Patients have been treated by peripheral blood stem-cell transplantation. After transplantation, patients are neutropenic. Occurrence of bloodstream infection during neutropenia is a severe complication. Analyze the impact of covariates (allo, sex) on the cause-specific hazards of event types (infection, end of neutropenia and death).

```
data(okiss,package="compeir")
# time : Time of neutropenia until first event in days
# status : Event status indicator.
# (1: infection, 2: end of neutropenia, 7: death,
#   11: censored observation)
# allo : Covariate transplant type indicator
# (0: autologous transplants, 1: allogeneic transplants)
```

```
# sex : Covariate sex indicator
# (f: female, m: male)
```

CHAPTER 9

# Joint models

In clinical studies, different types of outcomes (multivariate responses) can be observed from the same subject as we studied in Chapter 7. In this chapter, we consider data analysis for multivariate responses where at least one response is time-to-event. For example, we may have bivariate responses, where one outcome is repeated-measure response and the other outcome is time-to-event. Then, these outcomes are correlated because they are observed from the same subject. Joint modeling has been widely studied (Henderson et al., 2000; Ha et al., 2003; Rizopoulos, 2012) because a separated analysis, ignoring the inherent association between the outcomes from the subject, can lead to a biased result (Guo and Carlin, 2004). An unobserved random effect can be used to account for the association among multivariate outcomes. For the analysis of such dataset, the h-likelihood approach is very effective.

The R package jointdhglm supports a joint modeling framework based on the h-likelihood for analyzing jointly the repeated-measures data and event-time data including competing- risks events. The main function in the R package jointdhglm is `jmfit()`. For instance, the following code runs a joint model with Gaussian and event-time responses.

```
> jm1<-jointmodeling(Model="mean",
+   RespDist="gaussian",
+   Link="identity", LinPred=y1~x1+(1|id),
+   RandDist="gaussian")
> jm2<-jointmodeling(Model="mean",RespDist="FM",
+   Link="log",
+   LinPred=Surv(time,status)~x2+(1|id),
+   RandDist="gaussian")

> res<-jmfit(jm1,jm2,data1,data2)
+   Maxiter = 200, convergence = 10^-6)
```

The models for repeated-measure response and time-to-event `time` with censoring indicator `status` are defined by the `jointmodeling()` function by specifying their link functions, linear predictors, and distributions of random effects. The function `jmfit` runs joint models with two `jointmodeling` objects.

In the above example, the first response, `y1`, is specified to follow a Gaussian random-effect model with a fixed covariate `x1` and the subject identifier `id` specifying a random-effect model. The second time-to-event, `time` with censoring indicator `status`, is specified to follow a frailty model with a fixed covariate `x2` and the subject identifier `id` specifying a random-effect model. The `jmfit` function can fit the joint model with repeated-measure response and time-to-event. The `Maxiter` parameter specifies the maximum number of iterations and `convergence` specifies the tolerance of the convergence criterion.

## 9.1 Examples

### *9.1.1 Joint model for serum creatinine data*

In clinical trials, various response variables of interest are measured repeatedly over time on the same subject, which can be analyzed by using the mdhglm package. At the same time, an event time representing recurrent or terminating time is also obtained. For example, consider a clinical study to investigate the chronic renal allograft dysfunction in renal transplants (Sung et al., 1998). The renal function was evaluated from the serum creatinine (sCr) values. Since the time interval between the consecutive measurements differs from patient to patient, we focus on the mean creatinine levels over six months. In addition, a single terminating survival time (graft-loss time), measured by month, is observed from each patient. During the study period, there were 13 graft losses due to the kidney dysfunction. For other remaining patients, we assumed that the censoring occurred at the last follow-up time. Thus, the censoring rate is about 88%.

For each patient, sCr values and a single loss time are observed and we are interested in investigating the effects of covariates over these two types of responses. Ha et al. (2003) considered month, sex (=1 for male, =0 for female) and age as covariates for sCr, and sex and age for the loss time. They showed that the reciprocal of sCr levels tends to decrease linearly over time, having possibly constant variance. Thus, for the standard mixed linear model we use values 1/sCr as responses $y_{ij}$.

Consider a shared random effect as shown in Figure 9.1. For the 1/sCr values consider a linear mixed model

$$y_{ij} = x_{1ij}^t \beta + v_{1i} + e_{ij}, \tag{9.1}$$

where $x_{1ij}^t$ are covariates, $v_{1i} \sim \mathrm{N}(0, \sigma_{v1}^2)$ and $e_{ij} \sim \mathrm{N}(0, \sigma_e^2)$. For graft-loss time $t_i$, consider a frailty model with the conditional hazard function

$$\lambda(t_i|v_{1i}) = \lambda_0(t_i) \exp(x_{2i}^t \delta + \gamma v_{1i}),$$

where $\lambda_0(t)$ is the baseline hazard function, $x_{2i}^t$ are between-subject covariates and $\gamma$ is the shared parameter. Ha et al. (2003) considered a Weibull model for the baseline hazard function where $\lambda_0(t_i) = \tau t^{\tau-1}$ with a shape parameter $\tau$. Now we can fit the non-parametric baseline hazard model.

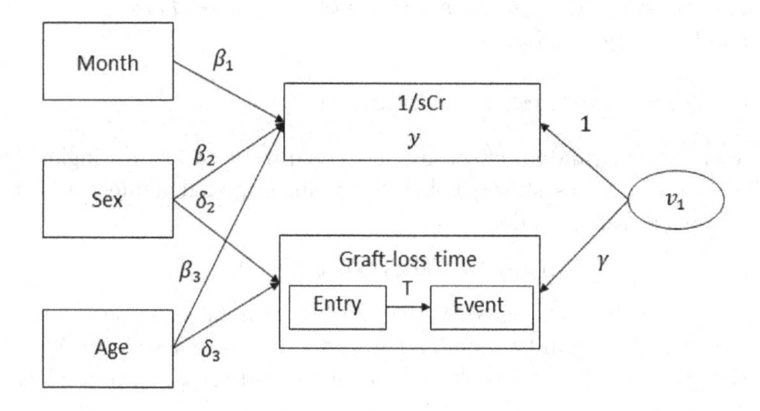


Figure 9.1 *Path diagram for the joint model for repeated measures and survival time.*

The joint model can be fitted by the jointdhglm package below. The results from joint models of linear mixed model for 1/sCr and frailty model with Weibull and non-parametric baseline hazard functions for graft-loss time are in Tables 9.1 and 9.2, respectively. The values of cAIC show that the non-parametric baseline hazard model is preferred to the Weibull baseline hazard model.

```
## Weibull baseline hazard model
> data(scr, package="jointdhglm")
> scr_surv<-subset(scr,first==1)
> jm1<-jointmodeling(Model="mean",RespDist="gaussian",
+  Link="identity",LinPred=icr~month+sex+age+(1|id),
+  RandDist="gaussian")
```

```
> jm2<-jointmodeling(Model="mean",RespDist="Weibull",
+  Link="log",
+  LinPred=Surv(sur_time,status==1)~sex+age+(1|id),
+  RandDist="gaussian")

> res1<-jmfit(jm1,jm2,scr,scr_surv)

## Non-parametric baseline hazard model
> jm1<-jointmodeling(Model="mean",RespDist="gaussian",
+  Link="identity",LinPred=icr~month+sex+age+(1|id),
+  RandDist="gaussian")
> jm2<-jointmodeling(Model="mean",RespDist="FM",
+  Link="log",
+  LinPred=Surv(sur_time,status==1)~sex+age+(1|id),
+  RandDist="gaussian")

> res2<-jmfit(jm1,jm2,scr,scr_surv)
```

We can fit two random-effect models separately by using dhglm and frailtyHL packages as shown below for a linear mixed model (9.1) and the following frailty model,

$$\lambda(t_i|v_{2i}) = \lambda_0(t_i)\exp(x_{2i}^t\delta + v_{2i}),$$

where $v_{2i} \sim \mathrm{N}(0, \sigma_{v2}^2)$. The cAIC for separated models can be computed by adding cAICs from two models for 1/sCr and graft-loss time. We can see that joint models are preferred to corresponding separate models.

```
## Linear mixed model
> model_mu<-DHGLMMODELING(Model="mean",Link="identity",
+  LinPred=icr~month+sex+age+(1|id),
+  RandDist="gaussian")
> model_phi<-DHGLMMODELING(Model="dispersion")

> fit1<-dhglmfit(RespDist="gaussian",DataMain=scr,
+  MeanModel=model_mu,DispersionModel=model_phi)

## Non-parametric baseline hazard model
> fit2<-frailtyHL(Surv(sur_time,status)~
+ sex+age+(1|id),RandDist="Normal",scr_surv)

## Weibull baseline hazard model
> fit2<-frailtyHL(Surv(sur_time,status)~
+ sex+age+(1|id),RandDist="Normal",scr_surv,
+ RespDist="Weibull")
```

Thus, the joint model with a non-parametric baseline hazard model is the final model to be selected. In Table 9.2, $\hat{\gamma} = -15.7$ indicates a negative correlation between the two responses, 1/sCr values and graft-loss time. Thus, a higher value of sCr has a higher risk for graft-loss time. For a linear mixed model, effects of month and sex are negatively significant and effect of age is positively significant. For the frailty model, the effect of age is negatively significant. Thus, patients tend to have lower expectation for sCr and lower risk graft-loss time as the age increases.

For the linear mixed model, the results from joint and separate models are similar. However, for the frailty model, the estimates for regression coefficients $\delta$ are quite different. Note that effect of age is very negatively significant in the joint model, but not significant in the separate model. Therefore, the joint model gives a more informative analysis.

Table 9.1 *Fitting results of joint model for linear mixed and Weibull baseline hazard model on serum creatinine data*

| | joint model | | | separate model | | |
|---|---|---|---|---|---|---|
| Parameter | Estimate | SE | t-value | Estimate | SE | t-value |
| | | | Linear mixed model for 1/sCr | | | |
| Intercept | 0.3508 | 0.1355 | 2.59 | 0.3509 | 0.1362 | 2.58 |
| month | −0.0012 | 0.0003 | −3.86 | −0.0012 | 0.0003 | −3.75 |
| sex (male) | −0.1113 | 0.0471 | −2.27 | −0.1112 | 0.0487 | −2.28 |
| age | 0.0105 | 0.0027 | 3.89 | 0.0104 | 0.0031 | 3.30 |
| $\log(\sigma_e^2)$ | −4.420 | | | −4.321 | | |
| $\log(\sigma_{v1}^2)$ | −3.501 | | | −3.517 | | |
| | | | Weibull baseline hazard model for graft loss time | | | |
| Intercept | −10.2357 | 3.0654 | −3.33 | −8.8715 | 2.7612 | −3.21 |
| sex (male) | −1.0205 | 0.8151 | −1.25 | −0.0814 | 0.8232 | −0.10 |
| age | −0.1136 | 0.0457 | −2.49 | −0.0413 | 0.0365 | −1.13 |
| $\log(\tau)$ | 2.765 | | | 2.340 | | |
| $\log(\sigma_{v2}^2)$ | | | | −1.977 | | |
| $\gamma$ | −11.536 | | | | | |
| cAIC | −1272.7 | | | −1205.5 | | |

### *9.1.2 Joint model for AIDS data*

Both longitudinal and survival data were collected in a recent clinical trial to compare the efficacy and safety of two antiretroviral drugs in treating patients who had failed or were intolerant of zidovudine (AZT)

Table 9.2 *Fitting results of joint model for linear mixed and non-parametric baseline hazard model on serum creatinine data*

| | joint model | | | separate model | | |
|---|---|---|---|---|---|---|
| Parameter | Estimate | SE | t-value | Estimate | SE | t-value |
| | Linear mixed model for 1/sCr | | | | | |
| Intercept | 0.3454 | 0.1374 | 2.51 | 0.3509 | 0.1362 | 2.58 |
| month | −0.0012 | 0.0003 | −3.90 | −0.0012 | 0.0003 | −3.75 |
| sex (male) | −0.1114 | 0.0491 | −2.27 | −0.1112 | 0.0487 | −2.28 |
| age | 0.0105 | 0.0032 | 3.32 | 0.0104 | 0.0031 | 3.30 |
| $\log(\sigma_e^2)$ | −4.320 | | | −4.321 | | |
| $\log(\sigma_{v1}^2)$ | −3.500 | | | −3.517 | | |
| | Non-parametric baseline hazard model for graft loss time | | | | | |
| sex (male) | −0.8566 | 1.5359 | −0.56 | −0.6136 | 0.9141 | −0.67 |
| age | 0.93486 | 0.1148 | −3.04 | −0.1646 | 0.0859 | −1.92 |
| $\log(\sigma_{v2}^2)$ | | | | 0.984 | | |
| $\gamma$ | −15.7485 | | | | | |
| cAIC | −1286.9 | | | −1211.5 | | |

therapy (Rizopoulos, 2015). In this trial, 467 HIV-infected patients were enrolled and randomly assigned to receive either didanosine (ddI) or zalcitabine (ddC). The number of CD4 cells per cubic millimeter of blood were recorded at study entry, and again at the 2, 6, 12, and 18 month visits. Times to death were also recorded with a 40% censoring rate.

Rizopoulos (2015) considered a joint model for the square root of CD4 value $y_{ij}$ for the $j$th visit and the time to death $t_i$ of the $i$th patient. As shown in Figure 9.2, visit month (coded as obstime) and drug (=1 for DDI (didanosine), =0 for DDC (zalcitabine)) are used as covariates for $y_{ij}$ and drug for $t_i$, respectively. For the response $y_{ij}$ consider a linear mixed model

$$y_{ij} = x_{1ij}^T \beta + v_{1i} + e_{ij}, \tag{9.2}$$

where $x_{1ij}^T$ are covariates, $v_{1i} \sim \mathrm{N}(0, \sigma_{v1}^2)$ and $e_{ij} \sim \mathrm{N}(0, \sigma_e^2)$. For death time $t_i$, consider a frailty model with the conditional hazard function

$$\lambda(t_i|v_{1i}) = \lambda_0(t_i) \exp(x_{2i}^t \delta + \gamma v_{1i}),$$

where $\lambda_0(t)$ is the baseline hazard function, $x_{2i}^T$ are between-subject covariates and $\gamma$ is the shared parameter. Rizopoulos (2015) considered a Weibull model for the baseline hazard function where $\lambda_0(t_i) = \tau t^{\tau-1}$ with a shape parameter $\tau$. Now we can also fit a non-parametric baseline hazard model.

The joint model can be fitted by the jointdhglm package below. The

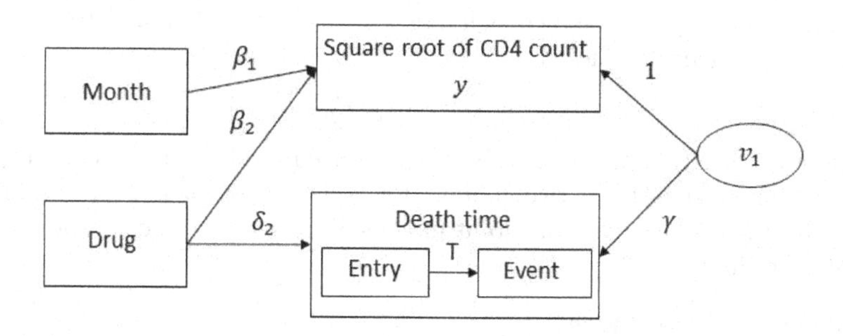


Figure 9.2 *Path diagram for the joint model for repeated measures and survival time on AIDS data.*

results from joint models of linear mixed model for square root of CD4 value and frailty model with Weibull and non-parametric baseline hazard functions for death time are in Tables 9.3 and 9.4, respectively. The values of cAIC show that the Weibull baseline hazard model is preferred to the non-parametric baseline hazard model.

```
## Weibull baseline hazard model
> data(aids, package="JM")
> data(aids.id, package="JM")
> jm1<-jointmodeling(Model="mean",RespDist="gaussian",
+  Link="identity",
+  LinPred=sqrt(CD4)~obstime+drug+(1|patient),
+  RandDist="gaussian")
> jm2<-jointmodeling(Model="mean",RespDist="Weibull",
+  Link="log",
+  LinPred=Surv(Time,death)~drug+(1|patient),
+  RandDist="gaussian")
> res1<-jmfit(jm1,jm2,aids,aids.id)
## Non-parametric baseline hazard model
> jm1<-jointmodeling(Model="mean",RespDist="gaussian",
+  Link="identity",
+  LinPred=sqrt(CD4)~obstime+drug+(1|patient),
+  RandDist="gaussian")
> jm2<-jointmodeling(Model="mean",RespDist="FM",
+  Link="log",
+  LinPred=Surv(Time,death)~drug+(1|patient),
+  RandDist="gaussian")
> res2<-jmfit(jm1,jm2,aids,aids.id)
```

We can fit two random-effect models separately by using dhglm and frailtyHL packages as shown below for a linear mixed model (9.2) and the following frailty model,

$$\lambda(t_i|v_{2i}) = \lambda_0(t_i)\exp(x_{2i}^t\delta + v_{2i}),$$

where $v_{2i} \sim \mathrm{N}(0, \sigma_{v2}^2)$. The cAIC for separated models can be computed by adding cAICs from two models for square root of CD4 value and death time. We can see that joint models are preferred to corresponding separate models.

```
## Linear mixed model
> model_mu<-DHGLMMODELING(Model="mean",Link="identity",
+  LinPred=sqrt(CD4)~obstime+drug+(1|patient),
+  RandDist="gaussian")
> model_phi<-DHGLMMODELING(Model="dispersion")
> fit1<-dhglmfit(RespDist="gaussian",DataMain=aids,
+  MeanModel=model_mu,DispersionModel=model_phi)
## Non-parametric baseline hazard model
> fit2<-frailtyHL(Surv(Time,death)~drug+(1|patient),
+  RandDist="Normal",aids.id)
## Weibull baseline hazard model
> fit2<-frailtyHL(Surv(Time,death)~drug+(1|patient),
+  RandDist="Normal",aids.id,
+  RespDist="Weibull")
```

Thus, the joint model with the Weibull baseline hazard model is the final model to be selected. In Table 9.3, $\hat{\gamma} = -1.082$ indicates the negative correlation between the two responses, CD4 values and death time. Thus, a higher value of CD4 counts has a lower risk for death time. For a linear mixed model, effects of month are negatively significant and effect of ddC drug is positive but non-significant. For the frailty model, the effect of ddC drug is positively significant. Thus, patients with treatment of ddC drug have higher risk than ddI drug patients.

For the linear mixed model, the results from joint and separate models are similar. However, for the frailty model, the estimates for regression coefficients $\delta$ are quite different. Note that the effect of ddC drug is significant in the joint model, but not significant in the separate model. Therefore, the joint model gives a more informative analysis.

### *9.1.3 Joint model for primary biliary cirrhosis data continued*

We show a joint model for repeated measures and competing event time. Consider PBC data again available in the R package JM (Rizopoulos,

Table 9.3 *Fitting results of joint model for linear mixed and Weibull baseline hazard model on AIDS data*

| | joint model | | | separate model | | |
|---|---|---|---|---|---|---|
| Parameter | Estimate | SE | t-value | Estimate | SE | t-value |
| | Linear mixed model for square root of CD4 count | | | | | |
| Intercept | 2.4584 | 0.0602 | 40.84 | 2.4601 | 0.0598 | 41.14 |
| month | −0.0334 | 0.0025 | −13.16 | −0.0311 | 0.0025 | −12.22 |
| drug (ddC) | 0.0948 | 0.0848 | 1.12 | 0.0983 | 0.0843 | 1.17 |
| $\log(\sigma_e^2)$ | −1.785 | | | −1.780 | | |
| $\log(\sigma_{v1}^2)$ | −0.258 | | | −0.274 | | |
| | Weibull baseline hazard model for death time | | | | | |
| Intercept | −2.2460 | 3.0654 | −3.33 | −8.8715 | 2.7612 | −3.21 |
| drug (ddC) | 0.3370 | 0.1502 | 2.24 | 0.2102 | 0.1460 | 1.44 |
| $\log(\tau)$ | 0.271 | | | 0.311 | | |
| $\log(\sigma_{v2}^2)$ | | | | −12.136 | | |
| $\gamma$ | −1.082 | | | | | |
| cAIC | 4260.3 | | | 4375.8 | | |

Table 9.4 *Fitting results of joint model for linear mixed and non-parametric baseline hazard model on AIDS data*

| | joint model | | | separate model | | |
|---|---|---|---|---|---|---|
| Parameter | Estimate | SE | t-value | Estimate | SE | t-value |
| | Linear mixed model for square root of CD4 count | | | | | |
| Intercept | 2.4139 | 0.0880 | 27.42 | 2.4601 | 0.0598 | 41.14 |
| month | −0.0336 | 0.0027 | −12.57 | −0.0311 | 0.0025 | −12.22 |
| drug (ddC) | 0.1255 | 0.1068 | 1.17 | 0.0983 | 0.0843 | 1.17 |
| $\log(\sigma_e^2)$ | −1.711 | | | −1.780 | | |
| $\log(\sigma_{v1}^2)$ | −0.249 | | | −0.274 | | |
| | Non-parametric baseline hazard model for death time | | | | | |
| drug (ddC) | 0.3427 | 0.1447 | 2.36 | 0.2150 | 0.1495 | 1.44 |
| $\log(\sigma_{v2}^2)$ | | | | −2.368 | | |
| $\gamma$ | −1.089 | | | | | |
| cAIC | 4789.6 | | | 4896.0 | | |

2015). We fit joint model for the logarithm of serum bilirubin in mg/dl $y_{ij}$ for the $j$th visit and the time to event $t_i$ of the $i$th patient. The event time in year is coded as `years` with status, `status`. Here, the `status` is coded as "dead" for the interesting event (140 observations), "transplanted" for the competing event (29 observations) and "alive" for censoring (143 observations).

As shown in Figure 9.3, visiting year, sex (=1 for male, =0 for female) and drug (=1 for D-penicil, =0 for placebo) are used as covariates for $y_{ij}$ and sex and drug for $t_i$, respectively. For $y_{ij}$ consider a linear mixed model

$$y_{ij} = x_{1ij}^t \beta + v_i + e_{ij},$$

where $x_{1ij}^t$ are covariates, $v_i \sim \mathrm{N}(0, \sigma_v^2)$ and $e_{ij} \sim \mathrm{N}(0, \sigma_e^2)$. For the time to event $t_i$, consider the cause-specific hazard frailty model as shown in Figure 8.3. Given a shared log-frailty $v_{1i}$, the conditional hazard function $\lambda_{ik}$ for the $i$th patient that failed from cause $k$ ($k = 1, 2$) can be expressed as

$$\begin{aligned}\lambda_{i1}(t|v_i) &= \lambda_{01}(t_i)\exp(x_{2i}^t\delta_1 + \gamma_1 v_i),\\ \lambda_{i2}(t|v_i) &= \lambda_{02}(t_i)\exp(x_{2i}^t\delta_2 + \gamma_2 v_i),\end{aligned}$$

where $\lambda_{0k}(t)$ is an unspecified baseline hazard function for cause $k$, $\delta_k$ is regression parameters for cause $k$ .

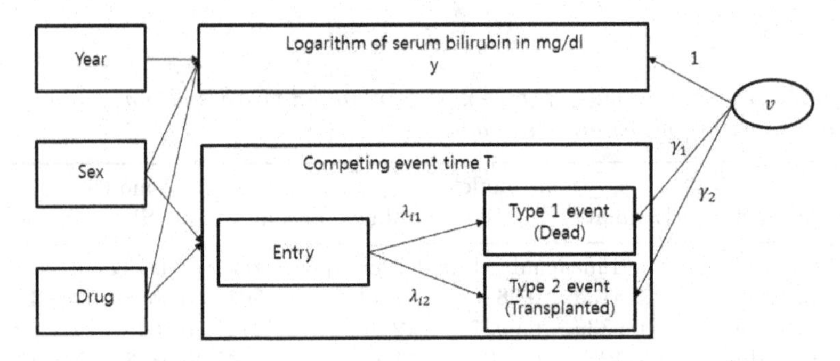


Figure 9.3 *Path diagram for the joint model of repeated measure and competing event time on PBC data.*

This model can be fitted by using the `jmfit` function in the jointdhglm package. In the R output, the estimates of shared parameters $\widehat{\gamma}_1 = 1.271$ and $\widehat{\gamma}_2 = 1.189$ show a positive associations between $y_{ij}$ and two (death and transplanted) events. This shows a positive association between two events. The visiting year effect for $y_{ij}$ is positively very significant (t-value = 22.462 with p-value < 0.001). The effect of drug is not significant for $y_{ij}$ (t-value = -0.971 with p-value = 0.317) and for death event (t-value = -0.539 with p-value 0.590), but it is negatively significant for transplanted event (t-value = -6.255 with p-value < 0.001). The effect of sex is positively significant for $y_{ij}$ (t-value = 2.113 with p-value = 0.035) and for death event (t-value = 2.295 with p-value = 0.022), but it is not significant for transplanted event (t-value = 0.535 with p-value = 0.593). However, when we fit the competing risk model for $t_i$ removing response $y_{ij}$, the effect of drug is not significant for transplanted event (t-value = -1.027 with p-value = 0.304).

```
# joint model for repeated measures and competing even time
> data(pbc2,package="JM")
> data(pbc2.id,package="JM")
> pbc2$sex<-if(pbc2$sex=="male",1,0)
> pbc2.id$sex<-if(pbc2.id$sex=="male",1,0)
> pbc2$drug<-if(pbc2$drug=="D-penicil",1,0)
> pbc2.id$drug<-if(pbc2.id$drug=="D-penicil",1,0)

> jm1<-jointmodeling(Model="mean",RespDist="gaussian",
+ Link="identity",
+ LinPred=log(serBilir)~year+drug+sex+(1|id),
+ RandDist="gaussian")
> jm2<-jointmodeling(Model="mean",RespDist="FM",
+ Link="log",
+ LinPred=Surv(years,status=="dead")~drug+sex+(1|id),
+ RandDist="gaussian")
> jm3<-jointmodeling(Model="mean",RespDist="FM",
+ Link="log",
+ LinPred=Surv(years,status=="transplanted")
+ ~drug+sex+(1|id),
+ RandDist="gaussian")

> data_conti<-pbc2
> data_surv<-pbc2.id
> res<-jmfit3(jm1,jm2,jm3,pbc2,pbc2.id,pbc2.id)

    "Estimates from the mean model"
[1] LinPred=log(serBilir)~year+drug+sex+(1|id)
              Estimate Std. Error t-value p-value
(intercept) 0.5909    0.0931      6.3464  0.0000
year         0.0970   0.0043     22.4619  0.0000
drug        -0.1239   0.1276     -0.9708  0.3166
sex          0.4221   0.1997      2.1133  0.0346

[1] Surv(years,status=="dead")~drug+sex+(1|id)
       Estimate Std. Error t-value p-value
drug  -0.1152   0.2138    -0.5389  0.5900
sex    0.6890   0.3002     2.2949  0.0217
```

```
[1] Surv(years,status=="transplanted")~drug+sex+(1|id)
      Estimate Std. Error t-value p-value
drug  -4.3272   0.6918   -6.2553  0.0000
sex    0.2915   0.5450    0.5348  0.5928

[1] "Estimates for logarithm of variance of random effect"
    Estimate
id    0.1858

[1] "Estimates for logarithm of residual variance"
Estimate : -1.4190

[1] "Estimates for the shared parameter"
        Estimate
gamma1   1.2705
gamma2   1.1888

# joint model for competing even time
> jm22<-jointmodeling(Model="mean",
+   RespDist="FM",Link="log",
+   LinPred=Surv(years,status=="dead")
+   ~drug+sex+(1|id),
+   RandDist="gaussian")
> jm32<-jointmodeling(Model="mean",
+  RespDist="FM",Link="log",
+  LinPred=Surv(years,status=="transplanted")~drug+sex+(1|id),
+  RandDist="gaussian")
res<-jmfit2(jm22,jm32,pbc2.id,pbc2.id)

[1] Surv(years,status=="dead")~drug+sex+(1|id)
      Estimate Std. Error t-value p-value
drug  -0.0226   0.1782   -0.1266  0.8993
sex    0.6891   0.2354    2.9274  0.0034

[1] Surv(years,status=="transplanted")~drug+sex+(1|id)
      Estimate Std. Error t-value p-value
drug  -0.3961   0.3855   -1.0273  0.3043
sex    0.2007   0.6250    0.3212  0.7481

[1] "Estimates for logarithm of variance of random effect"
    Estimate
id   -1.7168
```

```
[1] "Estimates for the shared parameter"
        Estimate
gamma1    1.4302
```

## 9.2 H-likelihood approach to joint models

### *9.2.1 H-likelihood construction*

Suppose that we are interested in the effects of covariates over the two types of responses, namely repeated-measure outcome $y(=y_1)$ and event time $T(=y_2)$. We consider a joint model which consists of two linked submodels, LMM for $y$ and frailty model for $T$. Let $y_{ij}$ be the $j$th repeated response of $i$th subject at time point $t$ ($i = 1, \ldots, q; j = 1, \ldots, n_i; n = \sum_i n_i$), and let $T_i$ be a single event time of $i$th subject and let $C_i$ be the corresponding censoring time. Denote by $v_i$ a shared random effect of the $i$th subject. Given $v_i$ suppose that $y_i = (y_{i1}, \cdots y_{in_i})^T$ and $T_i$ are conditionally independent, and $T_i$ and $C_i$ are also conditionally independent. Suppose the following joint hierarchical model with shared random effects, specified in terms of the two linked submodels,

$$\text{(i) } y_{ij} = x_{ij1}^T \beta_1 + v_i + \epsilon_{ij},$$

where $\epsilon_{ij} \sim \mathrm{N}(0, \phi)$, and

$$\text{(ii) } \lambda_i(t|v_i) = \lambda_0(t) \exp(x_{i2}^T \beta_2 + \gamma v_i),$$

where $v_i \sim \mathrm{N}(0, \alpha)$, $v_i$ and $\epsilon_{ij}$ are independent, and $\lambda_0(\cdot)$ is an unspecified baseline hazard function. Here $\beta_1$ and $\beta_2$ are $p_1 \times 1$ and $p_2 \times 1$ regression-parameter vectors corresponding to the vectors of covariates $x_{ij1}$ and $x_{i2}$, respectively. Here $\gamma$ is a real-valued association parameter that allows the magnitude of the associations to be different between two outcomes, $y_{ij}$ and $T_i$; if $\gamma > 0$ [$\gamma < 0$], $y_{ij}$ and hazard rate tend to be positively [negatively] correlated, and if $\gamma = 0$ they are not associated.

All observable random variables are repeated-measure responses $y_{ij}$ and time-to-event data $(t_i^*, \delta_i)$, defined by

$$t_i^* = \min(T_i, C_i) \text{ and } \delta_i = I(T_i \leq C_i).$$

Then, the h-likelihood of this joint model is immediately constructed, by taking $y_1 = y$ and $y_2 = (t^*, \delta)$ in (9.3). Because of the conditional independency between $y_i$ and $T_i$ and the non-informative censoring, the h-likelihood becomes

$$h = \sum_{ij} \ell_{1ij} + \sum_i \ell_{2i} + \sum_i \ell_{3i}, \tag{9.3}$$

where

$$\begin{aligned}\ell_{1ij} &= \ell_{1ij}(\beta_1, \phi; y_{ij}|v_i) = \log f_{\beta_1,\phi}(y_{ij}|v_i) \\ &= -\log(2\pi\phi)/2 - (y_{ij} - \eta_{1ij})^2/(2\phi), \\ \ell_{2i} &= \ell_{2i}(\beta_2, \lambda_0; t_i^*, \delta_i|v_i) = \log f_{\beta_2,\lambda_0}(t_i^*, \delta_i|v_i) \\ &= \delta_i(\log\lambda_0(t_i^*) + \eta_{2i}) - \Lambda_0(t_i^*)\exp(\eta_{2i}), \\ \ell_{3i} &= \ell_{3i}(\alpha; v_i) = \log f_\alpha(v_i) = -\log(2\pi\alpha)/2 - v_i^2/(2\alpha),\end{aligned}$$

$\ell_{1ij}$ is the conditional log-likelihoods for $y_{ij}$ given $v_i$, $\ell_{2i}$ is that for $(t_i^*, \delta_i)$ given $v_i$, $\ell_{3i}$ is the log-likelihood for $v_i$ and

$$\eta_{1ij} = x_{ij1}^T\beta_1 + v_i \text{ and } \eta_{2i} = x_{i2}^T\beta_2 + \gamma v_i$$

are linear predictors. This shows how to make h-likelihood for multivariate DHGLMs. Then, the use of DHGLM procedure in Chapter 6 leads to fitting algorithms. Now we describe those for joint models.

### *9.2.2 Iterative least squares equations*

The functional form of $\lambda_0(\cdot)$ in $\ell_{2i}$ of (9.3) is unknown. Following Breslow (1972), we define the baseline cumulative hazard function $\Lambda_0(t) = \int_0^t \lambda_0(u)du$ to be a step function with jumps $\lambda_{0r}$ at the observed event times $t_{(r)}$:

$$\Lambda_0(t) = \sum_{r:t_{(r)}\le t} \lambda_{0k}, \tag{9.4}$$

where $t_{(r)}$ is the $r$th ($r = 1, \ldots, D$) smallest distinct event time among the $t_i^*$'s and $\lambda_{0r} = \lambda_0(t_{(r)})$. By substituting (9.4) into (9.3), the second term $\sum_i \ell_{2i}$ in (9.3) becomes

$$\sum_i \ell_{2i} = \sum_r d_{(r)} \log\lambda_{0r} + \sum_i \delta_i\eta_{2i} - \sum_r \lambda_{0r}\Big\{\sum_{i\in R_{(r)}} \exp(\eta_{2i})\Big\},$$

where $d_{(r)}$ is the number of events at $t_{(r)}$ and $R_{(r)} = \{i : t_i^* \ge t_{(r)}\}$ is the risk set at $t_{(r)}$. As the number of $\lambda_{0r}$'s in $\sum_i \ell_{2i}$ above increases with the number of events, the function $\lambda_0(t)$ is potentially of high dimension. Following Ha et al. (2001), we use the profile h-likelihood $h^*$, given by

$$h^* = h|_{\lambda_0=\widehat{\lambda}_0} = \sum_{ij} \ell_{1ij} + \sum_i \ell_{2i}^* + \sum_i \ell_{3i},$$

where

$$\sum_i \ell_{2i}^* = \sum_i \ell_{2i}|_{\lambda_0=\widehat{\lambda}_0} = \sum_r d_{(r)} \log\widehat{\lambda}_{0r} + \sum_i \delta_i\eta_{2i} - \sum_r d_{(r)},$$

$$\widehat{\lambda}_{0r} = \widehat{\lambda}_{0r}(\beta_2, v) = \frac{d_{(r)}}{\sum_{i \in R_{(r)}} \exp(\eta_{2i})}$$

are the solutions of the estimating equations, $\partial h/\partial\lambda_{0r} = 0$, for $r = 1, \ldots, D$. This is proportional to the penalized partial h-likelihood $h_p$, given by

$$h_p = \sum_{ij} \ell_{1ij} + \sum_i \delta_i \eta_{2i} - \sum_r d_{(r)} \log\left\{ \sum_{i \in R_{(r)}} \exp(\eta_{2i}) \right\} + \sum_i \ell_{3i}.$$

Below we show how the h-likelihood procedure can be derived via $h_p$. Let $X_1$, $X_2$, and $Z$ be model matrices for vectors $\beta_1$, $\beta_2$, and $v = (v_1, \ldots, v_q)^T$, respectively. The score equations for fixed and random effects $(\beta_1, \beta_2, v)$ given dispersion parameters $\psi = (\phi, \alpha, \gamma)^T$ are

$$\begin{aligned}
\partial h_p/\partial\beta_1 &= X_1^T(y - \mu_1)/\phi, \\
\partial h_p/\partial\beta_2 &= X_2^T(\delta - \hat{\mu}_2), \\
\partial h_p/\partial v &= {Z_1}^T(y - \mu_1)/\phi + \gamma {Z_2}^T(\delta - \hat{\mu}_2) - v/\alpha,
\end{aligned}$$

where $\mu_1 = X_1\beta_1 + Z_1 v = \eta_1$, and $\hat{\mu}_2 = \exp(\log\hat{\Lambda}_0(t^*) + \eta_2)$ with $\eta_2 = X_2\beta_2 + \gamma Z_2 v$, and $Z_1$ is $n \times q$ group indicator matrix whose $ijk$th element $z_{ijk}$ is $\partial\eta_{1ij}/\partial v_k$ and $Z_2 = I_q$ which denotes a $q \times q$ identity matrix. Here $\hat{\Lambda}_0(t) = \sum_{r:t_{(r)} \leq t} \hat{\lambda}_{0r}$ is the estimator of cumulative baseline hazard.

This leads to the iterative least squares (ILS; see Ha et al. (2017)) joint equations for $\tau = (\beta_1^T, \beta_2^T, v^T)^T$, given by

$$\begin{pmatrix} {X_1}^T W_1 X_1 & 0 & X_1^T W_1 Z_1 \\ 0 & X_2^T W_2 X_2 & X_2^T(\gamma W_2) Z_2 \\ Z_1^T W_1 X_1 & {Z_2}^T(\gamma W_2) X_2 & \mathbf{Z}^T\mathbf{W}\mathbf{Z} + Q \end{pmatrix} \begin{pmatrix} \hat{\beta}_1 \\ \hat{\beta}_2 \\ \hat{v} \end{pmatrix} = \begin{pmatrix} {X_1}^T W_1 w_1 \\ {X_2}^T w_2 \\ \mathbf{Z}^T\mathbf{w}^* \end{pmatrix}, \tag{9.5}$$

where $W_1 = -\partial^2 h_p/\partial\eta_1\partial\eta_1^T = \phi^{-1} I_n$, and $W_2 = -\partial^2 h_p/\partial\eta_2\partial\eta_2^T$, $Q = -\partial^2\ell_3/\partial v\partial v^T = \alpha^{-1} I_q$, $w_1 = y$, $w_2 = W_2\eta_2 + (\delta - \hat{\mu}_2)$, and

$$\mathbf{Z} = \begin{pmatrix} Z_1 \\ \gamma Z_2 \end{pmatrix}, \quad \mathbf{W} = \begin{pmatrix} W_1 & \mathbf{0} \\ \mathbf{0} & W_2 \end{pmatrix} \quad \text{and} \quad \mathbf{w}^* = \begin{pmatrix} W_1 w_1 \\ w_2 \end{pmatrix}.$$

Note here that $\mathbf{Z}^T\mathbf{W}\mathbf{Z} = Z_1^T W_1 Z_1 + Z_2^T(\gamma^2 W_2) Z_2$ and that $\mathbf{Z}^T\mathbf{w}^* = Z_1^T W_1 w_1 + \gamma {Z_2}^T w_2$.

### *9.2.3 Fitting procedure*

The fitting procedure consists of the following two steps:

**(S1)** Estimation of fixed and random effects $\tau = (\beta_1^T, \beta_2^T, v^T)^T$ via the ILS equations (9.5).

**(S2)** Estimation of dispersion parameters $\psi = (\phi, \alpha, \gamma)^T$ as follows. For the estimation of $\psi$ we use the adjusted profile h-likelihood, given by

$$p_\tau(h_p) = [h_p - \frac{1}{2}\log\det\{H(h_p,\tau)/(2\pi)\}]|_{\tau=\widehat{\tau}},$$

where $\widehat{\tau} = \widehat{\tau}(\psi)$ are solutions of $\partial h_p/\partial\tau = 0$ for given $\psi$, and

$$H(h_p,\tau) = -\partial^2 h_p/\partial\tau\partial\tau^T$$

is observed information matrix for $\tau$. The estimating equations of $\psi$ are given by

$$\partial p_\tau(h_p)/\partial\psi = 0,$$

leading to the estimating equations

$$\widehat{\phi} = \frac{(y-\widehat{\mu}_1)^T(y-\widehat{\mu}_1)}{n-\kappa_0} \quad \text{and} \quad \widehat{\alpha} = \frac{\widehat{v}^T\widehat{v}}{q-\kappa_1},$$

where $\kappa_0 = -\phi\mathrm{tr}\{\widehat{H}^{-1}(\partial\widehat{H}/\partial\phi)\}$, $\kappa_1 = -\alpha\mathrm{tr}\{\widehat{H}^{-1}(\partial\widehat{H}/\partial\alpha)\}$, and $\widehat{H} = H(h_p,\tau)|_{\tau=\widehat{\tau}(\psi)}$. The estimator of $\gamma$ is also easily implemented via the Newton-Raphson method using the first and second derivatives, $\partial p_\tau(h_p)/\partial\gamma$ and $\partial^2 p_\tau(h_p)/\partial\gamma^2$. This approach can easily extended to a joint model with competing-risks data (Ha et al., 2017).

## 9.3 Exercises

1. Consider a simulated dataset used in the `jointModel` package (Kim, 2016) to illustrate the joint model. This dataset was generated under settings mimicking the prostate cancer study. For longitudinal outcomes, the `prostate` dataset has 697 observations (n = 100 patients) on the four variables. For survival data, the `dropout` dataset has 100 observations (n = 100 patients) on the four variables. For repeated-measured response `logPSA.postRT` with the covariate `logPSA.base` and time-to-event response `DropTime` with the censoring indicator `Status` and the covariate `logPSA.base2`, fit the joint model by using the `jointdhglm` package.

```
data(prostate,package="JointModel")

# ID: a numeric vector of patient ID.
# logPSA.postRT: a numeric vector
# containing Prostate-specific Antigen (PSA) levels
# after radiation
# therapy, i.e., log(PSA(t)+0.1) observed at time t.
# VisitTime: a numeric vector of visiting time.
# logPSA.base: a numeric vector of log(baseline PSA+0.1).
```

```
data(dropout,package="JointModel")

# ID2: a numeric vector of patient ID.
# Status: a numeric (binary) vector indicating
# whether the study drop-out time
# (end of follow-up) was informative.
# DropTime: a numeric vector of the end of follow-up.
# logPSA.base2: a numeric vector of log(baseline PSA+0.1).
```

2. The dataset `liver` in the `joineR` package (Philipson et al., 2012) gives the longitudinal observations of prothrombin index, a measure of liver function, for patients from a controlled trial into prednisone treatment of liver cirrhosis. Time-to-event information in the form of the event time and associated censoring indicator is also recorded along with a solitary baseline covariate - the allocated treatment arm in this instance. This is a subset of the full data where a number of variables were recorded both at entry and during the course of the trial. For repeated-measured response `prothrombin` and time-to-event response `survival` with the censoring indicator `cens` and the covariate `treatment`, fit the joint model by using the `jointdhglm` package.

```
data(liver, package="joineR")

# id : number for patient identification
# prothrombin : prothrombin index measurement
# time : time of prothrombin index measurement
# treatment : patient treatment indicator
# (0 = placebo, 1 = prednisone)
# survival : patient survival time
(in years)
# cens : censoring indicator
# (1 = died and 0 = censored)
```

CHAPTER 10

# Further topics

Up to now we have studied various applications of HGLMs for the analysis of data. In this chapter, we show the use of random-effect models for variable selection and hypothesis testing, covering a recent development in multiple testing. In some DHGLMs, the h-likelihood estimators become zero. This leads to a variable selection procedure that have advantages over existing procedures.

In this chapter, we also study an HGLM with discrete random effects. Up to now, we have investigated models with continuous random effects, in which the choice of random effects is crucial in defining the h-likelihood because of the Jacobian term (see Chapter 3). For the transformation of discrete random effects, such a Jacobian term does not appear in the extended likelihood. Thus, all the extended likelihoods, defined in any scale of the discrete random effect, are the h-likelihood. In this chapter we study how to use the h-likelihood for the inferences about discrete random effects; we focus on hypothesis testing, especially multiple testing of recent interest. Even though R codes for methods of this chapter have not been well developed yet, but it is still instructive to understand how random-effect models can be applied to these areas. We hope that readers of this book can contribute to the future development of R codes in these areas.

## 10.1 Examples

### *10.1.1 Diabetes data*

We first analyzed dataset on the disease progression of diabetes in Efron (2004) which is available in the lars package (Hastie and Efron, 2013). There are 10 predictive variables, including age, sex, body mass index (BMI), blood pressure (bp), and six serum measurements (tc, ldl, hdl, tch, ltg, glu), obtained from 442 diabetes patients. The response of interest is a quantitative measure of disease progression one year after baseline. We considered a quadratic model having $p = 64$ predictive variables, including 10 original predictive variables, 45 interactions, and 9 squares, where all predictive variables were standardized. H-likelihood (HL) estimates for variable selection in Section 10.2 can be obtained from the following R code and output.

```
> library(glmnet); library(ncvreg); library(lars)
> Fam="Linear"
> setwd("F:/dhglm_Rcode/hl-example/functions")
> filename <- list.files(pattern="*.R")
> for(i in 1:length(filename)) source(filename[i])
> data(diabetes,package="lars")
> X <- diabetes$x; Y <- diabetes$y
> N <- length(Y); P <- ncol(X)
> X <- scale(X)
> # Construct the quadratic terms
> del.id <- which(lapply(apply(X,2,unique),length)<=2)
> if(length(del.id)!=0){
> QQ <- X[,-del.id]^2
> colnames(QQ) <- paste("Q",colnames(X[,-del.id]),sep="")
> } else{
> QQ <- X^2
> colnames(QQ) <- paste("Q",colnames(X),sep="")
> }
> # Construct the interaction terms
> Comb <- combn(P,2)
> tmp <- matrix(0,nrow=N,ncol=ncol(Comb))
> colnames(tmp) <- 1:ncol(Comb)
> for(i in 1:ncol(Comb)){
> id <- Comb[,i]
> tmp[,i] <- X[,id[1]]*X[,id[2]]
> colnames(tmp)[i]<- paste("X",id[1],id[2],sep="")
> colnames(tmp)[i]<-paste(colnames(X)[id[1]],":",
+ colnames(X)[id[2]],sep="") }
# Construct the full design matrix
# with quadratic and interaction terms
> wX <- cbind(X,QQ,tmp); wX0 <- cbind(1,wX)
> set.seed(122); nF=10
> index <- split(sample(1:N),1:nF)
> L.vec <- exp(seq(log(10),log(0.001),length.out=100))
> ## 1. fit the LASSO
> Err <- c()
> for(k in 1:nF){
> ts.id <- index[[k]]
> GLM0 <- glmnet(x=wX[-ts.id,],y=Y[-ts.id],
```

```
+ family="gaussian",lambda=L.vec)
> B.mat <- rbind(GLM0$a0,as.matrix(GLM0$beta))
> cv <- c()
> for(i in 1:ncol(B.mat)){
> cv <- c(cv,mean((Y[ts.id]-wX0[ts.id,]%*%B.mat[,i])^2))}
> Err <- rbind(Err,cv)}
> Cv.err <- colMeans(Err)
> opt <- which.min(Cv.err); opt.L <- L.vec[opt]
> nL.cv <- min(Cv.err)
> GLM0 <- glmnet(x=wX,y=Y,family="gaussian",lambda=opt.L)
> nL <- c(GLM0$a0,as.matrix(GLM0$beta))
> ## 2. fit the SCAD
> Err <- c()
> for(k in 1:nF){
> ts.id <- index[[k]]
> SD0 <- ncvreg(X=wX[-ts.id,],y=Y[-ts.id],
+ family="gaussian",penalty="SCAD",eps=.005,lambda=L.vec)
> B.mat <- SD0$beta; cv <- c()
> for(i in 1:ncol(B.mat)){
> cv <- c(cv,mean((Y[ts.id]-wX0[ts.id,]%*%B.mat[,i])^2))}
> Err <- rbind(Err,cv)}
> Cv.err <- colMeans(Err); opt <- which.min(Cv.err)
> opt.L <- L.vec[opt]; sD.cv <- min(Cv.err)
> SD0 <- ncvreg(X=wX,y=Y,family="gaussian",
+ penalty="SCAD",eps=.005,lambda=c(opt.L,opt.L))
> sD <- SD0$beta[,1]
## 3. fit the HL
> Bc <- nL[-1]; S.vec <- 10^seq(-3.5,1,length.out=20)
> A=30; Err <- c()
> for(k in 1:nF){
> ts.id <- index[[k]]; B.mat <- SSS <- c()
> for(S in S.vec){
> h0 <- hl.glm.fun(Y=Y[-ts.id],X=wX[-ts.id,],Bc=Bc,A=A,S=S)
> SSS <- c(SSS,S); B.mat <-  cbind(B.mat,h0$Coef)}
> cv <- c()
> for(i in 1:ncol(B.mat)){
> cv <- c(cv,mean((Y[ts.id]-wX0[ts.id,]%*%B.mat[,i])^2))}
> Err <- rbind(Err,cv)}
> Cv.err <- colMeans(Err); opt <- which.min(Cv.err)
> S <- S.vec[opt]; HL.cv <- min(Cv.err)
> HL <- hl.glm.fun(Y=Y,X=wX,Bc=Bc,A=A,S=S)$Coef
> HL
```

```
### OUTPUT
                   age         sex         bmi         map
138.912687    0.000000  -10.860677   23.627749   15.170091
        tc         ldl         hdl         tch         ltg
  0.000000    0.000000  -12.522188    0.000000   22.890028
       glu        Qage        Qbmi        Qmap         Qtc
  2.926967    2.755124    2.125523    0.000000    0.000000
      Qldl        Qhdl        Qtch        Qltg        Qglu
  0.000000    0.000000    0.000000    0.000000    3.507128
   age:sex     age:bmi     age:map      age:tc     age:ldl
  7.456674    0.000000    1.686770    0.000000    0.000000
   age:hdl     age:tch     age:ltg     age:glu     sex:bmi
  0.000000    0.000000    1.554890    0.000000    0.000000
   sex:map      sex:tc     sex:ldl     sex:hdl     sex:tch
  2.287031    0.000000    0.000000    0.000000    0.000000
   sex:ltg     sex:glu     bmi:map      bmi:tc     bmi:ldl
  0.000000    0.000000    5.132527    0.000000    0.000000
   bmi:hdl     bmi:tch     bmi:ltg     bmi:glu      map:tc
  0.000000    0.000000    0.000000    0.000000    0.000000
   map:ldl     map:hdl     map:tch     map:ltg     map:glu
  0.000000    0.000000    0.000000    0.000000    0.000000
    tc:ldl      tc:hdl      tc:tch      tc:ltg      tc:glu
  0.000000    0.000000    0.000000    0.000000    0.000000
   ldl:hdl     ldl:tch     ldl:ltg     ldl:glu     hdl:tch
  0.000000    0.000000    0.000000    0.000000    0.000000
   hdl:ltg     hdl:glu     tch:ltg     tch:glu     ltg:glu
  0.000000    0.000000    0.000000    0.000000    0.000000
```

### *10.1.2 Finite mixture model*

For an example of discrete random effects, we may consider a finite mixture model. As an example, we use the dataset Npreg in the flexmix package (Gruen et al., 2015).

This dataset is generated from a two-components model:

$$\begin{aligned} \text{Component 1} &: \quad y_i = 5x_i + \epsilon_i \text{ for } i = 1, \cdots, 100 \\ \text{Component 2} &: \quad y_i = 15 + 10x_i^2 + \epsilon_i \text{ for } i = 101, \cdots, 200 \end{aligned}$$

where $y_i$ is the $i(= 1, \cdots, n)$th response variable, $x_i$ was generated from uniform distribution on $[0, 10]$ and $\epsilon_i \sim \text{N}(0, 9)$ with $n = 200$. To fit this dataset, we consider a finite mixture model with $K = 2$ components.

Let the unobserved random-effect $o_i = (o_{i1}, \cdots, o_{iK})$ follow multinomial distribution where $o_{ik} = 1$ if $y_i$ belongs to the $k$th component and $o_{ik} = 0$ otherwise with $p_k = P(o_{ik} = 1)$ satisfying $\sum p_k = 1$. If $o_i$ were known, MLEs for parameters can be obtained easily by using a simple regression model.

Here, the hierarchical likelihood is $f_\theta(y_i, o_i|x_i)$. From the h-likelihood, the log-likelihood of the $i$th observation $y_i$ is obtained by summing out all possible values of $o_i$

$$
\begin{aligned}
\log L &= \log f_\theta(y_i|x_i) \\
&= \log \sum_{o_i} f_\theta(y_i, o_i|x_i) \\
&= \log \sum_{k=1}^{K} P(o_{ik} = 1) f_\theta(y_i|x_i, o_{ik} = 1) \\
&= \log \sum_{k=1}^{K} p_k f_{\theta_k}(y_i|x_i)
\end{aligned}
$$

where $f_{\theta_k}(y_i|x_i)$ is the density from normal distribution $\mathrm{N}(\mu_{ik}, \phi_k)$ for the $k$th component with regression model $\mu_{ik} = \beta_{0k} + \beta_{1k} x_i + \beta_{2k} x_i^2$. $\theta_k = (\beta_{0k}, \beta_{1k}, \beta_{2k}, \phi_k, p_k)$ is fixed parameters for the $k$th component and $\theta = (\theta_1, \cdots, \theta_K)$. Then, the log-likelihood based on $n$ observations $(y_1, \cdots, y_n)$ is given by

$$
\sum_{i=1}^{n} \log\{\sum_{k=1}^{K} p_k f_{\theta_k}(y_i|x_i)\}.
$$

However, it is often difficult to maximize the likelihood directly. For this model a popular method for MLEs for parameters is the expectation-maximization (EM) algorithm (Demster et al., 1977). Note that the EM algorithm below can be viewed as a way of obtaining the MLEs from the h-likelihood, where the h-likelihood is the so-called complete-data $(y_i, o_i)$ likelihood. Thus, the hierarchical likelihood is

$$
f_\theta(y_i, o_i|x_i) = \prod_{k=1}^{K} \{p_k f_{\theta_k}(y_i|x_i)\}^{o_{ik}}
$$

and the h-likelihood (log-hierarchical likelihood) is

$$
\sum_{i=1}^{n} \sum_{k=1}^{K} o_{ik} \log\{p_k f_{\theta_k}(y_i|x_i)\}.
$$

1) E-step: Given, $p_k$ and $\theta_k$, the random-effect $o_{ik}$ is estimated by

$$\widehat{o}_{ik} = P(o_{ik} = 1|y_i, x_i) = \frac{f_\theta(y_i, o_{ik} = 1|x_i)}{\sum_{k=1}^K p_k f_{\theta_k}(y_i|x_i)} = \frac{p_k f_{\theta_k}(y_i|x_i)}{\sum_{k=1}^K p_k f_{\theta_k}(y_i|x_i)},$$

2) M-step: We maximize the h-likelihood for parameters $p_k, \theta_k$ by using $\widehat{o}_{ik}$

$$\max_{p_k,\theta_k} \sum_{i=1}^n \sum_{k=1}^K \widehat{o}_{ik} \log\{p_k f_{\theta_k}(y_i|x_i)\}.$$

The E- and M-steps are repeated until convergence has occurred. We can fit the above dataset using the flexmix package.

```
> library(flexmix)
> data(NPreg)
> m1 <- flexmix(yn ~ x + I(x^2), data = NPreg, k = 2)
> m1
Call:
flexmix(formula = yn ~ x + I(x^2), data = NPreg, k = 2)
Cluster sizes:
1 2
100 100
convergence after 15 iterations
and get a first look at the estimated parameters
of mixture component 1 by

> parameters(m1, component = 1)
Comp.1
coef.(Intercept) -0.20866478
coef.x 4.81612095
coef.I(x^2) 0.03632578
sigma 3.47494722
and

> parameters(m1, component = 2)
Comp.2
coef.(Intercept) 14.7175699
coef.x 9.8455831
coef.I(x^2) -0.9682393
sigma 3.4808477
for component 2.
```

The parameter estimates of both components are close to the true values. A cross-tabulation of true classes and cluster memberships can be obtained as follows.

```
> summary(m1)
Call:
flexmix(formula = yn ~ x + I(x^2), data = NPreg, k = 2)
prior size post>0 ratio
Comp.1 0.494 100 145 0.690
Comp.2 0.506 100 141 0.709
log Lik. -642.5451 (df=9)
AIC: 1303.09 BIC: 1332.775
```

Output gives the estimated probabilities of components $p_k$: $p_1 = 0.454$ and $p_2 = 0.506$.

## 10.2 Variable selections

With more variables we can potentially explain more of the systematic variation, but we potentially also bring in more noise. Keeping only the relevant variables in the model is a crucial step with several potential goals: better estimation, better prediction and better interpretation. When there are many potential predictors with equal status, i.e., no prior preferences among them, then having as few predictors as possible in the model would often help interpretation. When we have a large number of potential predictors, "overfitting" also becomes a serious problem.

Prediction accuracy is often improved by shrinking or simply setting some coefficients to zero by thresholding. In this chapter we discuss a general approach to simultaneously perform variable selection and estimation of the regression coefficients via random-effect models. The problem of too many predictors leads to overfitting, so variable selection can be seen as a regularization problem. Introducing random effects to deal with this problem provides a flexible framework and a likelihood-based solution.

### *10.2.1 Penalized least-squares methods*

In statistical analysis, variable selection is the process of choosing which variables to keep and which to exclude in the final model. Consider the regression model

$$y_i = x_i^T \beta + e_i, \quad i = 1, \ldots, n, \tag{10.1}$$

where $\beta$ is a $p \times 1$ vector of fixed unknown parameters and the $e_i$'s are i.i.d. with mean 0 and variance $\phi$.

Many classical subset selection methods have been proposed: stepwise forward-selection, stepwise backward-elimination or best-subset selection. The stepwise methods are fast and convenient, but have inferior performance compared to the best-subset method. The latter is preferable, but very quickly becomes impractical because, with $p$ predictors, we need to compare $2^p$ models. In general the subset selection methods are a discrete and highly variable process; furthermore, they cannot be easily adapted to applications where the number of variables $p$ is much greater than the sample size.

Subset selection is of course not the only way to perform variable selection. With the $L_1$-penalty, the penalized least squares (PLS) estimator, minimizing

$$Q_\lambda(\beta) = \frac{1}{2}\sum_{i=1}^{n}(y_i - x_i^T\beta)^2 + \lambda\sum_{j=1}^{p}|\beta_j|,$$

becomes the least absolute shrinkage and selection operator (LASSO): which automatically sets to zero those predictors with small estimated OLS coefficients, thus performing simultaneous estimation and variable selection. Tibshirani (1996) introduced and gave a comprehensive overview of LASSO as a method of PLS. LASSO has been criticized on the grounds that it typically ends up selecting too many variables to prevent over-shrinkage of the regression coefficients (Radchenko and James, 2008); otherwise, regression coefficients of selected variables are often over-shrunken. To improve LASSO, various other penalties have been proposed.

Fan and Li (2001) proposed the smoothly clipped absolute deviation (SCAD) penalty for oracle estimators, and Zou (2006) proposed the adaptive LASSO. Zou and Hastie (2005) noted that the prediction performances of the LASSO can be poor in cases where variable selection is ineffective; they proposed the elastic net penalty, which improves the prediction of LASSO.

With the $L_2$-penalty, we have

$$Q_\lambda(\beta) = \frac{1}{2}\sum_{i=1}^{n}(y_i - x_i^T\beta)^2 + \lambda\sum_{j=1}^{p}|\beta_j|^2,$$

and the PLS estimator becomes the ridge regression. In this case, all variables are kept in the model, but the resulting estimates are the shrunken version of the OLS estimates. Ridge regression often achieves good prediction performance, but it cannot produce a parsimonious model. The ridge estimator is the same as random-effect estimator, where $\beta_j$ are i.i.d. normal random effects.

The general version of the PLS is the penalized likelihood criterion

$$Q_\lambda(\beta) = \ell(\beta) - p_\lambda(\beta),$$

where $\ell(\beta) = \sum_{i=1}^n \log f_\phi(y_i|\beta)$ is the data log-likelihood, and $p_\lambda(\beta)$ is the sparseness penalty. We can in general put variable selection of any GLM-based regression model in this framework.

*Implied penalty functions*

Given fixed parameters $(w, \phi, \theta)$, the estimator of $\beta$ is obtained by maximizing the profile $h$-likelihood

$$h_p = (h_1 + h_2)|_{u=\hat{u}},$$

where $\hat{u}$ solves $dh/du = 0$. Since $h_1$ is the classical log-likelihood, the procedure corresponds to a penalized log-likelihood with implied penalty

$$p_\lambda(\beta) = -\phi h_2|_{u=\widehat{u}}.$$

Specifically, for fixed $w$, taking only terms that involve $\beta_j$ and $\widehat{u}_j$, the $j$th term of the penalty function is

$$p_\lambda(\beta_j) = \frac{\phi}{2\theta}\frac{\beta_j^2}{\widehat{u}_j} + \frac{\phi(w-2)}{2w}\log \widehat{u}_j + \frac{\phi}{w}\widehat{u}_j. \qquad (10.2)$$

Thus the random-effect model leads to a family of penalty functions $p_\lambda(\beta)$ that is indexed by $w$, which includes (i) the ridge ($w = 0$), (ii) LASSO penalties ($w = 2$) and (iii) the unbounded penalty with infinite value and derivative at the origin ($w > 2$).

As $w \to 0$, from (10.4) and (10.2) we get $\hat{u}_j \to 1$ and obtain the $L_2$-penalty

$$p_\lambda(\beta) \to \phi/(2\theta)\sum_{j=1}^p \beta_j^2 + \text{constant}.$$

For $w = 2$, again using (10.4) and (10.2) we have $\widehat{u}_j = |\beta_j|/\sqrt{\theta}$ to give the $L_1$-penalty

$$p_\lambda(\beta) = (\phi/\sqrt{\theta})\sum_{j=1}^p |\beta_j| + \text{constant}.$$

The penalty functions $p_\lambda(\cdot)$ at $w = 0$, 2, 10, and 30 with $\lambda = 1$ (solid),

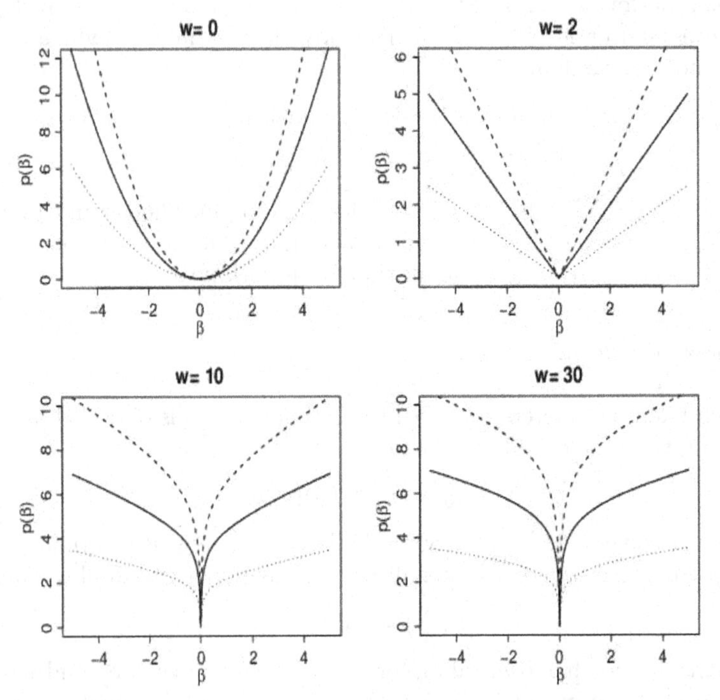


Figure 10.1 *Penalty function $p_\lambda(\beta)$ at different values of $w$, for $\lambda = 1$ (solid), $\lambda = 1.5$ (dashed) and $\lambda = 0.5$ (dotted). In general, larger values of $\lambda$ are associated with larger penalty, hence more shrinkage and more sparseness.*

1.5 (dashed) and 0.5 (dotted) are shown in Figure 10.1. As the concavity near the origin increases, the sparsity of local solutions increases, and as the slope becomes flat, the amount of shrinkage lessens. From Figure 10.1, we see that the DHGLM in Box 9 (referred to as HL below) controls the sparsity and shrinkage amount by choosing the values of $w$ and $\lambda$ simultaneously. The form of the penalty changes from a quadratic shape ($w = 0$) for ridge regression, to a cusped form ($w = 2$) for LASSO, and then to an unbounded form ($w > 2$) at the origin. Quadratic penalties correspond to ridge/shrinkage estimates, which often lead to better prediction, while cusped ones lead to simultaneous variable selection and estimation of LASSO and the smoothly-clipped absolute deviation (SCAD) penalty. Given $w > 2$, the amount of shrinkage becomes larger as $\lambda$ increases. Given $\lambda > 0$, the sparsity increases as $w$ becomes larger.

**Box 9: Random effect variable selection**

We now describe a random-effect model that generates a family of penalties, including the ridge-type (bell-shaped $L_2$), LASSO-type (cusped $L_1$), and a new unbounded penalty at the origin. In regression model (10.1), suppose $\beta$ are random effects; conditional on $u_j$, we have

$$\beta_j|u_j \sim \mathrm{N}(0, u_j\theta), \qquad (10.3)$$

where $\theta$ is a fixed dispersion parameter, and $u_j$'s are i.i.d. random variables. Because $\theta u_j = (a\theta)(u_j/a)$ for any $a > 0$, $\theta$ and $u_j$ are not separately identifiable. Thus, as is in HGLMs, we constrain $\mathrm{E}(u_j) = 1$. This imposes a constraint on random-effect estimates such that $\sum_{j=1}^{p} \hat{u}_j/p = 1$. Assume that $u_j$'s are from the gamma distribution with a parameter $w$ such that

$$f_w(u_j) = (1/w)^{1/w}\frac{1}{\Gamma(1/w)}u_j^{1/w-1}e^{-u_j/w}$$

with $\mathrm{E}(u_j) = 1$ and $\mathrm{Var}(u_j) = w$. In this random-effect model sparseness or selection is achieved in a transparent way, since if $\hat{u}_j = 0$ then $\hat{\beta}_j = 0$. This model (10.3) can be re-written as $\beta_j = \sqrt{\tau_j}e_j$, with $e_j \sim \mathrm{N}(0, 1)$ and

$$\log \tau_j = \log \theta + v_j,$$

and $v_j \equiv \log u_j$, which together with (10.1) defines a DHGLM, with h-likelilihood

$$h = h_1 + h_2,$$

where

$$h_1 = \sum_{i=1}^{n} \log f_\phi(y_i|\beta) = -\frac{n}{2}\log(2\pi\phi) - \frac{1}{2\phi}\sum_{i=1}^{n}(y_i - x_i^t\beta)^2,$$

$$h_2 = \sum_{j=1}^{p}\{\log f_\theta(\beta_j|u_j) + \log f_w(v_j)\},$$

$\log f_\theta(\beta_j|u_j) = -\frac{1}{2}\{\log(2\pi\theta) + \log u_j + \beta_j^2/(\theta u_j)\}$ and $\log f_w(v_j) = -\log(w)/w - \log\Gamma(1/w) + v_j/w - \exp(v_j)/w$. Here $\partial h/\partial u = 0$ gives the random-effect estimator

$$\widehat{u}_j \equiv \widehat{u}_j(\beta) = [\{8w\beta_j^2/\theta + (2-w)^2\}^{1/2} + (2-w)]/4. \qquad (10.4)$$

For $w = 0$ we get ridge regression and for $w = 2$ LASSO.

By controlling the amount of sparsity and shrinkage simultaneously, the HL has much higher chances of selecting the correct models without losing prediction accuracy than the other methods (Kwon et al., 2017). Ng et al. (2006) showed the consistency of all local solutions of the HL method, which implies the uniqueness of HL solution under certain conditions. Recently, Ng et al. (2017) showed that the HL estimator achieves consistent estimation of the number of change points, their locations, and their sizes, while LASSO and SCAD may not. Advantage of the HL method is to achieve asymptotic selection consistency without loosing prediction accuracy in finite samples.

*IWLS algorithm*

Given $\widehat{u}$, Lee and Oh (2014) proposed to update $\beta$ based on model (10.1) with $\beta$ satisfying (10.3). This is a purely random-effect model

$$Y = X\beta + e,$$

where $e \sim \mathrm{N}(0, \Sigma \equiv \mathrm{diag}\{\phi\})$, and $\beta \sim \mathrm{N}(0, D \equiv \mathrm{diag}\{\widehat{u}_j\theta\})$. From the mixed-model (Box 5) in Chapter 3 we update $\beta$ by solving

$$(X^T X + W_\lambda)\beta = Xy, \tag{10.5}$$

where $W_\lambda \equiv \mathrm{diag}\{\lambda/\widehat{u}_j\}$ and $\lambda = \phi/\theta$.

It is really fruitful to see the estimation in the simplest case with a single parameter $\beta$, where $z$ is the OLS estimate of $\beta$. Even more specifically, think of $\beta$ as the population mean and $z$ the sample mean. Here we can draw the penalized criterion, and illustrate various variable selection procedures in their effect on thresholding and shrinkage of the OLS estimate to zero. The IWLS step (10.5) gives

$$\hat{\beta} = z/[1 + \lambda/\hat{u}]. \tag{10.6}$$

The corresponding PLS criterion is

$$Q_\lambda(\beta) = \frac{1}{2}(z - \beta)^2 + p_\lambda(\beta). \tag{10.7}$$

The implied penalized likelihood $Q_\lambda(\beta)$ is nonconvex, but we use IWLSs for interconnected GLMs: (i) $y_i|\beta$ is normal, and (ii) $\beta_j|u_j$ is normal with (iii) gamma $u_j$. All three GLMs allow IWLSs for convex optimizations. Thus, the proposed IWLS algorithm overcomes the difficulties of a nonconvex optimization by solving three-interlinked convex optimizations.

Equalizing the score equations for $\beta$ from (10.6) and from the PLS (10.7) we have

$$\beta(1 + \lambda/\widehat{u}) - z = \partial Q_\lambda/\partial\beta = -(z - \beta) + p'_\lambda(\beta),$$

and we get a useful general formula

$$\widehat{u}(\beta) = \lambda\beta/p'_\lambda(\beta).$$

This formula allows us to obtain LASSO, SCAD, or the so-called adaptive LASSO (Zou, 2006) by using different random-effect estimates $\widehat{u}$ in the IWLS of (10.6). For the LASSO $p_\lambda(\beta) = \lambda|\beta|$, so $\widehat{u} = |\beta|$. The adaptive LASSO corresponds to $p_\lambda(\beta) = 2\lambda|\beta|/|z|$, so the estimate can be obtained by IWLS using

$$\widehat{u} = |\beta||z|/2$$

The SCAD estimate can be computed using

$$\widehat{u} = |\beta|/\left\{I(|\beta| \leq \lambda) + \frac{(a\lambda - |\beta|)_+}{(a-1)\lambda} I(|\beta| > \lambda)\right\},$$

for some $a > 2$.

From the model (10.3) it is clear that $\widehat{\beta}_j = 0$ when $\widehat{u}_j = 0$, which is how we achieve sparseness. We can allow thresholding by setting small $\widehat{u}_j$ to zero, but then the corresponding weight $1/\widehat{u}_j$ in $W_\lambda$ is undefined. We could exclude the corresponding predictors from (10.5), but instead we employ a perturbed random-effect estimate $\hat{u}_{\delta,k} = \lambda(|\beta_k| + \delta)/p'_\lambda(|\beta_k|)$ for a small positive $\delta = 10^{-8}$. Then, the weight is always defined. As long as $\delta$ is small, the solution is nearly identical to the original IWLS. Note that this algorithm is identical to that of Hunter and Li (2005) for the improvement of local quadratic approximation. For an alternative algorithm for the HL estimator see Kwon et al. (2017).

In random-effect models, we can use the ML or REML estimates for $(w, \phi, \theta)$, and compute the tuning parameter $\lambda$ as the ratio $\phi/\theta$. However, in the variable selection literature it is common to estimate the tuning parameter $\lambda$ by using the $K$-fold cross-validation method because in the PLS procedures $\lambda$ is not a model parameter.

### *10.2.2 Structured variable selection*

In many regression problems, the explanatory variables often possess a natural structure, so that we need variable selection, respecting these structures. In this section, we show a key advantage of the modeling approach, as they show how the variable selection is achieved more transparently, respecting the structures.

*Group selection*

The explanatory variables often possess a natural group structure. For example, (i) categorical factors are often represented by a group of indicator variables, and (ii) to capture flexible functional shapes, continuous factors can be represented by a linear combination of basis functions such as splines or polynomials. In these situations, the problem of selecting relevant variables is that of selecting groups rather than selecting individual variables. Depending on the situation, the individual variables in a group may or may not be meaningful scientifically. If they are not, we are typically not interested in selecting individual variables and the interest is limited to *group selection*. However, if the individual variables are meaningful, then we would be interested in selecting individual variables within each selected group; we refer to this as *bi-level selection* (Huang et al., 2012).

In this section we show how the group and bi-level selections can be achieved using the random-effect model approach. Suppose that the explanatory variables can be divided into $K$ groups, and the outcome $y = (y_1, \ldots, y_n)^T$ has mean $\mu = (\mu_1, \ldots, \mu_n)^T$, which follows a GLM with link function $\eta_i \equiv h(\mu_i)$, such that we have linear predictor

$$\eta = X\beta \equiv X_1\beta_1 + \cdots + X_K\beta_K,$$

where $\eta = (\eta_1, \ldots, \eta_n)^T$ is the vector of linear predictors;

$X \equiv (X_1, \cdots, X_K)$ is the collection of design matrices of the predictors; $\beta = (\beta_1, \ldots, \beta_K)^t$ is the vector of regression coefficients. Here, $X_k$ is $n \times p_k$ matrix and $\beta_k$ is a vector of length $p_k$; they are the design matrix and coefficients vector for the $k$th group, respectively.

For group variable selection, Lee et al. (2015b) considered a random-effect model

$$\beta_{kj}|u_k \sim \mathrm{N}(0, u_k\theta), \quad k = 1, \ldots, K, \; j = 1, \ldots, p_k, \tag{10.8}$$

and

$$u_k \sim \mathrm{gamma}(w_k), \quad k = 1, \ldots, K,$$

where, as before, $\theta$ and $w_k$ are regularization parameters that control the degree of shrinkage and sparseness of the estimates. Group selection is achieved as follows: if $\widehat{u}_k = 0$, then $\hat{\beta}_{kj} = 0$ for all $j = 1, \ldots, p_k$, whereas if $\hat{u}_k > 0$, then $\hat{\beta}_{kj} \neq 0$ for all $j = 1, \ldots, p_k$. This means that the model (10.8) is limited to group-only selection, as it does not impose sparsity within the selected groups.

Bi-level selection can be done by extending the model (10.8) as follows:

$$\beta_{kj}|u_k, v_{kj} \sim \mathrm{N}(0, u_k v_{kj}\theta), \quad k = 1, \ldots, K, \; j = 1, \ldots, p_k$$

and

$$u_k \sim \text{gamma}(w_k) \text{ and } v_{kj} \sim \text{gamma}(\tau).$$

Here $u_k$ is the random effect corresponding to the $k$th group and $v_{kj}$ is the random effect corresponding to the $j$th variable in the $k$th group. Hence this model selects variables at both the group level and the individual variable level within selected groups. For example, if $\widehat{u}_k = 0$, then $\hat{\beta}_{kj} = 0$ for all $j = 1, \ldots, p_k$. However, even if $\widehat{u}_k > 0$, $\hat{\beta}_{kj} = 0$ whenever $\hat{v}_{kj} = 0$.

*Interaction and hierarchy constraints*

Interaction terms in regression models form a natural hierarchy with the main effects, so their selection requires special considerations. For example, it is common practice that the presence of an interaction term requires both of the corresponding main effects in the model. This may be called a strong hierarchy constraint, while the weak version only requires one of the main effects to be present. This notion has also been called the "heredity principle" (Hamada and Wu, 1992). In these cases, it is sensible to select variables that respect the hierarchy. Accounting for hierarchy complicates the selection procedure, but it maintains the interpretability of the model. Using a random-effect model, we show how to impose sparse selection of interaction terms under the hierarchy constraints.

Consider a $p$-predictor regression model with both main and interaction terms. The outcome $y$ has mean $\mu$, and it follows a GLM with a linear predictor

$$\eta_i = \beta_0 + \sum_{j=1}^{p} x_{ij}\beta_j + \sum_{j<k} x_{ij}x_{ik}\delta_{jk}, \quad i = 1, \ldots, n,$$

which we shall write in matrix form as

$$\eta = X\beta + Z\delta$$

where $\eta \equiv (\eta_1, \ldots, \eta_n)$ is the vector of linear predictors, $\beta = (\beta_1, \ldots, \beta_p)$ and $\delta \equiv (\delta_{12}, \ldots, \delta_{p-1p})$ are the vectors of the corresponding regression coefficients for main and interaction terms, respectively. Similarly, $X$ is the design matrix of the intercept and linear terms for the main effects, and $Z$ is that of the cross-product terms for the interactions.

We now describe the random-effect model approach for sparse estimation of interaction terms with hierarchy constraint. The key advantage is that it is obvious how the constraint is achieved. Under the *strong hierarchy constraint,* it is required that if $\hat{\delta}_{kj} \neq 0$, then $\hat{\beta}_k \neq 0$ and $\hat{\beta}_j \neq 0$. To impose this constraint, Lee et al. (2015b) proposed the use of the random-effect model

$$\beta_j|u_j \sim \text{N}(0, u_j\theta), \quad \delta_{kj}|u_k, u_j, v_{kj} \sim \text{N}(0, u_k u_j v_{kj}\theta) \text{ for } k > j$$

and

$$u_j \sim \text{gamma}(w_1) \text{ and } v_{kj} \sim \text{gamma}(w_2).$$

Under this model, if $\hat{u}_j = 0$ or $\hat{u}_k = 0$ then $\hat{\delta}_{jk} = 0$. Conversely, if $\hat{\delta}_{jk} \neq 0$ then $\hat{\beta}_j \neq 0$ and $\hat{\beta}_k \neq 0$, and thus strong hierarchy holds; see Figure 10.2. Furthermore, sparsity is achieved if these random-effect estimates are zero with high probability. So, we have a conceptually transparent framework to impose the hierarchy constraint.

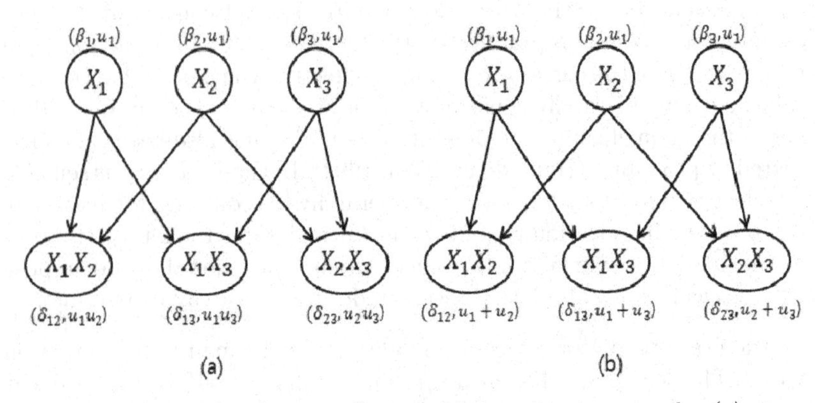


Figure 10.2 *A model with main effects and interaction terms under (a) strong hierarchy and (b) weak hierarchy constraints.*

Under the *weak-hierarchy constraint,* it is required that if $\hat{\delta}_{kj} \neq 0$ then either $\hat{\beta}_k \neq 0$ or $\hat{\beta}_j \neq 0$. To achieve this, one may use the random effect model

$$\beta_j|u_j \sim \text{N}(0, u_j\theta),$$
$$\delta_{kj}|u_k, u_j, v_{kj} \sim \text{N}(0, (u_k + u_j)v_{kj}\theta)$$

and

$$u_j \sim \text{gamma}(w_1) \text{ and } v_{kj} \sim \text{gamma}(w_2)$$

With this model the weak hierarchy constraint holds: If $\hat{u}_k = 0$ and $\hat{u}_j = 0$ then $\hat{\delta}_{kj} = 0$. This means that if $\hat{\delta}_{kj} \neq 0$ then either $\hat{\beta}_k \neq 0$ or $\hat{\beta}_j \neq 0$, i.e., an interaction can be included in the model if at least one corresponding main effect is included: see Figure 10.2.

*10.2.3 Functional marginality and general graph structure*

For completeness we describe here other statistical models where the notion of hierarchy applies, and show how to model them using the random effects approach. Suppose that we are interested to fit the second-order mixed polynomial model

$$\eta = X_1\beta_1 + \cdots + X_p\beta_p + X_1^2\delta_{11} + X_1X_2\delta_{12} \cdots + X_p^2\delta_{pp},$$

where $X_kX_j$ denote the componentwise product between two column vectors $X_k$ and $X_j$ of $X$. To maintain the functional marginality rule, if the second-order terms $X^2$ and $X_1X_2$ are in the model, then the first-order term $X$, and $X_1$ and $X_2$ should also be in the model. Lee et al. (2015b) proposed the use of the random-effect model

$$\begin{aligned} \beta_j|u_j &\sim \mathrm{N}(0, u_j\theta), \\ \delta_{jj}|u_j, v_{jj} &\sim \mathrm{N}(0, u_jv_{jj}\theta), \\ \delta_{kj}|u_k, u_j, v_{kj} &\sim \mathrm{N}(0, u_ku_jv_{kj}\theta) \end{aligned} \tag{10.9}$$

and

$$u_j \sim \mathrm{gamma}(w_1) \quad \text{and} \quad v_{kj} \sim \mathrm{gamma}(w_2).$$

With this model the functional marginality rule holds. If $\hat{u}_k = 0$ then $\hat{\delta}_{kj} = 0$ for all $j$. For example, if $\hat{\delta}_{kj} \neq 0$ then $\hat{\beta}_k \neq 0$ and $\hat{\beta}_j \neq 0$. This model is analogous to the strong hierarchy in the previous model, but now we include $\delta_{jj}$. It is not difficult to extend the model (10.9) to general higher-order models.

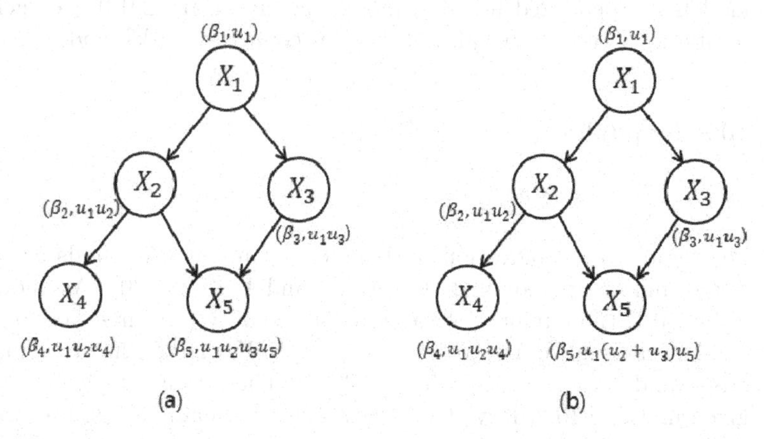


Figure 10.3 *The directed graph structure representing hierarchy of variables under (a) strong hierarchy and (b) weak hierarchy constraints.*

Various hierarchical structures can be represented by a directed graph as in Figure 10.3. We could consider graphs with either a strong or weak hierarchy constraint. In Figure 10.3(a) for strong hierarchy, $X_5$ can be included if $(X_1, X_2, X_3)$ are included in the model. This directed graph can be modeled by the following random-effect model

$$\begin{aligned} \beta_1|u_1 &\sim \mathrm{N}(0, u_1\theta), \\ \beta_2|u_1, u_2 &\sim \mathrm{N}(0, u_1u_2\theta), \\ \beta_3|u_1, u_3 &\sim \mathrm{N}(0, u_1u_3\theta), \\ \beta_4|u_1, u_2, u_4 &\sim \mathrm{N}(0, u_1u_2u_4\theta), \\ \beta_5|u_1, u_2, u_3, u_5 &\sim \mathrm{N}(0, u_1u_2u_3u_5\theta) \end{aligned}$$

and

$$u_j \sim \mathrm{gamma}(w), \quad j = 1, \ldots, 5.$$

In Figure 10.3(b) for weak hierarchy, $X_5$ can be included if the model includes, besides $X_1$, at least one of $X_2$ and $X_3$. This directed graph can be modeled by

$$\beta_5|u_1, u_2, u_3, u_5 \sim \mathrm{N}(0, u_1(u_2 + u_3)u_5\theta).$$

This illustrates how the random-effect model can be adapted to describe various hierarchical structures in the covariates.

The HL method can be easily applied to produce sparse versions of classical multivariate techniques, such as the principle component analysis, canonical covariance analysis, partial-least squares for Gaussian and that for survival outcomes (Lee et al., 2010, 2011a,b, 2013). Furthermore, it is straightforward to apply the HL method to various classes of HGLM models via penalized h-loglikelihood; see Ha et al. (2014b) for general frailty model and Ha et al. (2014a) for competing risks models.

## 10.3 Examples

### *10.3.1 Diabetes data*

For the disease progression of diabetes again, three methods are computed and compared: LASSO, SCAD and HL($w$ = 30). As shown in Table 10.1, the numbers of variables selected by the three methods are similar, varying from 10 to 15, though the HL method has the smallest cross-validated error (Kwon et al., 2016). The estimated coefficients are given in Table 10.2. If we look at estimates of main effects, the LASSO estimators are shrunk the most and the SCAD estimators the least. We see that all methods include the age:sex interaction in their final model, consistent with the known result that diabetes progression behaves differently in women after menopause.

Table 10.1 *Analysis of the diabetes data ($n = 442$, $p = 64$)*

| Method | LASSO | SCAD | HL |
|---|---|---|---|
| Number of variables | 15 | 12 | 14 |
| CV error | 2988.69 | 2982.85 | 2891.76 |

Table 10.2 *Estimated coefficients of the diabetes data*

| | LASSO | SCAD | HL |
|---|---|---|---|
| sex | −5.43 | −11.07 | −10.86 |
| bmi | 23.89 | 25.14 | 23.63 |
| map | 12.04 | 15.16 | 15.17 |
| hdl | −9.00 | −12.98 | −12.52 |
| ltg | 22.28 | 23.49 | 22.89 |
| glu | 0.89 | | 2.93 |
| age | | | |
| age$^2$ | 0.35 | 0.95 | 2.76 |
| bmi$^2$ | 1.29 | 0.06 | 2.13 |
| glu$^2$ | 2.25 | 2.31 | 3.51 |
| age:sex | 5.26 | 7.33 | 7.46 |
| age:map | 1.53 | 0.68 | 1.69 |
| age:ltg | 0.43 | 0.01 | 1.55 |
| age:glu | 0.58 | | |
| sex:map | 0.03 | | 2.29 |
| bmi:map | 3.87 | 5.23 | 5.13 |

As seen in Table 10.2, the HL solution chose interaction terms of age and a quadratic term of age, but age itself has not been chosen. This violates the functional marginality rule discussed in Section 2.1. The same problem is encountered with the other methods. With an automatic variable selection method with large $p$, the marginality rule will be easily violated, i.e., higher-order terms can appear without necessary lower-order interactions. A systematic way of handling such a problem is grouped model selection as we shall show.

### *10.3.2 Gene-gene interaction*

As an illustration we analyze gene-gene interaction in a cohort study called ULSAM (Uppsala Longitudinal Study of Adult Men). This is an ongoing population-based study of all available men in Uppsala County, Sweden, born between 1920 and 1924. We shall analyze a subset of $n =$ 1 179 subjects for which we have genetic data. These subjects were in average 71 years old, ranging from 70 to 74. The primary outcome is body-mass index (BMI), a major risk factor for many cardiovascular diseases; in this cohort, the BMI has an average of 26.3 and ranges from 16.7 to 46.3. Based on several criteria, we selected 10 single-nucleotide polymorphisms (SNPs) as the predictor variables; see Lee et al. (2015a) for details and for the exact identities of the SNPs.

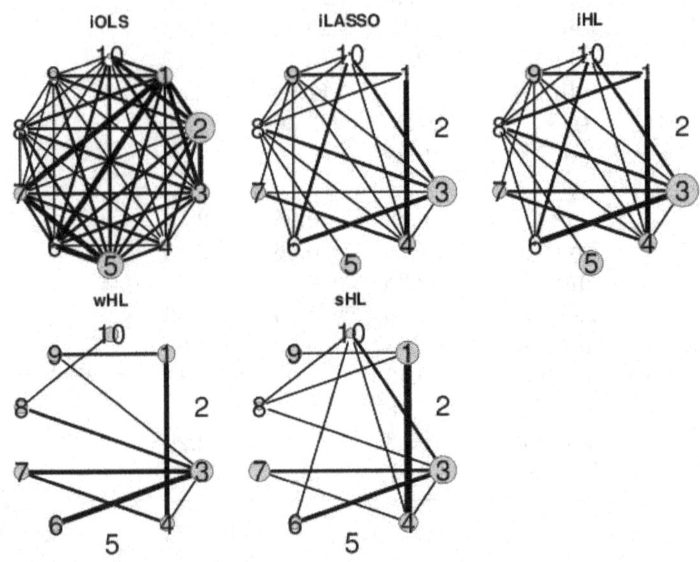


Figure 10.4 *Results from various methods applied to ULSAM data. In each graph, each node represents an SNP, the size of the main effect is represented by the circle size, and the interaction by the thickness of the line between two nodes.*

Figure 10.4 shows the results from various methods based on the interaction model; we put the prefix $i$ to indicate that all pairwise interactions are included in the model. The ordinary least-squares (iOLS) method estimates all the interaction terms, which are not informative: particularly, it cannot recognize the linkage disequilibrium between SNPs 1, 2, and 5. The largest interactions are between SNP pairs (1,6), (1,7), and (5,7). In contrast, all the sparse methods select (1,4) and (3,6) as the most interesting pairs. The hierarchy-constrained $w$HL and $s$HL have comparable sparsity, and they both select only one of the linked SNPs

1, 2, and 5. As expected, the unconstrained iLASSO and iHL methods select interaction terms without main effects, e.g., (6,10) and (1,8). Including an interaction term without main effects can lead to misleading conclusions, since for example the interpretation depends crucially on the coding of the variables, which is why it is not recommended in practice. If strong hierarchy is desired, the *s*HL method provides a sensibly sparse solution in this case.

## 10.4 Hypothesis testing

Recently, technological developments in areas such as genomics and neuroimaging have led to situations where thousands or more hypotheses need to be tested simultaneously. The problem of such a large-scale multiple testing can be expressed as a prediction problem for finding the true discoveries in an optimal way (Lee and Bjørnstad, 2013).

In this chapter, we formulate hypothesis testing as an HGLM with discrete random effects. Here all the extended likelihoods, defined in any scale of the discrete random effect, are the h-likelihood. In this chapter we study how to use the h-likelihood for the inferences about discrete random effects in multiple testing. We first study this by reviewing classical single hypothesis testing.

### *10.4.1 Classical single hypothesis testing*

Suppose that we are interested in a simple hypothesis testing problem on the mean of a random variable $Y \sim \mathrm{N}(\mu, 1)$

$$H_0 : \mu = \mu_0 \text{ vs. } H_1 : \mu = \mu_1.$$

The classical Neyman-Pearson likelihood ratio is

$$L = \frac{f(y|H_1)}{f(y|H_0)}.$$

Now, we introduce the idea of presenting the alternative hyptheses as a random effect, and will then give the h-likelihood ratio, $R$.

Let a discrete random effect be $o = 0$ if $H_0$ is true and $o = 1$ if $H_1$ is true, then the h-likelihood is

$$f(y, o) = f(y|o)P(o)$$

where $f(y|H_0) = f(y|o = 0)$ and $f(y|H_1) = f(y|o = 1)$. Hence, the h-likelihood ratio is

$$R = \frac{f(y, o = 1)}{f(y, o = 0)}.$$

For single hypothesis testing, $L$ and $R$ are proportional to each other, because

$$R = \frac{f(y, o = 1)}{f(y, o = 0)} = \frac{f(y|H_1)P(o = 1)}{f(y|H_0)P(o = 0)} = \frac{1 - p_0}{p_0} L$$

where $p_0 = P(o = 0)$.

When $p_0 = 0$ or $1$ $R$ is infinity or zero regardless of the values of $L$. Thus, for the test depending on the value of $L$, $p_0$ should be strictly between 0 and 1. However, with single hypothesis testing, $p_0$ is not estimable strictly between 0 and 1. Consequently, both $L$ and $R$ give equivalent optimal tests (see section below *Optimal likelihood ratio testing*), but the h-likelihood ratio, with the alternative hypotheses given as a discrete random effect, opens up a way for testing multiple hypotheses where $p_0$ is a constant between 0 and 1.

*Ratio of predictive probabilities*

Above we showed that the h-likelihood ratio is the product of the odds of the hypothesis, $(1 - p_0)/p_0$, and the classical likelihood ratio $L$. But it can also be interpreted as a ratio of predictive probabilities.

Let

$$f(y) = f(y|H_0)P(o = 0) + f(y|H_1)P(o = 1)$$

be the marginal distribution of $y$ with the random effects eliminated from the h-likelihood, and let

$$P(o = i|y) = \frac{f(y|H_i)P(o = i)}{f(y)}$$

be the predictive probability satisfying

$$P(o = 0|y) + P(o = 1|y) = 1.$$

The h-likelihood ratio is then the ratio of predictive probabilities, because

$$R = \frac{f(y, o = 1)}{f(y, o = 0)} = \frac{P(o = 1|y)f(y)}{P(o = 0|y)f(y)} = \frac{P(o = 1|y)}{P(o = 0|y)}.$$

Table 10.3 *Outcome of single hypothesis problem*

| | $\delta = 0$ | $\delta = 1$ |
|---|---|---|
| $o = 0$ | OK | Type I error |
| $o = 1$ | Type II error | OK |

*Optimal likelihood ratio testing*

Let $\delta$ be a predictor (test) for the true state $o$, i.e., $\delta = 1$ (discovery) if we reject $H_0$ and $\delta = 0$ (non-discovery) if we do not reject $H_0$. Now we can show that the optimal test is determined by the ratio of predictive probabilities $R$, equivalent to the h-likelihood-ratio. The latter is much easier to compute than the former. Note that

$$\text{type I error is } P(\delta = 1|H_0),$$
$$\text{type II error is } P(\delta = 0|H_1),$$
$$\text{the power is } P(\delta = 1|H_1).$$

The outcome of the test can be summarized as in Table 10.3.

To derive an optimal test we introduce the loss function that depends on a constant $\lambda$. With the loss

$$o(1 - \delta) + \lambda(1 - o)\delta,$$

given the data, we have the risk,

$$\begin{aligned} E\{o(1 - \delta) + \lambda(1 - o)\delta|y\} &= E\{o + \lambda(1 - o)\delta - o\delta|y\} \\ &= P(o = 1|y) + P(o = 0y)\{\lambda - R\delta\}. \end{aligned}$$

Thus, for each value of $\lambda$, the optimal test $\delta^{\lambda}$, minimizing the risk, is determined by the h-likelihood-ratio,

$$\delta^{\lambda} = I\{R > \lambda\}.$$

Note, however, that the h-likelihood-ratio

$$R = \frac{(1 - p_0)}{p_0} L,$$

is the product of the odds for the hypotheses, $(1 - p_0)/p_0$, and the Neyman-Pearson likelihood-ratio

$$L = \frac{\mathrm{N}(\mu_1, 1)}{\mathrm{N}(\mu_0, 1)} = \frac{f(y|H_1)}{f(y|H_0)}.$$

With a single hypothesis testing, $p_0$ (and therefore predictive probability) may not be estimable, so that we need to define the optimal test $\delta^\lambda$ without using estimator of $p_0$. In the single hypothesis testing, we define the optimal $\delta^\lambda$ as

$$\delta^\lambda = I\{R > \lambda\} = I\{L > \lambda^*\},$$

by determining $\lambda^* = \lambda p_0/(1-p_0)$ for some given $p_0$ with $0 < p_0 < 1$; we choose $\lambda^*$ to satisfy

$$P(\delta^{\lambda^*} = 1|H_0) \leq \alpha,$$

for a pre-specified significant level $\alpha = 0.05$ or $0.01$. It is well established that this Neyman-Pearson likelihood-ratio test $\delta^\lambda$ is generally the most powerful (efficient). Under our random-effect model, the optimal test $\delta^\lambda$ is a predictor for random effect $o$, so that we can view the likelihood-ratio test as a prediction of discrete random effect, controlling a prediction error (here type I error) in a prespecified significant level $\alpha$.

**Box 10: False discovery rate**

Suppose that we have $N$ null hypotheses $H_1, \cdots, H_N$ to test simultaneously. Table 10.4 summarizes the possible outcomes of a multiple testing, where $V_{01}$ is the number of false discoveries (positives) and $V_{10}$ is the number of false non-discoveries (nulls). In multiple testing, we observe $M_0$, $M_1$, and $N$ while all remaining cells such as $N_0$, $N_1, V_{00}, V_{01}, V_{10}$, and $V_{11}$, in Table 10.4 are unobserved random variables.

In multiple testing, Benjamini and Hochberg (1995) proposed to control the false discovery rate (FDR) among declared discoveries. They define FDR as the expected proportion of errors among rejected hypotheses (discoveries). The proportion of errors among the discoveries is $V_{01}/M_1$ and Benjamini and Hochberg (1995) suggested controlling the false discovery rate $E(V_{01}/M_1)$. Following Efron (2004), in this chapter we use the (marginal) FDR

$$FDR = E(V_{01})/E(M_1)$$

and the false non-discovery rate (FNDR),

$$FNDR = E(V_{10})/E(M_0).$$

Note that $E(V_{01})/E(N_0)$ and $E(V_{10})/E(N_1)$ are type I and type II error rates, respectively. The type I error rate is the false discovery rate under the null.

Table 10.4 *Outcomes of multiple testing*

| | $\delta = 0$ | $\delta = 1$ | Total |
|---|---|---|---|
| $o = 0$ | $V_{00}$ | $V_{01}$(Type I error) | $N_0$ |
| $o = 1$ | $V_{10}$ (Type II error) | $V_{11}$ | $N_1$ |
| Total | $M_0$ | $M_1$ | $N$ |

### *10.4.2 Multiple testing*

Most literature on multiple testing has focused on the error control of the test without paying much attention on the power of the test. Here, we model multiple testing as an HGLM with discrete random effects and show that the h-likelihood gives the optimal test maximizing the power of the test. The h-likelihood leads to an extension of the Neyman-Pearson likelihood-ratio test, which is often the most powerful in single hypothesis testing.

### *10.4.3 Random-effect model for multiple testing*

Suppose that a response $y_{ij1}$ for the $i$th site of the $j$th individual in the control group can be modeled for $i = 1, \ldots, N$ and $j = 1, \ldots, n_1$ as

$$y_{ij1} = \xi_i + \epsilon_{ij1},$$

while the response $y_{ij2}$ in the treatment group is modeled for $j = 1, \ldots, n_2$ as

$$y_{ij2} = \xi_i + w_i + \epsilon_{ij2},$$

where $\xi_i$ is the site effect, $w_i$ is the random treatment effect for the $i$th site, and $\epsilon_{ijm}$ is the error with $E(\epsilon_{ijm}) = 0$ and $\text{var}(\epsilon_{ijm}) = \phi_{im}$, for $m = 1, 2$. We shall express the model in terms of the difference in means:

$$d_i = \bar{y}_{i2} - \bar{y}_{i1},$$

where $\bar{y}_{im} = \sum_j y_{ijm}/n_m$ and $\psi_i = \phi_{i1}/n_1 + \phi_{i2}/n_2$. Then, we have

$$E(d_i|w_i) = w_i \quad \text{and} \quad \text{var}(d_i|w_i) = \psi_i.$$

The formal null hypotheses are of the form $H_{0i}$: $w_i = 0$. The inferential problem of interest is to identify only those cases with large effect sizes $w_i$, not those with $w_i$ close to 0. To achieve this, it is natural to assume that for the "Uninteresting" (null) cases, the $w_i$'s are independent with

$$E(w_i|H_{0i}) = 0 \quad \text{and} \quad \text{var}(w_i|H_{0i}) = \sigma^2,$$

with typically $0 < \sigma \ll 1$. The "Interesting" (alternative) cases are assumed to be independent with

$$E(w_i|H_{1i}) = \mu \neq 0 \text{ and } \text{var}(w_i|H_{1i}) = \tau^2.$$

Since this multiple testing problem has two states, let $o_i = 1$ if case $i$ is Interesting ($H_{1i}$), and $o_i = 0$ if case $i$ is Uninteresting ($H_{0i}$). Suppose that $P(o_i = 0) = p_0$ and $P(o_i = 1) = 1 - p_0$.

For simplicity of arguments in this book, we consider the case that $o_i$ are independent. Suppose that we have the following hierarchical model:

$$\begin{array}{l} \text{Conditional on } w_i \text{ and } o_i, \;\; E(d_i|w_i, o_i) = w_i \text{ and } \text{var}(d_i|w_i, o_i) = \psi_i \\ \text{Conditional on } o_i = 0, \; E(w_i|H_{0i}) = 0 \text{ and } \text{var}(w_i|H_{0i}) = \sigma^2 \\ \text{Conditional on } \; o_i = 1, \; E(w_i|H_{1i}) = \mu \text{ and } \text{var}(w_i|H_{1i}) = \tau^2. \end{array}$$

In this model we have $2N$ unobservables, $v = (w, o)$, where $w$ and $o$ are, respectively, the vector of $w_i$ and $o_i$. Let $y$ be the set of all observations. The h-likelihood for the unknown quantities $(v, \theta)$ is defined to be

$$L(v, \theta; y, v) = f_\theta(y, v) = f_\theta(y) P_\theta(v|y).$$

Suppose that effect sizes $w_i$ are not of inferential interest, we can eliminate them by integration. This leads to a hierarchical model for $d = (d_1, \cdots, d_N)$ with independent $d_i$:

$$\begin{aligned} \text{Given } o_i &= 0, \; E(d_i|H_{0i}) = 0 \text{ and } \text{var}(d_i|H_{0i}) = \psi_i + \sigma^2, \\ \text{given } o_i &= 1, \; E(d_i|H_{1i}) = \mu \text{ and } \text{var}(d_i|H_{1i}) = \psi_i + \tau^2. \end{aligned}$$

Here, the h-likelihood is given by

$$L(o) = L(o, \theta; d, o) = f_\theta(d, o) = \prod_{i=1}^{N} f_\theta(d_i, o_i) = \prod_{i=1}^{N} L(o_i),$$

where $L(o_i) = L(o_i, \theta; d_i, o) = f_\theta(d_i, o_i)$. Furthermore, we have

$$\begin{aligned} L(o_i &= 1) = P(o_i = 1) f_\theta(d_i|o_i = 1) = (1 - p_0) f_\theta(d_i|H_{1i}), \\ L(o_i &= 0) = P(o_i = 0) f_\theta(d_i|o_i = 0) = p_0 f_\theta(d_i|H_{0i}). \end{aligned}$$

**Box 11: h-likelihood-ratio test for multiple testing**
Consider the loss

$$\sum\{o_i(1-\delta_i)+\lambda(1-o_i)\delta_i\}.$$

Then, as is in the single hypothesis testing, given $\lambda$, the optimal rule $\delta^\lambda=\{\delta_1^\lambda,\cdots,\delta_N^\lambda\}$ becomes

$$\delta_i^\lambda=I(R_i>\lambda). \quad (10.10)$$

This shows that the likelihood-ratio test in single hypothesis testing can be straightforwardly extended to multiple testing via the h-likelihood. In multiple testing,

$$p_0=\frac{E(N_0)}{N}$$

is estimable, so that $R_i$ can be used directly to control the error rate.
For controlling the FDR, we have

$$FDR(\lambda)=\frac{E(V_{01})}{E(M_1)}=\frac{\sum P(o_i=0,\delta_i^\lambda=1)}{E(\sum\delta_i^\lambda)},$$

where $V_{01}=\sum\delta_i^\lambda(1-o_i)$. It follows that an estimated FDR is given by

$$\widehat{FDR(\lambda)}=\frac{\hat{p}_0\sum P(R_i>\lambda|H_{0i})}{\sum I(R_i>\lambda)},$$

and FDR can be controlled by using $\widehat{FDR(\lambda)}$ at a specific level by varying $\lambda$. Parameters can be estimated by maximizing the marginal likelihood (Lee and Bjørnstad, 2013). Thus, if the $MLE$ for $\hat{\theta}$ is consistent the likelihood-ratio test is asymptotically optimal (the most powerful, having the least FNDR). Lee and Lee (2017) showed that this likelihood-ratio test can be easily extended for correlated cases.

*Optimal test*

The h-likelihood-ratio of $o_i=1$ vs. $o_i=0$ is

$$R_i=\frac{L(o_i=1)}{L(o_i=0)}=\frac{P(o_i=1|d_i)f_\theta(d_i)}{P(o_i=0|d_i)f_\theta(d_i)}=\frac{P(o_i=1|d_i)}{P(o_i=0|d_i)}=\frac{(1-p_0)}{p_0}L_i,$$

where $f_\theta(d_i)=f_\theta(d_i|H_{0i})p_0+f_\theta(d_i|H_{1i})(1-p_0)$ is the marginal likelihood component from the $i$th difference $d_i$ and

$$L_i=f_\theta(d_i|H_{1i})/f_\theta(d_i|H_{0i})$$

is the Neyman-Pearson likelihood-ratio test for the $i$th hypothesis. Let $\delta_i$ be a test for the $i$th hypothesis $H_i$, i.e., $\delta_i = 1$ (discovery) if $H_i$ is rejected and 0 (non-discovery) if not. The optimal test, h-likelihood-ratio test, is in Box 11, which shows how to control FDR error rate. For further developments, see Lee et al. (2017).

### *10.4.4 Neuroimaging data*

Not much is known about gender-related differences in behavioral performance or the functional activation of brain regions during the resting state. During specific cognitive tasks such as language or visuospatial tasks, there is clearly a gender difference (Bell et al., 2006). Males have significantly greater glucose metabolism than females in bilateral motor cortices and the right temporal region, including the hippocampus, which may explain an increased visuospatial and motor functionality often found among males. In contrast, females tend to have greater glucose metabolism in the left frontal cortex, including the inferior and orbitofrontal cortices, which could be associated with better language and emotional processing for females. Lee and Lee (2017) considered positron emission tomography (PET) data from the study of the Korean standard template by Lee et al. (2003). The data consist of scans of 28 healthy males and 22 healthy females. Each image has $N$= 189,201 voxels. For detailed descriptions of the data, refer to Lee et al. (2003).

The goal is to identify the significantly different voxels of the brain between males and females. Figure 10.5 shows the significantly different areas in the brain by the likelihood-ratio test. Previous methods have not identified any voxel in the brain to be significant for these data and we conclude that our method based on the likelihood-ratio test is the most powerful one. Lee and Lee (2017) showed that a larger number of significant voxels can be identified by modeling the spatial correlation structure of the brain images.

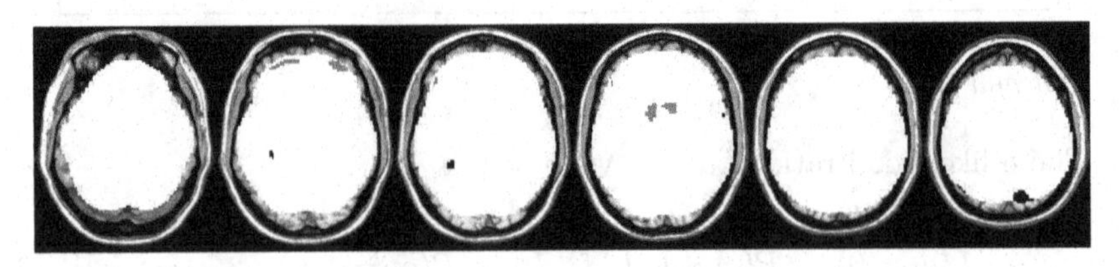

Figure 10.5 *Multiple testing for the neuroimage data by using likelihood-ratio testing. The gray-colored (black-colored) regions are negatively (positively) activated.*

# References


Agresti, A. (2007). *An Introduction to Categorical Data Analysis*. John Wiley & Sons, New York, 2nd edition.

Alam, M., Rönnegård, L., and Shen, X. (2015). *hglm: Hierarchical Generalized Linear Models*. R package version 2.1-1.

Asar, O. and Ilk, O. (2013). mmm: An R package for analyzing multivariate longitudinal data with multivariate marginal models. *Computer Methods and Programs in Biomedicine*, 112:649–654.

Asar, O. and Ilk, O. (2014). *mmm: an R Package for Analyzing Multivariate Longitudinal Data with Multivariate Marginal Models*. R package version 1.4.

Azzalini, A., Bowman, A. W., and Hardle, W. (1989). On the use of nonparametric regression for model checking. *Biometrika*, 76:1–11.

Bates, D. and Maechler, M. (2009). *lme4: Linear Mixed-Effects Models Using S4 Classes*. R package version 0.999375-32.

Bell, E. C., Willson, M. C., Wilman, A. H., and Dave, S. (2006). Males and females differ in brain activation during cognitive tasks. *Neuroimage*, 30:529–538.

Benjamini, Y. and Hochberg, Y. (1995). Controlling the false discovery rate: A practical and powerful approach to multiple testing. *Journal of the Royal Statistical Society: Series B (Statistical Methodology)*, 57:289–300.

Birnbaum, A. (1962). On the foundations of statistical inference. *Journal of the American Statistical Association*, 57:269–306.

Bissell, A. F. (1972). A negative binomial model with varying element sizes. *Biometrika*, 59:435–441.

Bjørnstad, J. F. (1990). Predictive likelihood principle: A review (with discussion). *Statistical Science*, 5:242–265.

Bjørnstad, J. F. (1996). On the generalization of the likelihood function and the likelihood principle. *Journal of the American Statistical Association*, 91:791–806.

Bock, R. D. and Lieberman, M. (1970). Fitting a reponse model for n dichotomously scored items. *Psychometrika*, 35:179–197.

Bollerslev, T. (1986). Generalized autoregressive conditional heteroskedasticity. *Journal of Econometrics*, 31:307–327.

Breslow, N. E. (1972). Discussion of professor Cox's paper. *Journal of the Royal Statistical Society: Series B (Statistical Methodology)*, 34:216–217.

Breslow, N. E. (1974). Covariance analysis of censored survival data. *Biometrics*, 30:89–99.

Breslow, N. E. (1984). Extra-poisson variation in log-linear models. *Applied Statistics*, 33:38–44.

Breslow, N. E. and Clayton, D. G. (1993). Approximate inference in generalized linear mixed models. *Journal of the American Statistical Association*, 88:9–25.

Cao, C., Shi, J. Q., and Lee, Y. (2017). Robust functional regression model for marginal mean and subject-specific inferences. *Statistical Methods in Medical Research*, in press.

Carlin, B. P. and Louis, T. A. (2000). Empirical Bayes: Past, present and future. *Journal of the American Statistical Association*, 95:1286–1289.

Christian, N. J., Ha, I. D., and Jeong, J. (2016). Hierarchical likelihood inference on clustered competing risks data. *Statistics in Medicine*, 35:251–267.

Clayton, D. G. and Kaldor, J. (1987). Empirical Bayes estimates of age-standardized relative risks for use in disease mapping. *Biometrics*, 43:671–681.

Cochran, W. G. and Cox, G. M. (1957). *Experimental Designs*. John Wiley & Sons, New York, 2nd edition.

Crowder, M. J. (1978). Beta-binomial anova for proportions. *Applied Statistics*, 27:34–37.

Demster, A., Laird, N., and Rubin, D. (1977). Maximum likelihood from incomplete data via the EM-alogrithm. *Journal of the Royal Statistical Society: Series B (Statistical Methodology)*, 39:1–38.

Diggle, P., Liang, K. Y., and Zeger, S. L. (1994). *Analysis of Longitudinal Data*. Clarendon Press, Oxford.

Donohue, M., Overholser, R., Xu, R., and Vaida, F. (2011). Conditional Akaike information under generalized linear and proportional hazards mixed models. *Biometrika*, 98:685–700.

Donohue, M. and Xu, R. (2013). *phmm: Proportional Hazards Mixed-Effects Model (PHMM)*. R package version 0.7-5.

Duchateau, L. and Janssen, P. (2008). *The Frailty Model.* Springer-Verlag, New York, 1st edition.

Durbin, J. and Koopman, S. J. (2000). Time series analysis of non-gaussian observations based on state space models from both classical and Bayesian perspectives (with discussion). *Journal of the Royal Statistical Society: Series B (Statistical Methodology)*, 62:3–56.

Efron, B. (2004). Large scale simultaneous hypothesis testing: The choice of a null hypothesis. *Journal of the American Statistical Association*, 99:96–104.

Elston, D. A., Moss, R., Boulinier, T., Arrowsmith, C., and Lambin, X. (2001). Analysis of aggregation, a worked example: Numbers of ticks on red grouse chicks. *Parasitology*, 122:563–569.

Engel, J. (1992). Modelling variation in industrial experiments. *Applied Statistics*, 41:579–593.

Engle, R. F. (1982). Autoregressive conditional heteroscedasticity with estimates of the variance of United Kingdom inflation. *Econometrica: Journal of the Econometric Society*, pages 987–1007.

Fan, J. and Li, R. (2001). Variable selection via nonconcave penalized likelihood and its oracle properties. *Journal of the American Statistical Association*, 96:1348–1360.

Felleki, M., Lee, Y., Lee, D., Gilmour, A., and Rönnegård, L. (2012). Estimation of breeding values for mean and dispersion, their variance and correlation using double hierarchical generalized linear models. *Genetics Research*, 94:307–317.

Fieuws, S. and Verbeke, G. (2006). Pairwise fitting of mixed models for the joint modeling of multivariate longitudinal profiles. *Biometris*, 62:424–431.

Firth, D., Glosup, J., and Hinkley, D. V. (1991). Model checking with nonparametric curves. *Biometrika*, 78:245–252.

Fisher, R. A. (1922). On the mathematical foundations of theoretical statistics. *Philosophical Transaction of the Royal Statistical Society*, 222:309–368.

Fleming, T. R. and Harrington, D. P. (1991). *Counting Processes and Survival Analysis.* John Wiley & Sons, New York.

Fox, J., Weisberg, S., Adler, D., Bates, D., Baud-Bovy, G., Ellison, S., Firth, D., Friendly, M., Gorjanc, G., Graves, S., Heiberger, R., Monette, G., Murdoch, D., Nilsson, H., Ogle, D., Riply, B., Venables, W., Winsemius, D., and Zeileis, A. (2016). *car: Companion to Applied Regression.* R package version 2.1-2.

Gonzalez, J. R., Rondeau, V., Mazroui, Y., Mauguen, A., and Diakité, A. (2012). *frailtypack: Frailty Models Using a Semi-Parametrical Penalized Likelihood Estimation or a Parametrical Estimation.* R package version 2.2-23.

Goossens, H. H. L. M. and Opstal, A. J. V. (2000). Blink-perturbed saccades in monkey. i. behavioral analysis. *Journal of Neurophysiology*, 83:3411–3429.

Grambauer, N. and Neudecker, A. (2011). *compeir: Event-Specific Incidence Rates for Competing Risks Data.* R package version 1.0.

Green, P. J. and Silverman, B. W. (1994). *Nonparametric Regression and Generalized Linear Models: A Roughness Penalty Approach.* Chanpman and Hall, London, 1st edition.

Gruen, B., Leisch, F., Sarkar, D., Mortier, F., and Picard, N. (2015). *flexmix: Flexible Mixture Modeling.* R package version 2.3-13.

Guo, X. and Carlin, B. P. (2004). Separate and joint modeling of longitudinal and event time data using standard computer packages. *American Statistician*, 58:16–24.

Ha, I. D., Jeong, J., and Lee, Y. (2017). *Statistical Modelling of Survival Data with Random Effects: H-Likelihood Approach.* Springer.

Ha, I. D., Lee, M., Oh, S., Jeong, J., Sylvester, R., and Lee, Y. (2014a). Variable selection in subdistribution hazard frailty models with competing risks data. *Statistics in Medicine*, 33:4590–4604.

Ha, I. D. and Lee, Y. (2003). Estimating frailty models via poisson hierarchical generalized linear models. *Journal of Computational and Graphical Statistics*, 12:663–681.

Ha, I. D. and Lee, Y. (2005). Comparison of hierarchical likelihood versus orthodox best linear unbiased predictor approaches for frailty models. *Biometrika*, 92:717–723.

Ha, I. D., Lee, Y., and MacKenzie, G. (2007). Model selection for multicomponent frailty models. *Statistics in Medicine*, 26:4790–4807.

Ha, I. D., Lee, Y., and Noh, M. (2015). *jointdhglm: Joint Modelling for DHGLMs and Frailty Models.* R package version 1.1.

Ha, I. D., Lee, Y., and Song, J. K. (2001). Hierarchical likelihood approach for frailty models. *Biometrika*, 88:233–243.

Ha, I. D., Noh, M., and Lee, Y. (2010). Bias reduction of likelihood estimators in semi-parametric frailty models. *Scandinavian Journal of Statistics*, 37:307–320.

Ha, I. D., Noh, M., and Lee, Y. (2012). *frailtyHL: Frailty Models via H-Likelihood.* R package version 1.1.

Ha, I. D., Pan, J., Oh, S., and Lee, Y. (2014b). Variable selection in general frailty models using penalized h-likelihood. *Journal of Computational Graphical Statistics*, 23:1044–10607.

Ha, I. D., Park, T. S., and Lee, Y. (2003). Joint modeling of repeated measures and survival time data. *Biometrical Journal*, 45:647–658.

Ha, I. D., Sylvester, R., Legrand, C., and MacKenzie, G. (2011). Frailty modelling for survival data from multi-centre clinical trials. *Statistics in Medicine*, 30:2144–2159.

Hamada, M. and Wu, C. F. J. (1992). Analysis of designed experiments with complex aliasing. *Journal of Quality Technology*, 24:130–137.

Harvey, A. C., Ruiz, E., and Shephard, N. (1994). Multivariate stochastic variance models. *Review of Economic Studies*, 61:247–264.

Hastie, T. and Efron, B. (2013). *lars: Least Angle Regression, Lasso and Forward Stagewise.* R package version 1.2.

Hastie, T., Tibshirani, R., and Friedman, J. (2009). *The Elements of Statistical Learning: Data Mining, Inference, and Prediction.* Springer, Canada.

Henderson, C. R. (1975). Best linear unbiased estimation and prediction under a selection model. *Biometrics*, 31:423–447.

Henderson, R., Diggle, P., and Dobson, A. (2000). Joint modelling of longitudinal measurements and event time data. *Biostatistics*, 1:465–480.

Hilbe, J. M. (2014). *Modeling Count Data.* Cambridge University Press, 2nd edition.

Hougaard, P. (2000). *Analysis of Multivariate Survival Data.* Springer-Verlag, New York, 1st edition.

Huang, J., Breheny, P., and Ma, S. (2012). A selective review of group selection in high-dimensional models. *Statistical Science*, 27:481–499.

Huang, X. and Wolfe, R. A. (2002). A frailty model for informative censoring. *Biometrics*, 58:510–520.

Hudak, S. J., Saxena, A., Bucci, R. J., and Malcolm, R. C. (1978). Development of standard methods of testing and analyzing crack growth rate data. Technical report, AFML-TR-78-40. Westinghouse R&D Center, Westinghouse Electric Corp., Pittsburgh, PA.

Hunter, D. and Li, R. (2005). Variable selection using mm algorithms. *Annals of Statistics*, 33:1617–1642.

Ibanez, E. N., Garcia, M., and Sorensen, D. (2010). Gsevm v.2 mcmc software to analyze genetically structured environmental variance models. *Journal of Animal Breeding and Genetics*, 127:249–251.

Kalbfleisch, J. D. and Prentice, R. L. (1980). *The Statistical Analysis of Failure Time Data.* John Wiley & Sons, New York, 2nd edition.

Kim, S. (2016). *JointModel: Semiparametric Joint Models for Longitudinal and Counting Processes.* R package version 1.0.

Komarek, A. (2015). *mixAK: Multivariate Normal Mixture Models and Mixtures of Generalized Linear Mixed Models Including Model Based Clustering.* R package version 4.2.

Kwon, S., Oh, S., and Lee, Y. (2017). The use of random-effect models for high-dimensional variable selection problems. To appear at *Computational Statistics and Data Analyis.*

Lee, D., Kang, H., Jang, M., Cho, S. S., Kang, W. J., Lee, J. S., Kang, E., Lee, K. U., Woo, J. I., and Lee, M. C. (2003). Application of false discovery rate control in the assessment of decrease of fdg uptake in early Alzheimer dementia. *Korean Journal of the Nuclear Medicine*, 37:374–381.

Lee, D., Lee, W., Lee, Y., and Pawitan, Y. (2010). Super-sparse principal component analyses for high-throughput genomic data. *BMC Bioinformatics*, 11:296.

Lee, D., Lee, W., Lee, Y., and Pawitan, Y. (2011a). Sparse partial least-squares regression and its applications to high-throughput data analysis. *Chemometrics and Intelligent Laboratory Systems*, 109:1–8.

Lee, D. and Lee, Y. (2017). Extended likelihood approach to multiple test with directional error control under hidden Markov random field model. To appear at *Journal of Multivariate Analysis.*

Lee, D., Lee, Y., Pawian, Y., and Lee, W. (2013). Sparse partial least-squares regression for high-throughput survival data analysis. *Statistics in Medicine*, 32:5340–5352.

Lee, S., Pawitan, Y., Ingelsson, E., and Lee, Y. (2015a). Sparse estimation of gene-gene interactions in prediction models. *Statistical Methods in Medical Research.*

Lee, S., Pawitan, Y., and Lee, Y. (2015b). A random-effect model approach for group variable selection. *Computational Statistics and Data Analysis*, 89:147–157.

Lee, W., Lee, D., Lee, Y., and Pawitan, Y. (2011b). Sparse canonical covariance analysis for high-throughput data. *Statistical Applications in Genetics and Molecular Biology*, 10:Article 30.

Lee, W. and Lee, Y. (2012). Modifications of reml algorithm for hglms. *Statistics and Computing*, 22:959–966.

Lee, Y. and Bjørnstad, J. F. (2013). Extended likelihood approach to large-scale multiple testing. *Journal of the Royal Statistical Society: Series B (Statistical Methodology)*, 75:553–575.

Lee, Y. and Ha, I. D. (2010). Orthodox blup versus h-likelihood methods for inferences about random effects in tweedie mixed models. *Statistics and Computing*, 20:295–303.

Lee, Y., Jang, M., and Lee, W. (2011c). Pediction interval for disease mapping using hierarchical likelihood. *Computational Statistics*, 26:159–179.

Lee, Y. and Kim, G. (2016). H-likelihood predictive intervals for unobservables. *International Statistical Review*, 84:487–505.

Lee, Y., Molas, M., and Noh, M. (2016a). mdhglm: A package for fitting multivariate double hierarchical generalized linear models with h-likelihood. To appear at *the R Journal.*

Lee, Y., Molas, M., and Noh, M. (2016b). *mdhglm: Multivariate Double Hierarchical Generalized Linear Models.* R package version 1.6.

Lee, Y. and Nelder, J. A. (1996). Hierarchical generalized linear models (with discussion). *Journal of the Royal Statistical Society: Series B (Statistical Methodology)*, 58:619–678.

Lee, Y. and Nelder, J. A. (1997). Extended quasi-likelihood and estimating equations approach. *IMS Notes Monograph Series,* edited by Basawa, Godambe and Tayler. pages 139–148.

Lee, Y. and Nelder, J. A. (1998). Generalized linear models for the analysis of quality-improvement experiments. *Canadian Journal of Statistics*, 26:95–105.

Lee, Y. and Nelder, J. A. (2001a). Hierarchical generalised linear models: A synthesis of generalised linear models, random-effect models and structured dispersions. *Biometrika*, 88:987–1006.

Lee, Y. and Nelder, J. A. (2001b). Modelling and analysing correlated non-normal data. *Statistical Modelling*, 1:7–16.

Lee, Y. and Nelder, J. A. (2004). Conditional and marginal models: another view (with discussion). *Statistical Science*, 19:219–238.

Lee, Y. and Nelder, J. A. (2005). Likelihood for random-effect models (with discussion). *Statistical and Operational Research Transactions*, 29:141–182.

Lee, Y. and Nelder, J. A. (2006). Double hierarchical generalized linear models (with discussion). *Applied Statistics*, 55:139–185.

Lee, Y. and Nelder, J. A. (2009). Likelihood inference for models with unobservables: another view (with discussion). *Statistical Science*, 24:255–293.

Lee, Y., Nelder, J. A., and Pawitan, Y. (2006). *Generalized Linear Models with Random Effects.* Chapman & Hall, Boca Raton, 1st edition.

Lee, Y., Nelder, J. A., and Pawitan, Y. (2017). *Generalized Linear Models with Random Effects.* Chapman & Hall, Boca Raton, 2nd edition.

Lee, Y. and Noh, M. (2016). *dhglm: Double Hierarchical Generalized Linear Models.* R package version 1.6.

Lee, Y. and Oh, H. S. (2014). A new sparse variable selection via random-effect model. *Journal of Multivariate Analysis*, 125:89–99.

Lesaffre, E. and Lawson, A. B. (2012). *Bayesian Biostatistics.* John Wiley & Sons, Chichester.

Loehlin, J. C. and Nichols, R. C. (1976). *Heredity, Environments and Personality: A Study of 850 Twins.* University of Texas Press, Austin.

Lu, C. J. and Meeker, W. Q. (1993). Using degeneration measurements to estimate a time-to-failure distribution. *Technometrics*, 35:161–174.

Mantel, N., Bohidar, N. R., and Ciminera, J. L. (1977). Mantel-Haenszel analyses of litter-matched time-to-response data, with modifications for recovery of interlitter information. *Cancer Research*, 37:3863–3868.

McCullagh, P. and Nelder, J. A. (1989). *Generalized Linear Models.* Chapman & Hall, London.

McGilchrist, C. A. and Aisbett, C. W. (1991). Regression with frailty in survival analysis. *Biometrics*, 47:461–466.

Meng, X. (2009). Decoding the h-likelihood. *Statistical Science*, 24:280–293.

Meng, X. (2010). *What's the H in H-likelihood: A holy grail or an Achilles' heel? In Bayesian Statistics 9.* Clarendon Press, Oxford.

Molas, M. (2012). *Complex data modelling using likelihood and h-likelihood methods.* PhD thesis, Erasmus University Rotterdam.

Molas, M. and Lesaffre, E. (2010). Hurdle models for multilevel zero-inflated data via h-likelihood. *Statistics in Medicine*, 29:3294–3310.

Molas, M. and Lesaffre, E. (2011). Hierarchical generalized linear models: The R package hglmmm. *Journal of Statistical Software*, 39:1–20.

Myers, P. H., Montgomery, D. C., and Vining, G. G. (2002). *Generalized Linear Models with Applications in Engineering and the Sciences.* John Wiley & Sons, New York.

Nelder, J. A. (1990). Nearly parallel lines in residual plots. *The American Statistician*, 44:221–222.

Nelder, J. A. and Lee, Y. (1991). Generalized linear models for the analysis of Taguchi-type experiments. *Applied Stochastic Models and Data Analysis*, 7:107–120.

Nelder, J. A. and Lee, Y. (1992). Likelihood, quasi-likelihood and pseudo-likelihood: Some comparisons. *Journal of the Royal Statistical Society: Series B (Statistical Methodology)*, 54:273–294.

Nelder, J. A. and Lee, Y. (1998). Joint modeling of mean and dispersion. *Technometrics*, 40:168–171.

Nelder, J. A. and Wedderburn, R. W. M. (1972). Generalized linear models. *Journal of Royal Statistical Society: Series A*, 135:370–384.

Ng, T., Lee, W., and Lee, Y. (2017). Change-point estimators with true identification property. To appear at *Bernoulli Journal.*

Ng, T., Oh, S., and Lee, Y. (2006). Going beyond oracle property: Selection consistency of all local solutions of the generalized linear model. To appear at *Statistical Methodology.*

Noh, M. and Lee, Y. (2007a). REML estimation for binary data in GLMMs. *Journal of Multivariate Analysis*, 98:896–915.

Noh, M. and Lee, Y. (2007b). Robust modelling for inference from glm classes. *Journal of the American Statistical Association*, 102:1059–1072.

Noh, M. and Lee, Y. (2017). Joint random-effect models for mean and dispersion via the R package dhglm. To appear at *Journal of Statistical Software.*

Noh, M., Lee, Y., Oud, J. H., and Toharudin, T. (2017). Hierarchical likelihood approach to non-gaussian factor analysis. *Prepared for submission.*

Noh, M., Pawitan, Y., and Lee, Y. (2005). Robust ascertainment-adjusted parameter estimation. *Genetic Epidemiology*, 29:68–75.

Paik, M. C., Lee, Y., and Ha, I. D. (2015). Frequentist inference on random effects using summarizability. *Statistica Sinica*, 25:1107–1132.

Patterson, H. D. and Thompson, R. (1971). Recovery of inter-block information when block sizes are unequal. *Biometrika*, 58:545–554.

Pawitan, Y. (2001). *In All Likelihood: Statistical Modelling and Inference Using Likelihood.* Clarendon Press, Oxford.

Pearson, K. (1920). Notes on the history of correlation. *Biometrika*, 13:25–45.

Pepe, M. S. and Mori, M. (1993). Kaplan-Meier, marginal or conditional probability curves in summarizing competing risks failure time data? *Statistics in Medicine*, 12:737–751.

Phadke, M. S., Kacka, R. N., Speeney, D. V., and Grieco, M. J. (1983). Off-line quality control for integrated circuit fabrication using experimental design. *Bell System Technical Journal*, 62:1273–1309.

Philipson, P., Sousa, I., Diggle, P., Williamson, P., Kolamunnage-Dona, R., and Henderson, R. (2012). *joineR: Joint Modelling of Repeated Measurements and Time-to-Event Data.* R package version 1.0-3.

Pierce, D. A. and Schafer, D. W. (2008). Residuals in generalized linear models. *Journal of the American Statistical Association*, 81:977–986.

Pinheiro, J. C. and Bates, D. (2000). *Mixed-Effects Models in S and S-PLUS*. Springer, New York.

Price, C. J., Kimmel, C. A., Tyle, R. W., and Marr, M. C. (1985). The developmental toxicity of ethylene glycol in rats and mice. *Toxicological Applications in Pharmacology*, 81:113–127.

Radchenko, P. and James, G. (2008). Variable inclusion and shrinkage algorithms. *Journal of the American Statistical Association*, 103:1304–1315.

Ripatti, S. and Palmgren, J. (2000). Estimation of multivariate frailty models using penalized partial likelihood. *Biometrics*, 56:1016–1022.

Rizopoulos, D. (2012). *Joint Models for Longitudinal and Time-to-Event Data with Applications in R*. Chapman & Hall, London.

Rizopoulos, D. (2013). *ltm: Latent Models under IRT*. R package version 1.00.

Rizopoulos, D. (2015). *JM: Joint Modeling of Longitudinal and Survival Data*. R package version 1.4-2.

Robinson, G. K. (1991). That blup is a good thing: The estimation of random effects. *Statistical Science*, 6:15–32.

Rönnegård, L., Felleki, M., Fikse, F., Mulder, H. A., and Strandberg, E. (2010). Genetic heterogeneity of residual variance-estimation of variance components using double hierarchical generalized linear models. *Genetics Selection Evolution*, 42:8.

Rönnegård, L., Felleki, M., Fikse, F., Mulder, H. A., and Strandberg, E. (2013). Variance component and breeding value estimation for genetic heterogeneity of residual variance in Swedish holstein dairy cattle. *Journal of Dairy Science*, 96:2627–2636.

Rousset, F., Ferdy, J. B., and Courtiol, A. (2016). *spaMM: Mixed Models, Particularly Spatial GLMMs*. R package version 1.7.2.

Rubin, D. B. and Wu, Y. N. (1997). Modeling schizophrenic behavior using general mixture components. *Biometrics*, 53:243–261.

Rue, H., Martino, S., and Chopin, N. (2009). Approximate Bayesian inference for latent gaussian models by using integrated nested Laplace approximations. *Journal of the Royal Statistical Society: Series B (Statistical Methodology)*, 71:319–392.

Sakamoto, Y., Ishiguro, M., and Kitagawa, G. (1986). *Akaike Information Criterion Statistics*. KTK Scientific Publisher, Tokyo, 1st edition.

Self, S. G. and Liang, K. Y. (1987). Asymptotic properties of maximum likelihood estimators and likelihood ratio tests under nonstandard conditions. *Journal of the American Statistical Association*, 82:605–610.

Shephard, N. and Pitt, M. R. (1997). Likelihood analysis of non-gaussian measurement time series. *Biometrika*, 84:653–667.

Silverman, J. (1967). Variations in cognitive control and psychophysiological defense in schizophrenias. *Psychosomatic Medicine*, 29:225–251.

Smyth, G. K. and Verbyla, A. P. (1996). A conditional likelihood approach to residual maximum likelihood estimation in generalized linear models. *Journal of the Royal Statistical Society: Series B (Statistical Methodology)*, 58:565–572.

Sorensen, D. and Waagepetersen, R. (2003). Normal linear models with genetically structured residual variance heterogeneity: A case study. *Genetical Research*, 82:207–222.

Strokes, M. E., Davis, C. S., and Koch, G. G. (1995). *Categorical Data Analysis Using the SAS System.* SAS Institute, Cary, NC.

Sung, K. H., Kang, K. W., Kang, C. M., Kwak, J. Y., Park, T. S., and Lee, S. Y. (1998). Study on the factors affecting the chronic renal allograft dysfunction. *The Korean Journal of Nephrology*, 17:483–493.

Thall, P. F. and Vail, S. C. (1990). Some covariance models for longitudinal count data with overdispersion. *Biometrics*, 46:657–671.

Therneau, T. and Lumley, T. (2015). *Survival: Survival Analysis.* R package version 2.38-3.

Therneau, T. M. (2015). *coxme: Mixed Effects Cox Models.* R package version 2.2-5.

Therneau, T. M. and Grambsch, P. M. (2000). *Modelling Survival Data: Extending the Cox Model.* Springer-Verlag, New York.

Tibshirani, R. (1996). Regression shrinkage and selection via the lasso. *Journal of the Royal Statistical Society: Series B (Statistical Methodology)*, 58:267–288.

Trapletti, A., Hornik, K., and LeBaron, B. (2016). *tseries: Time Series Analysis and Computational Finance.* R package version 0.10-35.

Vaida, F. and Xu, R. (2000). Proportional hazards model with random effects. *Statistics in Medicine*, 19:3309–3324.

Waagepetersen, R., Ibanez, E. N., and Sorensen, D. (2008). A comparison of strategies for Markov chain Monte Carlo computation in quantitative genetics. *Genetics Selection Evolution*, 40:161–176.

Wei, L. J., Lin, D. Y., and Weissfeld, L. (1989). Regression analysis of multivariate incomplete failure time data by modeling marginal distributions. *Journal of the American Statistical Association*, 84:1065–1073.

Wolfinger, R. (1993). Laplace's approximation for nonlinear mixed models. *Biometrika*, 80:791–795.

Wolfinger, R. D. and Tobias, R. D. (1998). Joint estimation of location, dispersion, and random effects in robust design. *Technometrics*, 40:62–71.

Xue, X. and Brookmeyer, R. (1996). Bivariate frailty model for the analysis of multivariate survival time. *Lifetime Data Analysis*, 2:277–289.

Yau, K. K. W. (2001). Multilevel models for survival analysis with random effects. *Biometrics*, 57:96–102.

Yun, S. and Lee, Y. (2004). Comparison of hierarchical and marginal likelihood estimators for binary outcomes. *Computational Statistics and Data Analysis*, 45:639–650.

Yun, S. and Lee, Y. (2006). Robust estimation in mixed linear models with non-monotone missingness. *Statistics in Medicine*, 25:3877–3892.

Zhou, B., Fine, J., and Latouche, A. (2015). *crrSC: Competing Risks Regression for Stratified and Clustered Data.* R package version 1.1.

Zhou, B., Fine, J., Latouche, A., and Labopin, M. (2012). Competing risks regression for clustered data. *Biostatistics*, 13:371–383.

Zou, H. (2006). The adaptive lasso and its oracle properties. *Journal of the American Statistical Association*, 101:1418–1429.

Zou, H. and Hastie, T. (2005). Regularization and variable selection via the elastic net. *Journal of the Royal Statistical Society: Series B (Statistical Methodology)*, 67:301–320.

# Data Index

# Author Index

# Subject Index

Printed in the United States
by Baker & Taylor Publisher Services